An Overview of Supergravity

" Combining Supersymmetry & General Relativity "

Edited by Paul F. Kisak

Contents

Chapter 1

Supergravity

In theoretical physics, **supergravity** (**supergravity theory**; **SUGRA** for short) is a field theory that combines the principles of supersymmetry and general relativity. Together, these imply that, in supergravity, the supersymmetry is a local symmetry (in contrast to non-gravitational supersymmetric theories, such as the Minimal Supersymmetric Standard Model). Since the generators of supersymmetry (SUSY) are convoluted with the Poincaré group to form a super-Poincaré algebra, it can be seen that supergravity follows naturally from supersymmetry.[1] All traditional literature on supergravity is generally written in terms of Cartan connections.[2]

1.1 Gravitons

Like any field theory of gravity, a supergravity theory contains a spin-2 field whose quantum is the graviton. Supersymmetry requires the graviton field to have a superpartner. This field has spin 3/2 and its quantum is the gravitino. The number of gravitino fields is equal to the number of supersymmetries.

1.2 History

1.2.1 Gauge supersymmetry

The first theory[3] of local supersymmetry was proposed in 1975 by Dick Arnowitt and Pran Nath and was called gauge supersymmetry.

1.2.2 SUGRA

SUGRA, or supergravity, was discovered in 1976 by Dan Freedman, Sergio Ferrara and Peter van Nieuwenhuizen,[4] but was quickly generalized to many different theories in various numbers of dimensions and additional (N) supersymmetry charges. Supergravity theories with N>1 are usu-

ally referred to as extended supergravity (SUEGRA). Some supergravity theories were shown to be equivalent to certain higher-dimensional supergravity theories via dimensional reduction (e.g. $N = 1$ 11-dimensional supergravity is dimensionally reduced on S^7 to $N = 8$, $d = 4$ SUGRA). The resulting theories were sometimes referred to as Kaluza–Klein theories as Kaluza and Klein constructed in 1919 a 5-dimensional gravitational theory, that when dimensionally reduced on circle, its 4-dimensional non-massive modes describe electromagnetism coupled to gravity.

1.2.3 mSUGRA

mSUGRA means minimal SUper GRAvity. The construction of a realistic model of particle interactions within the $N = 1$ supergravity framework where supersymmetry (SUSY) is broken by a super Higgs mechanism was carried out by Ali Chamseddine, Richard Arnowitt and Pran Nath in 1982. In these classes of models collectively now known as minimal supergravity Grand Unification Theories (mSUGRA GUT), gravity mediates the breaking of SUSY through the existence of a hidden sector. mSUGRA naturally generates the Soft SUSY breaking terms which are a consequence of the Super Higgs effect. Radiative breaking of electroweak symmetry through Renormalization Group Equations (RGEs) follows as an immediate consequence. mSUGRA is one of the most widely investigated models of particle physics due to its predictive power—requiring only four input parameters and a sign to determine the low energy phenomenology from the scale of Grand Unification.

See also: Gravity-Mediated Supersymmetry Breaking in the MSSM

1.2.4 11d: the maximal SUGRA

One of these supergravities, the 11-dimensional theory, generated considerable excitement as the first potential candidate for the theory of everything. This excitement was built on four pillars, two of which have now been largely discredited:

- Werner Nahm showed[5] that 11 dimensions was the largest number of dimensions consistent with a single graviton, and that a theory with more dimensions would also have particles with spins greater than 2. These problems are avoided in 12 dimensions if two of these dimensions are timelike, as has been often emphasized by Itzhak Bars.

- In 1981, Ed Witten showed[6] that 11 was the smallest number of dimensions that was big enough to contain the gauge groups of the Standard Model, namely SU(3) for the strong interactions and SU(2) times U(1) for the electroweak interactions. Today many techniques exist to embed the standard model gauge group in supergravity in any number of dimensions. For example, in the mid and late 1980s, the obligatory gauge symmetry in type I and heterotic string theories was often used. In type II string theory they could also be obtained by compactifying on certain Calabi–Yau manifolds. Today one may also use D-branes to engineer gauge symmetries.

- In 1978, Eugène Cremmer, Bernard Julia and Joël Scherk (CJS) found[7] the classical action for an 11-dimensional supergravity theory. This remains today the only known classical 11-dimensional theory with local supersymmetry and no fields of spin higher than two. Other 11-dimensional theories are known that are quantum-mechanically inequivalent to the CJS theory, but classically equivalent (that is, they reduce to the CJS theory when one imposes the classical equations of motion). For example, in the mid 1980s Bernard de Wit and Hermann Nicolai found an alternate theory in D=11 Supergravity with Local SU(8) Invariance. This theory, while not manifestly Lorentz-invariant, is in many ways superior to the CJS theory in that, for example, it dimensionally-reduces to the 4-dimensional theory without recourse to the classical equations of motion.

- In 1980, Peter Freund and M. A. Rubin showed that compactification from 11 dimensions preserving all the SUSY generators could occur in two ways, leaving only 4 or 7 macroscopic dimensions (the other 7 or 4 being compact).[8] Unfortunately, the noncompact dimensions have to form an anti-de Sitter space. Today it is understood that there are many possible compactifications, but that the Freund-Rubin compactifications are invariant under all of the supersymmetry transformations that preserve the action.

Thus, the first two results appeared to establish 11 dimensions uniquely, the third result appeared to specify the theory, and the last result explained why the observed universe appears to be four-dimensional.

Many of the details of the theory were fleshed out by Peter van Nieuwenhuizen, Sergio Ferrara and Daniel Z. Freedman.

1.2.5 The end of the SUGRA era

The initial excitement over 11-dimensional supergravity soon waned, as various failings were discovered, and attempts to repair the model failed as well. Problems included:

- The compact manifolds which were known at the time and which contained the standard model were not compatible with supersymmetry, and could not hold quarks or leptons. One suggestion was to replace the compact dimensions with the 7-sphere, with the symmetry group SO(8), or the squashed 7-sphere, with symmetry group SO(5) times SU(2).

- Until recently, the physical neutrinos seen in experiments were believed to be massless, and appeared to be left-handed, a phenomenon referred to as the chirality of the Standard Model. It was very difficult to construct a chiral fermion from a compactification — the compactified manifold needed to have singularities, but physics near singularities did not begin to be understood until the advent of orbifold conformal field theories in the late 1980s.

- Supergravity models generically result in an unrealistically large cosmological constant in four dimensions, and that constant is difficult to remove, and so require fine-tuning. This is still a problem today.

- Quantization of the theory led to quantum field theory gauge anomalies rendering the theory inconsistent. In the intervening years physicists have learned how to cancel these anomalies.

Some of these difficulties could be avoided by moving to a 10-dimensional theory involving superstrings. However, by moving to 10 dimensions one loses the sense of uniqueness of the 11-dimensional theory.

The core breakthrough for the 10-dimensional theory, known as the first superstring revolution, was a demonstration by Michael B. Green, John H. Schwarz and David Gross that there are only three supergravity models in 10 dimensions which have gauge symmetries and in which all of the gauge and gravitational anomalies cancel. These were theories built on the groups $SO(32)$ and $E_8 \times E_8$, the direct product of two copies of E_8. Today we know that, using D-branes for example, gauge symmetries can be introduced in other 10-dimensional theories as well.[9]

1.2.6 The second superstring revolution

Initial excitement about the 10-dimensional theories, and the string theories that provide their quantum completion, died by the end of the 1980s. There were too many Calabi–Yaus to compactify on, many more than Yau had estimated, as he admitted in December 2005 at the 23rd International Solvay Conference in Physics. None quite gave the standard model, but it seemed as though one could get close with enough effort in many distinct ways. Plus no one understood the theory beyond the regime of applicability of string perturbation theory.

There was a comparatively quiet period at the beginning of the 1990s; however, several important tools were developed. For example, it became apparent that the various superstring theories were related by "string dualities", some of which relate weak string-coupling (i.e. perturbative) physics in one model with strong string-coupling (i.e. non-perturbative) in another.

Then it all changed, in what is known as the second superstring revolution. Joseph Polchinski realized that obscure string theory objects, called D-branes, which he had discovered six years earlier, are stringy versions of the p-branes that were known in supergravity theories. The treatment of these p-branes was not restricted by string perturbation theory; in fact, thanks to supersymmetry, p-branes in supergravity were understood well beyond the limits in which string theory was understood.

Armed with this new nonperturbative tool, Edward Witten and many others were able to show that all of the perturbative string theories were descriptions of different states in a single theory which Witten named M-theory. Furthermore, he argued that M-theory's long wavelength limit (i.e. when the quantum wavelength associated to objects in the theory are much larger than the size of the 11th dimension) should be described by the 11-dimensional supergravity that had fallen out of favor with the first superstring revolution 10 years earlier, accompanied by the 2- and 5-branes.

Historically, then, supergravity has come "full circle". It is a commonly used framework in understanding features of string theories, M-theory and their compactifications to lower spacetime dimensions.

1.3 Relation to superstrings

Particular 10-dimensional supergravity theories are considered "low energy limits" of the 10-dimensional superstring theories; more precisely, these arise as the massless, tree-level approximation of string theories. True effective field theories of string theories, rather than truncations, are rarely available. Due to string dualities, the conjectured 11-dimensional M-theory is required to have 11-dimensional supergravity as a "low energy limit". However, this doesn't necessarily mean that string theory/M-theory is the only possible UV completion of supergravity; supergravity research is useful independent of those relations.

1.4 4D $N = 1$ SUGRA

Before we move on to SUGRA proper, let's recapitulate some important details about general relativity. We have a 4D differentiable manifold M with a Spin(3,1) principal bundle over it. This principal bundle represents the local Lorentz symmetry. In addition, we have a vector bundle T over the manifold with the fiber having four real dimensions and transforming as a vector under Spin(3,1). We have an invertible linear map from the tangent bundle TM to T. This map is the vierbein. The local Lorentz symmetry has a gauge connection associated with it, the spin connection.

The following discussion will be in superspace notation, as opposed to the component notation, which isn't manifestly covariant under SUSY. There are actually *many* different versions of SUGRA out there which are inequivalent in the sense that their actions and constraints upon the torsion tensor are different, but ultimately equivalent in that we can always perform a field redefinition of the supervierbeins and spin connection to get from one version to another.

In 4D N=1 SUGRA, we have a 4|4 real differentiable supermanifold M, i.e. we have 4 real bosonic dimensions and 4 real fermionic dimensions. As in the nonsupersymmetric case, we have a Spin(3,1) principal bundle over M. We have an $\mathbf{R}^{4|4}$ vector bundle T over M. The fiber of T transforms under the local Lorentz group as follows; the four real bosonic dimensions transform as a vector and the four real fermionic dimensions transform as a Majorana spinor. This Majorana spinor can be reexpressed as a complex left-handed Weyl spinor and its complex conjugate right-handed Weyl spinor (they're not independent of each other). We also have a spin connection as before.

We will use the following conventions; the spatial (both bosonic and fermionic) indices will be indicated by M, N, ... The bosonic spatial indices will be indicated by μ, ν, ..., the left-handed Weyl spatial indices by α, β,..., and the right-handed Weyl spatial indices by $\dot{\alpha}$, $\dot{\beta}$, ... The indices for the fiber of T will follow a similar notation, except that they will be hatted like this: $\hat{M}$, $\hat{\alpha}$. See van der Waerden notation for more details. $M = (\mu, \alpha, \dot{\alpha})$. The supervierbein is denoted by $e_N^{\hat{M}}$, and the spin connection by $\omega_{\hat{M}\hat{N}P}$. The *inverse* supervierbein is denoted by $E_{\hat{M}}^{N}$.

The supervierbein and spin connection are real in the sense that they satisfy the reality conditions

$$e_N^{\hat{M}}(x,\overline{\theta},\theta)^* = e_N^{\hat{M}^*}(x,\theta,\overline{\theta}) \text{ where } \mu^* = \mu, \alpha^* = \dot{\alpha}, \text{ and } \dot{\alpha}^* = \alpha \text{ and } \omega(x,\overline{\theta},\theta)^* = \omega(x,\theta,\overline{\theta}).$$

The covariant derivative is defined as

$$D_{\hat{M}} f = E_{\hat{M}}^{N}\left(\partial_N f + \omega_N[f]\right)$$

The covariant exterior derivative as defined over supermanifolds needs to be super graded. This means that every time we interchange two fermionic indices, we pick up a +1 sign factor, instead of -1.

The presence or absence of R symmetries is optional, but if R-symmetry exists, the integrand over the full superspace has to have an R-charge of 0 and the integrand over chiral superspace has to have an R-charge of 2.

A chiral superfield X is a superfield which satisfies $\overline{D}_{\dot{\alpha}} X = 0$. In order for this constraint to be consistent, we require the integrability conditions that $\left\{\overline{D}_{\dot{\alpha}}, \overline{D}_{\dot{\beta}}\right\} = c_{\dot{\alpha}\dot{\beta}}^{\dot{\gamma}} \overline{D}_{\dot{\gamma}}$ for some coefficients c.

Unlike nonSUSY GR, the torsion has to be nonzero, at least with respect to the fermionic directions. Already, even in flat superspace, $D_{\dot{\alpha}} e_{\hat{\alpha}} + \overline{D}_{\dot{\alpha}} e_{\hat{\alpha}} \neq 0$. In one version of SUGRA (but certainly not the only one), we have the following constraints upon the torsion tensor:

$$T_{\hat{\alpha}\hat{\beta}}^{\hat{\gamma}} = 0$$

$$T_{\hat{\alpha}\hat{\beta}}^{\hat{\mu}} = 0$$

$$T_{\hat{\alpha}\hat{\beta}}^{\hat{\mu}} = 0$$

$$T_{\hat{\alpha}\hat{\beta}}^{\hat{\mu}} = 2i\sigma_{\hat{\alpha}\hat{\beta}}^{\hat{\mu}}$$

$$T_{\hat{\mu}\hat{\alpha}}^{\hat{\nu}} = 0$$

$$T_{\hat{\mu}\hat{\nu}}^{\hat{\rho}} = 0$$

Here, $\underline{\alpha}$ is a shorthand notation to mean the index runs over either the left or right Weyl spinors.

The superdeterminant of the supervierbein, $|e|$, gives us the volume factor for M. Equivalently, we have the volume 4|4-superform $e^{\hat{\mu}=0} \wedge \cdots \wedge e^{\hat{\mu}=3} \wedge e^{\hat{\alpha}=1} \wedge e^{\hat{\alpha}=2} \wedge e^{\dot{\hat{\alpha}}=1} \wedge e^{\dot{\hat{\alpha}}=2}$.

If we complexify the superdiffeomorphisms, there is a gauge where $E_{\dot{\alpha}}^{\mu} = 0$, $E_{\dot{\alpha}}^{\beta} = 0$ and $E_{\dot{\alpha}}^{\dot{\beta}} = \delta_{\dot{\alpha}}^{\dot{\beta}}$. The resulting chiral superspace has the coordinates x and Θ.

R is a scalar valued chiral superfield derivable from the supervielbeins and spin connection. If f is any superfield, $\left(\overline{D}^2 - 8R\right) f$ is always a chiral superfield.

The action for a SUGRA theory with chiral superfields X, is given by

$$S = \int d^4x d^2\Theta 2\mathcal{E}\left[\frac{3}{8}\left(\overline{D}^2 - 8R\right)e^{-K(X,X)/3} + W(X)\right] + c.c.$$

where K is the Kähler potential and W is the superpotential, and $\mathcal{E}$ is the chiral volume factor.

Unlike the case for flat superspace, adding a constant to either the Kähler or superpotential is now physical. A constant shift to the Kähler potential changes the effective Planck constant, while a constant shift to the superpotential changes the effective cosmological constant. As the effective Planck constant now depends upon the value of the chiral superfield X, we need to rescale the supervierbeins (a field redefinition) to get a constant Planck constant. This is called the **Einstein frame**.

1.5 N = 8 supergravity in 4 dimensions

N=8 Supergravity is the most symmetric quantum field theory which involves gravity and a finite number of fields. It can be found from a dimensional reduction of 11D supergravity by making the size of 7 of the dimensions go to zero. It has 8 supersymmetries which is the most any gravitational theory can have since there are 8 half-steps between spin 2 and spin -2. (A graviton has the highest spin in this theory which is a spin 2 particle). More supersymmetries would mean the particles would have superpartners with spins higher than 2. The only theories with spins higher than 2 which are consistent involve an infinite number of particles (such as String Theory and Higher-Spin Theories). Stephen Hawking in his *A Brief History of Time* speculated that this theory could be the Theory of Everything. However, in later years this was abandoned in favour of String Theory. There has been renewed interest in the 21st century with the possibility that this theory may be finite.

1.6 Higher-dimensional SUGRA

Main article: Higher-dimensional supergravity

Higher-dimensional SUGRA is the higher-dimensional, supersymmetric generalization of general relativity. Supergravity can be formulated in any number of dimensions up to eleven. Higher-dimensional SUGRA focuses upon supergravity in greater than four dimensions.

The number of supercharges in a spinor depends on the dimension and the signature of spacetime. The supercharges occur in spinors. Thus the limit on the number of supercharges cannot be satisfied in a spacetime of arbitrary dimension. Some theoretical examples in which this is satisfied are:

- 12-dimensional two-time theory

- 11-dimensional maximal SUGRA

- 10-dimensional SUGRA theories
  - Type IIA SUGRA: $N = (1, 1)$
  - IIA SUGRA from 11d SUGRA
  - Type IIB SUGRA: $N = (2, 0)$
  - Type I gauged SUGRA: $N = (1, 0)$

- 9d SUGRA theories
  - Maximal 9d SUGRA from 10d
  - T-duality
  - $N = 1$ Gauged SUGRA

The supergravity theories that have attracted the most interest contain no spins higher than two. This means, in particular, that they do not contain any fields that transform as symmetric tensors of rank higher than two under Lorentz transformations. The consistency of interacting higher spin field theories is, however, presently a field of very active interest.

1.7 See also

1.8 Notes

[1] P. van Nieuwenhuizen, Phys. Rep. 68, 189 (1981)

[2] "supergravity in nLab". *ncatlab.org*. Retrieved 2015-10-05.

[3] P. Nath and R. Arnowitt, "Generalized Super-Gauge Symmetry as a New Framework for Unified Gauge Theories", *Physics Letters B* **56** (1975) 177

[4] D.Z. Freedman, P. van Nieuwenhuizen and S. Ferrara, "Progress Toward A Theory Of Supergravity", *Physical Review* **D13** (1976) pp 3214–3218.

[5] Werner Nahm, "Supersymmetries and their representations". *Nuclear Physics B* **135** no 1 (1978) pp 149-166, doi:10.1016/0550-3213(78)90218-3

[6] Ed Witten, "Search for a realistic Kaluza-Klein theory". *Nuclear Physics B* **186** no 3 (1981) pp 412-428, doi:10.1016/0550-3213(81)90021-3

[7] E. Cremmer, B. Julia and J. Scherk, "Supergravity theory in eleven dimensions", *Physics Letters* **B76** (1978) pp 409-412,

[8] Peter G.O. Freund; Mark A. Rubin (1980). "Dynamics of dimensional reduction". *Physics Letters B* **97** (2): 233–235. Bibcode:1980PhLB...97..233F. doi:10.1016/0370-2693(80)90590-0.

[9] Blumenhagen, R.; Cvetic, M.; Langacker, P.; Shiu, G. (2005). "Toward Realistic Intersecting D-Brane Models". arXiv:hep-th/0502005 [hep-th].

1.9 References

1.9.1 Historical

- P. Nath and R. Arnowitt, "Generalized Super-Gauge Symmetry as a New Framework for Unified Gauge Theories", *Physics Letters B '*56 (1975) 177.

- D.Z. Freedman, P. van Nieuwenhuizen and S. Ferrara, "Progress Toward A Theory Of Supergravity", *Physical Review* **D13** (1976) pp 3214–3218.

- E. Cremmer, B. Julia and J. Scherk, "Supergravity theory in eleven dimensions", *Physics Letters* **B76** (1978) pp 409–412. scanned version

- P. Freund and M. Rubin, "Dynamics of dimensional reduction", *Physics Letters* **B97** (1980) pp 233–235.

- Ali H. Chamseddine, R. Arnowitt, Pran Nath, "Locally Supersymmetric Grand Unification", " Phys. Rev.Lett.49:970,1982"

- Michael B. Green, John H. Schwarz, "Anomaly Cancellation in Supersymmetric D=10 Gauge Theory and Superstring Theory", *Physics Letters* **B149** (1984) pp117–122.

1.9.2 General

- Bernard de Wit(2002) Supergravity

- A Supersymmetry Primer (1998); updated in (2006).

- Adel Bilal, Introduction to supersymmetry (2001) ArXiv hep-th/0101055, (*a comprehensive introduction to supersymmetry*).

- Friedemann Brandt, Lectures on supergravity (2002) ArXiv hep-th/0204035, (*an introduction to 4-dimensional N = 1 supergravity*).

- Wess, Julius; Bagger, Jonathan (1992). *Supersymmetry and Supergravity*. Princeton University Press. p. 260. ISBN 0-691-02530-4.

Chapter 2

Field (physics)

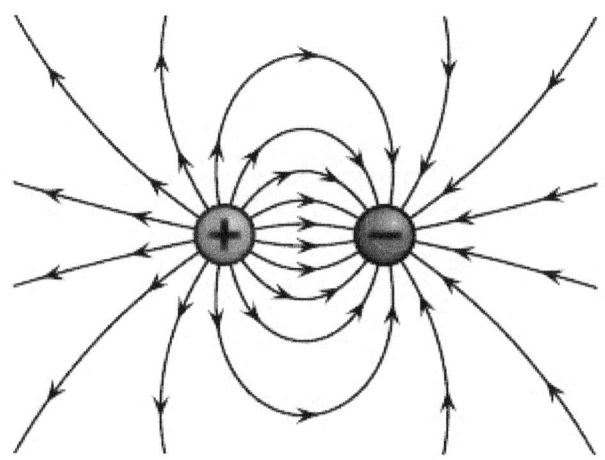

Illustration of the electric field surrounding a positive (red) and a negative (blue) charge.

In physics, a **field** is a physical quantity that has a value for each point in space and time.[1][2][3] For example, on a weather map, the surface wind velocity is described by assigning a vector to each point on a map. Each vector represents the speed and direction of the movement of air at that point. As another example, an electric field can be thought of as a "condition in space"[4] emanating from an electric charge and extending throughout the whole of space. When a test electric charge is placed in this electric field, the particle accelerates due to a force. Physicists have found the notion of a field to be of such practical utility for the analysis of forces that they have come to think of a force as due to a field.[5]

In the modern framework of the quantum theory of fields, even without referring to a test particle, a field occupies space, contains energy, and its presence eliminates a true vacuum.[6] This led physicists to consider electromagnetic fields to be a physical entity, making the field concept a supporting paradigm of the edifice of modern physics. "The fact that the electromagnetic field can possess momentum and energy makes it very real... a particle makes a field, and a field acts on another particle, and the field has such familiar properties as energy content and momentum, just as particles can have".[7] In practice, the strength of most fields has been found to diminish with distance to the point of being undetectable. For instance the strength of many relevant classical fields, such as the gravitational field in Newton's theory of gravity or the electrostatic field in classical electromagnetism, is inversely proportional to the square of the distance from the source (i.e. they follow the Gauss's law). One consequence is that the Earth's gravitational field quickly becomes undetectable on cosmic scales.

A field can be classified as a scalar field, a vector field, a spinor field or a tensor field according to whether the represented physical quantity is a scalar, a vector, a spinor or a tensor, respectively. A field has a unique tensorial character in every point where it is defined: i.e. a field cannot be a scalar field somewhere and a vector field somewhere else. For example, the Newtonian gravitational field is a vector field: specifying its value at a point in spacetime requires three numbers, the components of the gravitational field vector at that point. Moreover, within each category (scalar, vector, tensor), a field can be either a *classical field* or a *quantum field*, depending on whether it is characterized by numbers or quantum operators respectively. In fact in this theory an equivalent representation of field is a field particle, namely a boson.[8]

2.1 History

To Isaac Newton his law of universal gravitation simply expressed the gravitational force that acted between any pair of massive objects. When looking at the motion of many bodies all interacting with each other, such as the planets in the Solar System, dealing with the force between each pair of bodies separately rapidly becomes computationally inconvenient. In the eighteenth century, a new quantity was devised to simplify the bookkeeping of all these gravitational forces. This quantity, the gravitational field, gave at each point in space the total gravitational force which would be felt by an object with unit mass at that point. This did not change the physics in any way: it did not matter if you calculated all the gravitational forces on an object individually and then added them together, or if you first added all the contributions together as a gravitational field and then applied it to an object.[9]

The development of the independent concept of a field truly began in the nineteenth century with the development of the theory of electromagnetism. In the early stages, André-Marie Ampère and Charles-Augustin de Coulomb could manage with Newton-style laws that expressed the forces between pairs of electric charges or electric currents. However, it became much more natural to take the field approach and express these laws in terms of electric and magnetic fields; in 1849 Michael Faraday became the first to coin the term "field".[9]

The independent nature of the field became more apparent with James Clerk Maxwell's discovery that waves in these fields propagated at a finite speed. Consequently, the forces on charges and currents no longer just depended on the positions and velocities of other charges and currents at the same time, but also on their positions and velocities in the past.[9]

Maxwell, at first, did not adopt the modern concept of a field as fundamental quantity that could independently exist. Instead, he supposed that the electromagnetic field expressed the deformation of some underlying medium—the luminiferous aether—much like the tension in a rubber membrane. If that were the case, the observed velocity of the electromagnetic waves should depend upon the velocity of the observer with respect to the aether. Despite much effort, no experimental evidence of such an effect was ever found; the situation was resolved by the introduction of the special theory of relativity by Albert Einstein in 1905. This theory changed the way the viewpoints of moving observers should be related to each other in such a way that velocity of electromagnetic waves in Maxwell's theory would be the same for all observers. By doing away with the need for a background medium, this development opened the way for physicists to start thinking about fields as truly independent entities.[9]

In the late 1920s, the new rules of quantum mechanics were first applied to the electromagnetic fields. In 1927, Paul Dirac used quantum fields to successfully explain how the decay of an atom to lower quantum state lead to the spontaneous emission of a photon, the quantum of the electromagnetic field. This was soon followed by the realization (following the work of Pascual Jordan, Eugene Wigner, Werner Heisenberg, and Wolfgang Pauli) that all particles, including electrons and protons, could be understood as the quanta of some quantum field, elevating fields to the status of the most fundamental objects in nature.[9] That said, John Wheeler and Richard Feynman seriously considered Newton's pre-field concept of action at a distance (although they set it aside because of the ongoing utility of the field concept for research in general relativity and quantum electrodynamics).

2.2 Classical fields

Main article: Classical field theory

There are several examples of classical fields. Classical field theories remain useful wherever quantum properties do not arise, and can be active areas of research. Elasticity of materials, fluid dynamics and Maxwell's equations are cases in point.

Some of the simplest physical fields are vector force fields. Historically, the first time that fields were taken seriously was with Faraday's lines of force when describing the electric field. The gravitational field was then similarly described.

2.2.1   Newtonian gravitation

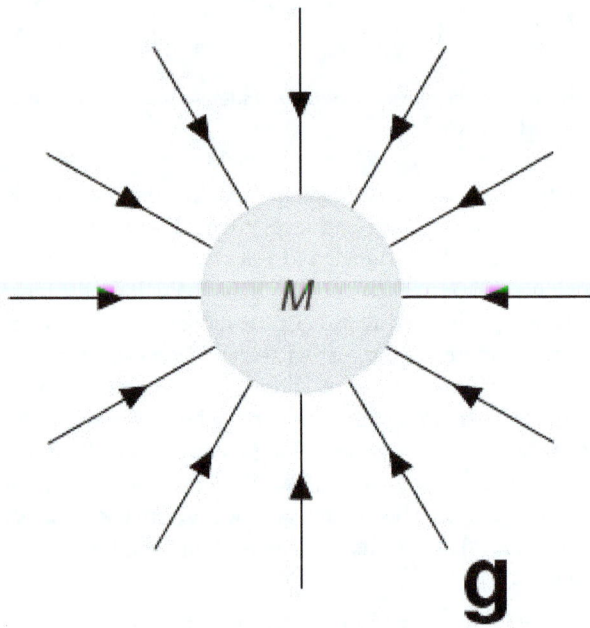

In classical gravitation, mass is the source of an attractive gravitational field **g**.

A classical field theory describing gravity is Newtonian gravitation, which describes the gravitational force as a mutual interaction between two masses.

Any body with mass M is associated with a gravitational field **g** which describes its influence on other bodies with mass. The gravitational field of M at a point $\mathbf{r}$ in space corresponds to the ratio between force $\mathbf{F}$ that M exerts on a small or negligible test mass m located at $\mathbf{r}$ and the test mass itself:[10]

$$\mathbf{g}(\mathbf{r}) = \frac{\mathbf{F}(\mathbf{r})}{m}.$$

Stipulating that m is much smaller than M ensures that the presence of m has a negligible influence on the behavior of M.

According to Newton's law of universal gravitation, $\mathbf{F}(\mathbf{r})$ is given by[10]

$$\mathbf{F}(\mathbf{r}) = -\frac{GMm}{r^2}\hat{\mathbf{r}},$$

where $\hat{\mathbf{r}}$ is a unit vector lying along the line joining M and m and pointing from m to M. Therefore, the gravitational field of $\mathbf{M}$ is[10]

$$\mathbf{g}(\mathbf{r}) = \frac{\mathbf{F}(\mathbf{r})}{m} = -\frac{GM}{r^2}\hat{\mathbf{r}}.$$

The experimental observation that inertial mass and gravitational mass are equal to an unprecedented level of accuracy leads to the identity that gravitational field strength is identical to the acceleration experienced by a particle. This is the starting point of the equivalence principle, which leads to general relativity.

Because the gravitational force $\mathbf{F}$ is conservative, the gravitational field $\mathbf{g}$ can be rewritten in terms of the gradient of a scalar function, the gravitational potential $\Phi(\mathbf{r})$:

$$\mathbf{g}(\mathbf{r}) = -\nabla\Phi(\mathbf{r}).$$

2.2.2 Electromagnetism

Main article: Electromagnetism

Michael Faraday first realized the importance of a field as a physical quantity, during his investigations into magnetism. He realized that electric and magnetic fields are not only fields of force which dictate the motion of particles, but also have an independent physical reality because they carry energy.

These ideas eventually led to the creation, by James Clerk Maxwell, of the first unified field theory in physics with the introduction of equations for the electromagnetic field. The modern version of these equations is called Maxwell's equations.

Electrostatics

Main article: Electrostatics

A charged test particle with charge q experiences a force $\mathbf{F}$ based solely on its charge. We can similarly describe the electric field $\mathbf{E}$ so that $\mathbf{F} = q\mathbf{E}$. Using this and Coulomb's law tells us that the electric field due to a single charged particle as

$$\mathbf{E} = \frac{1}{4\pi\epsilon_0}\frac{q}{r^2}\hat{\mathbf{r}}.$$

The electric field is conservative, and hence can be described by a scalar potential, $V(\mathbf{r})$:

$$\mathbf{E}(\mathbf{r}) = -\nabla V(\mathbf{r}).$$

Magnetostatics

Main article: Magnetostatics

A steady current I flowing along a path ℓ will exert a force on nearby moving charged particles that is quantitatively different from the electric field force described above. The force exerted by I on a nearby charge q with velocity $\mathbf{v}$ is

$$\mathbf{F}(\mathbf{r}) = q\mathbf{v} \times \mathbf{B}(\mathbf{r}),$$

where $\mathbf{B}(\mathbf{r})$ is the magnetic field, which is determined from I by the Biot–Savart law:

$$\mathbf{B}(\mathbf{r}) = \frac{\mu_0 I}{4\pi}\int\frac{d\boldsymbol{\ell} \times d\hat{\mathbf{r}}}{r^2}.$$

The magnetic field is not conservative in general, and hence cannot usually be written in terms of a scalar potential. However, it can be written in terms of a vector potential, $\mathbf{A}(\mathbf{r})$:

$$\mathbf{B}(\mathbf{r}) = \nabla \times \mathbf{A}(\mathbf{r})$$

Electrodynamics

Main article: Electrodynamics

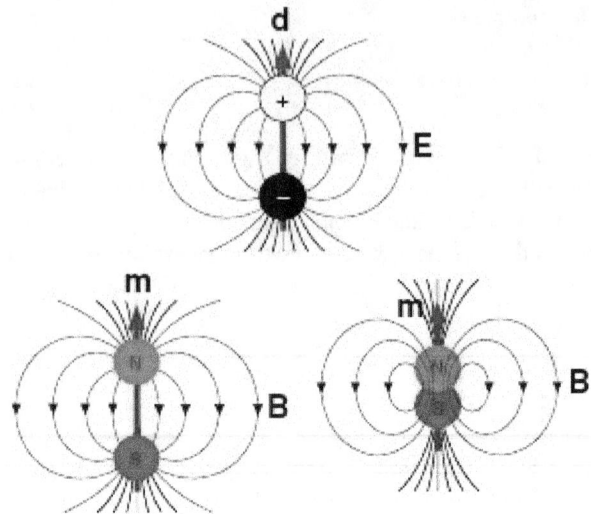

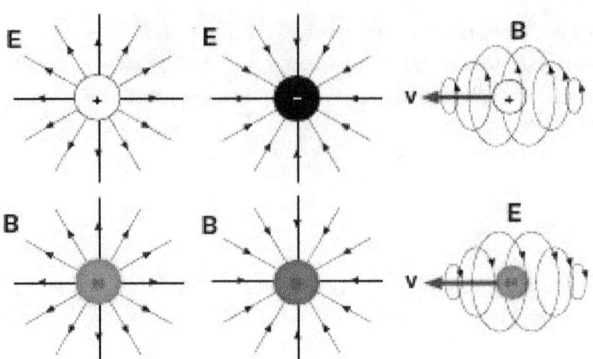

The E fields and B fields due to electric charges (black/white) and magnetic poles (red/blue).[11][12] **Top:** *E field due to an electric dipole moment **d**.* **Bottom left:** *B field due to a* mathematical *magnetic dipole **m** formed by two magnetic monopoles.* **Bottom right:** *B field due to a pure magnetic dipole moment **m** found in ordinary matter (not from monopoles).*

In general, in the presence of both a charge density $\rho(\mathbf{r}, t)$ and current density $\mathbf{J}(\mathbf{r}, t)$, there will be both an electric and a magnetic field, and both will vary in time. They are determined by Maxwell's equations, a set of differential equations which directly relate $\mathbf{E}$ and $\mathbf{B}$ to ρ and $\mathbf{J}$.[13]

Alternatively, one can describe the system in terms of its scalar and vector potentials V and $\mathbf{A}$. A set of integral equations known as *retarded potentials* allow one to calculate V and $\mathbf{A}$ from ρ and $\mathbf{J}$,[note 1] and from there the electric and magnetic fields are determined via the relations[14]

$$\mathbf{E} = -\nabla V - \frac{\partial \mathbf{A}}{\partial t}$$

$$\mathbf{B} = \nabla \times \mathbf{A}.$$

At the end of the 19th century, the electromagnetic field was understood as a collection of two vector fields in space. Nowadays, one recognizes this as a single antisymmetric 2nd-rank tensor field in spacetime.

2.2.3   Gravitation in general relativity

Einstein's theory of gravity, called general relativity, is another example of a field theory. Here the principal field is the metric tensor, a symmetric 2nd-rank tensor field in spacetime. This replaces Newton's law of universal gravitation.

The E fields and B fields due to electric charges (black/white) and magnetic poles (red/blue).[11][12] E *fields due to stationary electric charges and* B *fields due to stationary magnetic charges (note in nature N and S monopoles do not exist). In motion (velocity v), an electric charge induces a* B *field while a magnetic charge (not found in nature) would induce an* E *field. Conventional current is used.*

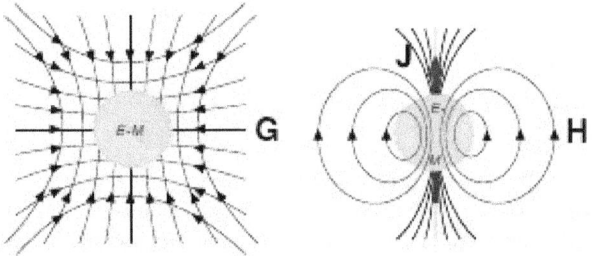

In general relativity, mass-energy warps space time (Einstein tensor G),[15] and rotating asymmetric mass-energy distributions with angular momentum J generate GEM fields H[16]

2.2.4   Waves as fields

Waves can be constructed as physical fields, due to their finite propagation speed and causal nature when a simplified physical model of an isolated closed system is set. They are also subject to the inverse-square law.

For electromagnetic waves, there are optical fields, and terms such as near- and far-field limits for diffraction. In practice, though the field theories of optics are superseded by the electromagnetic field theory of Maxwell.

2.3   Quantum fields

Main article: Quantum field theory

It is now believed that quantum mechanics should underlie all physical phenomena, so that a classical field theory should, at least in principle, permit a recasting in quantum mechanical terms; success yields the corresponding quantum field theory. For example, quantizing classical electrodynamics gives quantum electrodynamics. Quantum electrodynamics is arguably the most successful scientific theory; experimental data confirm its predictions to a higher precision (to more significant digits) than any other theory.[17] The two other fundamental quantum field theories are quantum chromodynamics and the electroweak theory.

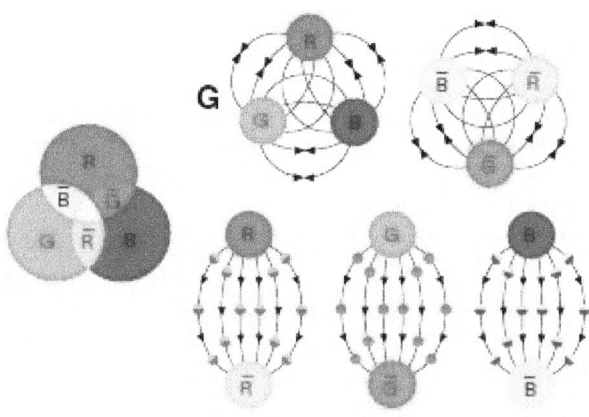

*Fields due to color charges, like in quarks (**G** is the gluon field strength tensor). These are "colorless" combinations. **Top:** Color charge has "ternary neutral states" as well as binary neutrality (analogous to electric charge). **Bottom:** The quark/antiquark combinations.[11][12]*

In quantum chromodynamics, the color field lines are coupled at short distances by gluons, which are polarized by the field and line up with it. This effect increases within a short distance (around 1 fm from the vicinity of the quarks) making the color force increase within a short distance, confining the quarks within hadrons. As the field lines are pulled together tightly by gluons, they do not "bow" outwards as much as an electric field between electric charges.[18]

These three quantum field theories can all be derived as special cases of the so-called standard model of particle physics. General relativity, the Einsteinian field theory of gravity, has yet to be successfully quantized. However an extension, thermal field theory, deals with quantum field theory at *finite temperatures*, something seldom considered in quantum field theory.

In BRST theory one deals with odd fields, e.g. Faddeev–Popov ghosts. There are different descriptions of odd classical fields both on graded manifolds and supermanifolds.

As above with classical fields, it is possible to approach their quantum counterparts from a purely mathematical view using similar techniques as before. The equations governing the quantum fields are in fact PDEs (specifically, relativistic wave equations (RWEs)). Thus one can speak of Yang–Mills, Dirac, Klein–Gordon and Schrödinger fields as being solutions to their respective equations. A possible problem is that these RWEs can deal with complicated mathematical objects with exotic algebraic properties (e.g. spinors are not tensors, so may need calculus over spinor fields), but these in theory can still be subjected to analytical methods given appropriate mathematical generalization.

2.4 Field theory

Field theory usually refers to a construction of the dynamics of a field, i.e. a specification of how a field changes with time or with respect to other independent physical variables on which the field depends. Usually this is done by writing a Lagrangian or a Hamiltonian of the field, and treating it as the classical mechanics (or quantum mechanics) of a system with an infinite number of degrees of freedom. The resulting field theories are referred to as classical or quantum field theories.

The dynamics of a classical field are usually specified by the Lagrangian density in terms of the field components; the dynamics can be obtained by using the action principle.

It is possible to construct simple fields without any a priori knowledge of physics using only mathematics from several variable calculus, potential theory and partial differential equations (PDEs). For example, scalar PDEs might consider quantities such as amplitude, density and pressure fields for the wave equation and fluid dynamics; temperature/concentration fields for the heat/diffusion equations. Outside of physics proper (e.g., radiometry and computer graphics), there are even light fields. All these previous examples are scalar fields. Similarly for vectors, there are vector PDEs for displacement, velocity and vorticity fields in (applied mathematical) fluid dynamics, but vector calculus may now be needed in addition, being calculus over vector fields (as are these three quantities, and those for vector PDEs in general). More generally problems in continuum mechanics may involve for example, directional elasticity (from which comes the term *tensor*, derived from the Latin word for stretch), complex fluid flows or anisotropic diffusion, which are framed as matrix-tensor PDEs, and then require matrices or tensor fields, hence matrix or tensor calculus. It should be noted that the scalars (and hence the vectors, matrices and tensors) can be real or complex as both are fields in the abstract-algebraic/ring-theoretic sense.

In a general setting, classical fields are described by sections of fiber bundles and their dynamics is formulated in the terms of jet manifolds (covariant classical field theory).[19]

In modern physics, the most often studied fields are those that model the four fundamental forces which one day may lead to the Unified Field Theory.

2.4.1 Symmetries of fields

A convenient way of classifying a field (classical or quantum) is by the symmetries it possesses. Physical symmetries are usually of two types:

Spacetime symmetries

Main articles: Global symmetry and Spacetime symmetries

Fields are often classified by their behaviour under transformations of spacetime. The terms used in this classification are:

- scalar fields (such as temperature) whose values are given by a single variable at each point of space. This value does not change under transformations of space.

- vector fields (such as the magnitude and direction of the force at each point in a magnetic field) which are specified by attaching a vector to each point of space. The components of this vector transform between themselves contravariantly under rotations in space. Similarly, a dual (or co-) vector field attaches a dual vector to each point of space, and the components of each dual vector transform covariantly.

- tensor fields, (such as the stress tensor of a crystal) specified by a tensor at each point of space. Under rotations in space, the components of the tensor transform in a more general way which depends on the number of covariant indices and contravariant indices.

- spinor fields (such as the Dirac spinor) arise in quantum field theory to describe particles with spin which transform like vectors except for the one of their component; in other words, when one rotates a vector field 360 degrees around a specific axis, the vector field turns to itself; however, spinors in same case turn to their negatives.

Internal symmetries

Main article: Internal symmetry

Fields may have internal symmetries in addition to spacetime symmetries. For example, in many situations one needs fields which are a list of space-time scalars: (φ_1, φ_2, ... φN). For example, in weather prediction these may be temperature, pressure, humidity, etc. In particle physics, the color symmetry of the interaction of quarks is an example of an internal symmetry of the strong interaction, as is the isospin or flavour symmetry.

If there is a symmetry of the problem, not involving spacetime, under which these components transform into each other, then this set of symmetries is called an *internal symmetry*. One may also make a classification of the charges of the fields under internal symmetries.

2.4.2 Statistical field theory

Main article: Statistical field theory

Statistical field theory attempts to extend the field-theoretic paradigm toward many-body systems and statistical mechanics. As above, it can be approached by the usual infinite number of degrees of freedom argument.

Much like statistical mechanics has some overlap between quantum and classical mechanics, statistical field theory has links to both quantum and classical field theories, especially the former with which it shares many methods. One important example is mean field theory.

2.4.3 Continuous random fields

Classical fields as above, such as the electromagnetic field, are usually infinitely differentiable functions, but they are in any case almost always twice differentiable. In contrast, generalized functions are not continuous. When dealing carefully with classical fields at finite temperature, the mathematical methods of continuous random fields are used, because thermally fluctuating classical fields are nowhere differentiable. Random fields are indexed sets of random variables; a continuous random field is a random field that has a set of functions as its index set. In particular, it is often mathematically convenient to take a continuous random field to have a Schwartz space of functions as its index set, in which case the continuous random field is a tempered distribution.

We can think about a continuous random field, in a (very) rough way, as an ordinary function that is $\pm\infty$ almost everywhere, but such that when we take a weighted average of all the infinities over any finite region, we get a finite result. The infinities are not well-defined; but the finite values can be associated with the functions used as the weight functions to get the finite values, and that can be well-defined. We can define a continuous random field well enough as a linear map from a space of functions into the real numbers.

2.5 See also

- Field strength

- Lagrangian and Eulerian specification of a field
- Covariant Hamiltonian field theory
- Scalar field theory

2.6 Notes

[1] This is contingent on the correct choice of gauge. V and $\mathbf{A}$ are not completely determined by ρ and $\mathbf{J}$; rather, they are only determined up to some scalar function $f(\mathbf{r}, t)$ known as the gauge. The retarded potential formalism requires one to choose the Lorenz gauge.

2.7 References

[1] John Gribbin (1998). *Q is for Quantum: Particle Physics from A to Z*. London: Weidenfeld & Nicolson. p. 138. ISBN 0-297-81752-3.

[2] Richard Feynman (1970). *The Feynman Lectures on Physics Vol II*. Addison Wesley Longman. ISBN 978-0-201-02115-8. A "field" is any physical quantity which takes on different values at different points in space.

[3] Ernan McMullin (2002). "The Origins of the Field Concept in Physics" (PDF). *Phys. Perspect.* **4**: 13–39.

[4] Richard P. Feynman (1970). *The Feynman Lectures on Physics Vol II*. Addison Wesley Longman.

[5] Richard P. Feynman (1970). *The Feynman Lectures on Physics Vol I*. Addison Wesley Longman.

[6] John Archibald Wheeler (1998). *Geons, Black Holes, and Quantum Foam: A Life in Physics*. London: Norton. p. 163.

[7] Richard P. Feynman (1970). *The Feynman Lectures on Physics Vol I*. Addison Wesley Longman.

[8] Steven Weinberg (November 7, 2013). "Physics: What We Do and Don't Know". *New York Review of Books*.

[9] Weinberg, Steven (1977). "The Search for Unity: Notes for a History of Quantum Field Theory". *Daedalus* **106** (4): 17–35. JSTOR 20024506.

[10] Kleppner, David; Kolenkow, Robert. *An Introduction to Mechanics*. p. 85.

[11] Parker, C.B. (1994). *McGraw Hill Encyclopaedia of Physics* (2nd ed.). Mc Graw Hill. ISBN 0-07-051400-3.

[12] M. Mansfield, C. O'Sullivan (2011). *Understanding Physics* (4th ed.). John Wiley & Sons. ISBN 978-0-47-0746370.

[13] Griffiths, David. *Introduction to Electrodynamics* (3rd ed.). p. 326.

[14] Wangsness, Roald. *Electromagnetic Fields* (2nd ed.). p. 469.

[15] J.A. Wheeler, C. Misner, K.S. Thorne (1973). *Gravitation*. W.H. Freeman & Co. ISBN 0-7167-0344-0.

[16] I. Ciufolini and J.A. Wheeler (1995). *Gravitation and Inertia*. Princeton Physics Series. ISBN 0-691-03323-4.

[17] Peskin, Michael E.; Schroeder, Daniel V. (1995). *An Introduction to Quantum Fields*. Westview Press. p. 198. ISBN 0-201-50397-2.. Also see precision tests of QED.

[18] R. Resnick, R. Eisberg (1985). *Quantum Physics of Atoms, Molecules, Solids, Nuclei and Particles* (2nd ed.). John Wiley & Sons. p. 684. ISBN 978-0-471-87373-0.

[19] Giachetta, G., Mangiarotti, L., Sardanashvily, G. (2009) *Advanced Classical Field Theory*. Singapore: World Scientific. ISBN 978-981-283-895-7 (arXiv: 0811.0331v2)

2.8 Further reading

- "Fields". *Principles of Physical Science. Encyclopaedia Britannica (Macropaedia)* **25** (fifteenth ed.). 1994. p. 815.
- Landau, Lev D. and Lifshitz, Evgeny M. (1971). *Classical Theory of Fields* (3rd ed.). London: Pergamon. ISBN 0-08-016019-0. Vol. 2 of the Course of Theoretical Physics.
- Jepsen, Kathryn (July 18, 2013). "Real talk: Everything is made of fields" (PDF). *Symmetry Magazine*.

2.9 External links

- Particle and Polymer Field Theories

Chapter 3

Supersymmetry

"SUSY" redirects here. For other uses, see Susy (disambiguation).
For the episode of the American TV series *Angel*, see Supersymmetry (Angel).

In particle physics, **Supersymmetry (SUSY)** is a proposed type of spacetime symmetry that relates two basic classes of elementary particles: bosons, which have an integer-valued spin, and fermions, which have a half-integer spin.[1] Each particle from one group is associated with a particle from the other, known as its superpartner, the spin of which differs by a half-integer. In a theory with perfectly "unbroken" supersymmetry, each pair of superpartners would share the same mass and internal quantum numbers besides spin. For example, there would be a "selectron" (superpartner electron), a bosonic version of the electron with the same mass as the electron, that would be easy to find in a laboratory. Thus, since no superpartners have been observed, if supersymmetry exists it must be a spontaneously broken symmetry so that superpartners may differ in mass.[2][3] Spontaneously-broken supersymmetry could solve many mysterious problems in particle physics including the hierarchy problem. The simplest realization of spontaneously-broken supersymmetry, the so-called Minimal Supersymmetric Standard Model, is one of the best studied candidates for physics beyond the Standard Model.

There is only indirect evidence and motivation for the existence of supersymmetry. Direct confirmation would entail production of superpartners in collider experiments, such as the Large Hadron Collider (LHC). The first run of the LHC found no evidence for supersymmetry (all results were consistent with the Standard Model), and thus set limits on superpartner masses in supersymmetric theories. While some remain enthusiastic about supersymmetry,[4] this first run at the LHC led some physicists to explore other ideas.[5] The LHC resumed its search for supersymmetry and other new physics in its second run.

3.1 Motivations

There are numerous phenomenological motivations for supersymmetry close to the electroweak scale, as well as technical motivations for supersymmetry at any scale.

3.1.1 The hierarchy problem

Supersymmetry close to the electroweak scale ameliorates the hierarchy problem that afflicts the Standard Model. In the Standard Model, the electroweak scale receives enormous Planck-scale quantum corrections. The observed hierarchy between the electroweak scale and the Planck scale must be achieved with extraordinary fine tuning. In a supersymmetric theory, on the other hand, Planck-scale quantum corrections cancel between partners and superpartners (owing to a minus sign associated with fermionic loops). The hierarchy between the electroweak scale and the Planck scale is achieved in a natural manner, without miraculous fine-tuning.

3.1.2 Gauge coupling unification

The idea that the gauge symmetry groups unify at high-energy is called Grand unification theory. In the Standard Model, however, the weak, strong and electromagnetic couplings fail to unify at high energy. In a supersymmetry theory, the running of the gauge couplings are modified, and precise high-energy unification of the gauge couplings is achieved. The modified running also provides a natural mechanism for radiative electroweak symmetry breaking.

3.1.3 Dark matter

TeV-scale supersymmetry (augmented with a discrete symmetry) typically provides a candidate dark matter particle at a mass scale consistent with thermal relic abundance calculations.[6][7]

3.1.4 Other technical motivations

Supersymmetry is also motivated by solutions to several theoretical problems, for generally providing many desirable mathematical properties, and for ensuring sensible behavior at high energies. Supersymmetric quantum field theory is often much easier to analyze, as many more problems become exactly solvable. When supersymmetry is imposed as a *local* symmetry, Einstein's theory of general relativity is included automatically, and the result is said to be a theory of supergravity. It is also a necessary feature of the most popular candidate for a theory of everything, superstring theory.

Another theoretically appealing property of supersymmetry is that it offers the only "loophole" to the Coleman–Mandula theorem, which prohibits spacetime and internal symmetries from being combined in any nontrivial way, for quantum field theories like the Standard Model with very general assumptions. The Haag-Lopuszanski-Sohnius theorem demonstrates that supersymmetry is the only way spacetime and internal symmetries can be combined consistently.[8]

3.2 History

A supersymmetry relating mesons and baryons was first proposed, in the context of hadronic physics, by Hironari Miyazawa during 1966. This supersymmetry did not involve spacetime, that is, it concerned internal symmetry, and was broken badly. Miyazawa's work was largely ignored at the time.[9][10][11][12]

J. L. Gervais and B. Sakita (during 1971),[13] Yu. A. Golfand and E. P. Likhtman (also during 1971), and D.V. Volkov and V.P. Akulov (1972),[14] independently rediscovered supersymmetry in the context of quantum field theory, a radically new type of symmetry of spacetime and fundamental fields, which establishes a relationship between elementary particles of different quantum nature, bosons and fermions, and unifies spacetime and internal symmetries of microscopic phenomena. Supersymmetry with a consistent Lie-algebraic graded structure on which the Gervais–Sakita rediscovery was based directly first arose during 1971[15] in the context of an early version of string theory by Pierre Ramond, John H. Schwarz and André Neveu.

Finally, Julius Wess and Bruno Zumino (during 1974)[16] identified the characteristic renormalization features of four-dimensional supersymmetric field theories, which identified them as remarkable QFTs, and they and Abdus Salam and their fellow researchers introduced early particle physics applications. The mathematical structure of supersymmetry (Graded Lie superalgebras) has subse-

quently been applied successfully to other topics of physics, ranging from nuclear physics,[17][18] critical phenomena,[19] quantum mechanics to statistical physics. It remains a vital part of many proposed theories of physics.

The first realistic supersymmetric version of the Standard Model was proposed during 1977 by Pierre Fayet and is known as the Minimal Supersymmetric Standard Model or MSSM for short. It was proposed to solve, amongst other things, the hierarchy problem.

3.3 Applications

3.3.1 Extension of possible symmetry groups

One reason that physicists explored supersymmetry is because it offers an extension to the more familiar symmetries of quantum field theory. These symmetries are grouped into the Poincaré group and internal symmetries and the Coleman–Mandula theorem showed that under certain assumptions, the symmetries of the S-matrix must be a direct product of the Poincaré group with a compact internal symmetry group or if there is not any mass gap, the conformal group with a compact internal symmetry group. During 1971 Golfand and Likhtman were the first to show that the Poincaré algebra can be extended through introduction of four anticommuting spinor generators (in four dimensions), which later became known as supercharges. During 1975 the Haag-Lopuszanski-Sohnius theorem analyzed all possible superalgebras in the general form, including those with an extended number of the supergenerators and central charges. This extended super-Poincaré algebra paved the way for obtaining a very large and important class of supersymmetric field theories.

The supersymmetry algebra

Main article: Supersymmetry algebra

Traditional symmetries of physics are generated by objects that transform by the tensor representations of the Poincaré group and internal symmetries. Supersymmetries, however, are generated by objects that transform by the spinor representations. According to the spin-statistics theorem, bosonic fields commute while fermionic fields anticommute. Combining the two kinds of fields into a single algebra requires the introduction of a Z_2-grading under which the bosons are the even elements and the fermions are the odd elements. Such an algebra is called a Lie superalgebra.

The simplest supersymmetric extension of the Poincaré algebra is the Super-Poincaré algebra. Expressed in terms of two Weyl spinors, has the following anti-commutation relation:

$$\{Q_\alpha, \bar{Q}\beta\} = 2(\sigma^\mu)_{\alpha\dot\beta} P_\mu$$

and all other anti-commutation relations between the Qs and commutation relations between the Qs and Ps vanish. In the above expression $P_\mu = -i\partial_\mu$ are the generators of translation and σ^μ are the Pauli matrices.

There are representations of a Lie superalgebra that are analogous to representations of a Lie algebra. Each Lie algebra has an associated Lie group and a Lie superalgebra can sometimes be extended into representations of a Lie supergroup.

3.3.2   The Supersymmetric Standard Model

Main article: Minimal Supersymmetric Standard Model

Incorporating supersymmetry into the Standard Model requires doubling the number of particles since there is no way that any of the particles in the Standard Model can be superpartners of each other. With the addition of new particles, there are many possible new interactions. The simplest possible supersymmetric model consistent with the Standard Model is the Minimal Supersymmetric Standard Model (MSSM) which can include the necessary additional new particles that are able to be superpartners of those in the Standard Model.

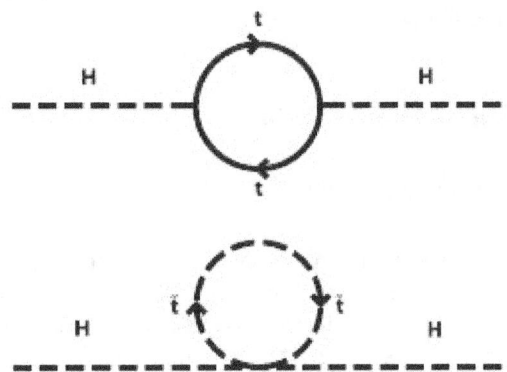

Cancellation of the Higgs boson quadratic mass renormalization between fermionic top quark loop and scalar stop squark tadpole Feynman diagrams in a supersymmetric extension of the Standard Model

One of the main motivations for SUSY comes from the quadratically divergent contributions to the Higgs mass squared. The quantum mechanical interactions of the Higgs boson causes a large renormalization of the Higgs mass and unless there is an accidental cancellation, the natural size of the Higgs mass is the greatest scale possible. This problem is known as the hierarchy problem. Supersymmetry reduces the size of the quantum corrections by having automatic cancellations between fermionic and bosonic Higgs interactions. If supersymmetry is restored at the weak scale, then the Higgs mass is related to supersymmetry breaking which can be induced from small non-perturbative effects explaining the vastly different scales in the weak interactions and gravitational interactions.

In many supersymmetric Standard Models there is a heavy stable particle (such as neutralino) which could serve as a weakly interacting massive particle (WIMP) dark matter candidate. The existence of a supersymmetric dark matter candidate is related closely to R-parity.

The standard paradigm for incorporating supersymmetry into a realistic theory is to have the underlying dynamics of the theory be supersymmetric, but the ground state of the theory does not respect the symmetry and supersymmetry is broken spontaneously. The supersymmetry break can not be done permanently by the particles of the MSSM as they currently appear. This means that there is a new sector of the theory that is responsible for the breaking. The only constraint on this new sector is that it must break supersymmetry permanently and must give superparticles TeV scale masses. There are many models that can do this and most of their details do not matter. In order to parameterize the relevant features of supersymmetry breaking, arbitrary soft SUSY breaking terms are added to the theory which temporarily break SUSY implicitly but could never arise from a complete theory of supersymmetry breaking.

Gauge-coupling unification

Main article: Minimal Supersymmetric Standard Model § Gauge-coupling unification

One piece of evidence for supersymmetry existing is gauge coupling unification. The renormalization group evolution of the three gauge coupling constants of the Standard Model is somewhat sensitive to the present particle content of the theory. These coupling constants do not quite meet together at a common energy scale if we run the renormalization group using the Standard Model.[20] With the addition of minimal SUSY joint convergence of the coupling constants is projected at approximately 10^{16} GeV.[20]

3.3.3 Supersymmetric quantum mechanics

Main article: Supersymmetric quantum mechanics

Supersymmetric quantum mechanics adds the SUSY super-algebra to quantum mechanics as opposed to quantum field theory. Supersymmetric quantum mechanics often becomes relevant when studying the dynamics of supersymmetric solitons, and due to the simplified nature of having fields which are only functions of time (rather than space-time), a great deal of progress has been made in this subject and it is now studied in its own right.

SUSY quantum mechanics involves pairs of Hamiltonians which share a particular mathematical relationship, which are called *partner Hamiltonians*. (The potential energy terms which occur in the Hamiltonians are then known as *partner potentials*.) An introductory theorem shows that for every eigenstate of one Hamiltonian, its partner Hamiltonian has a corresponding eigenstate with the same energy. This fact can be exploited to deduce many properties of the eigenstate spectrum. It is analogous to the original description of SUSY, which referred to bosons and fermions. We can imagine a "bosonic Hamiltonian", whose eigenstates are the various bosons of our theory. The SUSY partner of this Hamiltonian would be "fermionic", and its eigenstates would be the theory's fermions. Each boson would have a fermionic partner of equal energy.

3.3.4 Supersymmetry: Applications to condensed matter physics

SUSY concepts have provided useful extensions to the WKB approximation. Additionally, SUSY has been applied to disorder averaged systems both quantum and non-quantum (through statistical mechanics), the Fokker-Planck equation being an example of a non-quantum theory. The 'supersymmetry' in all these systems arises from the fact that one is modelling one particle and as such the 'statistics' don't matter. The use of the supersymmetry method provides a mathematical rigorous alternative to the replica trick, but only in non-interacting systems, which attempts to address the so-called 'problem of the denominator' under disorder averaging. For more on the applications of supersymmetry in condensed matter physics see the book[21]

3.3.5 Supersymmetry in optics

Integrated optics was recently found[22] to provide a fertile ground on which certain ramifications of SUSY can be explored in readily-accessible laboratory settings. Making use of the analogous mathematical structure of the quantum-mechanical Schrödinger equation and the wave equation governing the evolution of light in one-dimensional settings, one may interpret the refractive index distribution of a structure as a potential landscape in which optical wave packets propagate. In this manner, a new class of functional optical structures with possible applications in phase matching, mode conversion[23] and space-division multiplexing becomes possible. SUSY transformations have been also proposed as a way to address inverse scattering problems in optics and as a one-dimensional transformation optics [24]

3.3.6 Mathematics

SUSY is also sometimes studied mathematically for its intrinsic properties. This is because it describes complex fields satisfying a property known as holomorphy, which allows holomorphic quantities to be exactly computed. This makes supersymmetric models useful "toy models" of more realistic theories. A prime example of this has been the demonstration of S-duality in four-dimensional gauge theories[25] that interchanges particles and monopoles.

The proof of the Atiyah-Singer index theorem is much simplified by the use of supersymmetric quantum mechanics.

3.4 General supersymmetry

Supersymmetry appears in many related contexts of theoretical physics. It is possible to have multiple supersymmetries and also have supersymmetric extra dimensions.

3.4.1 Extended supersymmetry

Main article: Extended supersymmetry

It is possible to have more than one kind of supersymmetry transformation. Theories with more than one supersymmetry transformation are known as extended supersymmetric theories. The more supersymmetry a theory has, the more constrained are the field content and interactions. Typically the number of copies of a supersymmetry is a power of 2, i.e. 1, 2, 4, 8. In four dimensions, a spinor has four degrees of freedom and thus the minimal number of supersymmetry generators is four in four dimensions and having eight copies of supersymmetry means that there are 32 supersymmetry generators.

The maximal number of supersymmetry generators possible is 32. Theories with more than 32 supersymmetry generators automatically have massless fields with spin greater than 2. It is not known how to make massless fields with spin greater than two interact, so the maximal number of supersymmetry generators considered is 32. This is due to the Weinberg-Witten theorem. This corresponds to an $N = 8$ supersymmetry theory. Theories with 32 supersymmetries automatically have a graviton.

For four dimensions there are the following theories, with the corresponding multiplets[26](CPT adds a copy, whenever they are not invariant under such symmetry)

- $N = 1$

Chiral multiplet: $(0,\frac{1}{2})$ Vector multiplet: $(\frac{1}{2},1)$ Gravitino multiplet: $(1,\frac{3}{2})$ Graviton multiplet: $(\frac{3}{2},2)$

- $N = 2$

hypermultiplet: $(-\frac{1}{2},0^2,\frac{1}{2})$ vector multiplet: $(0,\frac{1}{2}^2,1)$ supergravity multiplet: $(1,\frac{3}{2}^2,2)$

- $N = 4$

Vector multiplet: $(-1,-\frac{1}{2}^4,0^6,\frac{1}{2}^4,1)$ Supergravity multiplet: $(0,\frac{1}{2}^4,1^6,\frac{3}{2}^4,2)$

- $N = 8$

Supergravity multiplet: $(-2,-\frac{3}{2}^8,-1^{28},-\frac{1}{2}^{56},0^{70},\frac{1}{2}^{56},1^{28},\frac{3}{2}^8,2)$

3.4.2 Supersymmetry in alternate numbers of dimensions

It is possible to have supersymmetry in dimensions other than four. Because the properties of spinors change drastically between different dimensions, each dimension has its characteristic. In d dimensions, the size of spinors is approximately $2^{d/2}$ or $2^{(d-1)/2}$. Since the maximum number of supersymmetries is 32, the greatest number of dimensions in which a supersymmetric theory can exist is eleven.

3.5 Supersymmetry in quantum gravity

Supersymmetry is part of a larger enterprise of theoretical physics to unify everything we know about the universe into a single consistent set of physical principles, known as the quest for a Theory of Everything (TOE). A significant part of this larger enterprise is the quest for a theory of quantum gravity, which would unify the classical theory of general relativity and the Standard Model, which explains the other three basic forces in physics (electromagnetism, the strong interaction, and the weak interaction), and provides a palette of fundamental particles upon which all four forces act. Two of the most active methods of forming a theory of quantum gravity are string theory and loop quantum gravity (LQG), although in theory, supersymmetry could be a component of other theories as well.

For string theory to be consistent, supersymmetry seems to be required at some level (although it may be a strongly broken symmetry). In particle theory, supersymmetry is recognized as a way to stabilize the hierarchy between the unification scale and the electroweak scale (or the Higgs boson mass), and can also provide a natural dark matter candidate. String theory also requires extra spatial dimensions which have to be compactified as in Kaluza–Klein theory.

Loop quantum gravity (LQG) predicts no additional spatial dimensions, nor anything else about particle physics. These theories can be formulated in three spatial dimensions and one dimension of time, although in some LQG theories dimensionality is an emergent property of the theory, rather than a fundamental assumption of the theory. Also, LQG is a theory of quantum gravity which does not require supersymmetry. Lee Smolin, one of the originators of LQG, has proposed that a loop quantum gravity theory incorporating either supersymmetry or extra dimensions, or both, be called "loop quantum gravity II".

If experimental evidence confirms supersymmetry in the form of supersymmetric particles such as the neutralino that is often believed to be the lightest superpartner, some people believe this would be a major boost to string theory. Since supersymmetry is a required component of string theory, any discovered supersymmetry would be consistent with string theory. If the Large Hadron Collider and other major particle physics experiments fail to detect supersymmetric partners or evidence of extra dimensions, many versions of string theory which had predicted certain low mass superpartners to existing particles may need to be significantly revised. The failure of experiments to discover either supersymmetric partners or extra spatial dimensions, as of 2013, has encouraged loop quantum gravity researchers.

3.6 Current status

Supersymmetric models are constrained by a variety of experiments, including measurements of low-energy observables – for example, the anomalous magnetic moment of the muon at Brookhaven; the WMAP dark matter density measurement and direct detection experiments – for example, XENON−100 and LUX; and by particle collider experiments, including B-physics, Higgs phenomenology and direct searches for superpartners (sparticles), at the Large Electron–Positron Collider, Tevatron and the LHC.

Historically, the tightest limits were from direct production at colliders. The first mass limits for squarks and gluinos were made at CERN by the UA1 experiment and the UA2 experiment at the Super Proton Synchrotron. LEP later set very strong limits.,[27] which in 2006 were extended by the D0 experiment at the Tevatron.[28][29] From 2003, WMAP's and Planck's dark matter density measurements have strongly constrained supersymmetry models, which, if they explain dark matter, have to be tuned to invoke a particular mechanism to sufficiently reduce the neutralino density.

Prior to the beginning of the LHC, in 2009 fits of available data to CMSSM and NUHM1 indicated that squarks and gluinos were most likely to have masses in the 500 to 800 GeV range, though values as high as 2.5 TeV were allowed with low probabilities. Neutralinos and sleptons were expected to be quite light, with the lightest neutralino and the lightest stau most likely to be found between 100 to 150 GeV.[30]

The first run of the LHC found no evidence for supersymmetry, and, as a result, surpassed existing experimental limits from the Large Electron–Positron Collider and Tevatron and partially excluded the aforementioned expected ranges.[31]

During 2011 and 2012, the LHC discovered a Higgs boson with a mass of about 125 GeV, and with couplings to fermions and bosons which are consistent with the Standard Model. The MSSM predicts that the mass of the lightest Higgs boson should not be much higher than the mass of the Z boson, and, in the absence of fine tuning (with the supersymmetry breaking scale on the order of 1 TeV), should not exceed 130 GeV. Furthermore, for values of the MSSM parameter *tan* $\beta \leq 3$, it predicts a Higgs mass below 114 GeV over most of the parameter space.[32] This region of Higgs mass was excluded by LEP by 2000. The LHC result is somewhat problematic for the minimal supersymmetric model, as the value of 125 GeV is relatively large for the model and can only be achieved with large radiative loop corrections from top squarks, which many theorists consider to be "unnatural" (see naturalness and fine tuning).[33] On the other hand, the lightest Higgs boson in the MSSM is Standard Model-like, which is consistent with measurements of the Higgs boson couplings at the LHC.

3.7   See also

- Supersymmetric gauge theory
- Wess–Zumino model
- Minimal Supersymmetric Standard Model
- Supersymmetry as a quantum group
- Quantum group
- Supercharge
- Superfield
- Supergeometry
- Supergravity
- Supergroup
- Superspace
- Superpartner

3.8   References

[1] Haber, Howie. "SUPERSYMMETRY, PART I (THEORY)" (PDF). *Reviews, Tables and Plots*. Particle Data Group (PDG). Retrieved 8 July 2015.

[2] Martin, Stephen P. (1997). "A Supersymmetry Primer". arXiv:hep-ph/9709356.

[3] Dine, Michael (2007). *Supersymmetry and String Theory: Beyond the Standard Model*. p. 169.

[4] Ellis, John. "The Physics Landscape after the Higgs Discovery at the LHC". *arXiv*. Invited plenary talk at SILAFAE 2014. Retrieved 8 July 2015.

[5] Wolchover, Natalie (November 20, 2012). "Supersymmetry Fails Test, Forcing Physics to Seek New Ideas". *Quanta Magazine*.

[6] Jonathan Feng: Supersymmetric Dark Matter *(pdf)*, University of California, Irvine, 11 May 2007

[7] Torsten Bringmann: The WIMP "Miracle" *(pdf)* University of Hamburg

[8] R. Haag, J. T. Lopuszanski and M. Sohnius, "All Possible Generators Of Supersymmetries Of The S Matrix", Nucl. Phys. B 88 (1975) 257

[9] H. Miyazawa (1966). "Baryon Number Changing Currents". *Prog. Theor. Phys.* **36** (6): 1266–1276. Bibcode:1966PThPh..36.1266M. doi:10.1143/PTP.36.1266.

[10] H. Miyazawa (1968). "Spinor Currents and Symmetries of Baryons and Mesons". *Phys. Rev.* **170** (5): 1586–1590. Bibcode:1968PhRv..170.1586M. doi:10.1103/PhysRev.170.1586.

[11] Michio Kaku, *Quantum Field Theory*, ISBN 0-19-509158-2, pg 663.

[12] Peter Freund, *Introduction to Supersymmetry*, ISBN 0-521-35675-X, pages 26-27, 138.

[13] Gervais, J. -L.; Sakita, B. (1971). "Field theory interpretation of supergauges in dual models". *Nuclear Physics B* **34** (2): 632–639. Bibcode:1971NuPhB..34..632G. doi:10.1016/0550-3213(71)90351-8.

[14] D.V. Volkov, V.P. Akulov, Pisma Zh.Eksp.Teor.Fiz. 16 (1972) 621; Phys.Lett. B46 (1973) 109; V.P. Akulov, D.V. Volkov, Teor.Mat.Fiz. 18 (1974) 39

[15] Ramond, P. (1971). "Dual Theory for Free Fermions". *Physical Review D* **3** (10): 2415–2418. Bibcode:1971PhRvD...3.2415R. doi:10.1103/PhysRevD.3.2415.

[16] Wess, J.; Zumino, B. (1974). "Supergauge transformations in four dimensions". *Nuclear Physics B* **70**: 39–50. Bibcode:1974NuPhB..70...39W. doi:10.1016/0550-3213(74)90355-1.

[17] http://users.physik.fu-berlin.de/~{}kleinert/kleinert/?p=supersym suggested here

[18] Iachello, F. (1980). "Dynamical Supersymmetries in Nuclei". *Physical Review Letters* **44** (12): 772–775. Bibcode:1980PhRvL..44..772I. doi:10.1103/PhysRevLett.44.772.

[19] Friedan, D.; Qiu, Z.; Shenker, S. (1984). "Conformal Invariance, Unitarity, and Critical Exponents in Two Dimensions". *Physical Review Letters* **52** (18): 1575–1578. Bibcode:1984PhRvL..52.1575F. doi:10.1103/PhysRevLett.52.1575.

[20] Gordon L. Kane, *The Dawn of Physics Beyond the Standard Model*, Scientific American, June 2003, page 60 and *The frontiers of physics*, special edition, Vol 15, #3, page 8

[21] *Supersymmetry in Disorder and Chaos*, Konstantin Efetov, Cambridge university press, 1997.

[22] Miri, M.-A.; Heinrich, M.; El-Ganainy, R.; Christodoulides, D. N. (2013). "Superymmetric optical structures". *Physical Review Letters* (APS) **110** (23): 233902. arXiv:1304.6646. Bibcode:2013PhRvL.110w3902M. doi:10.1103/PhysRevLett.110.233902. PMID 25167493. Retrieved April 2014.

[23] Heinrich, M.; Miri, M.-A.; Stützer, S.; El-Ganainy, R.; Nolte, S.; Szameit, A.; Christodoulides, D. N. (2014). "Superymmetric mode converters". *Nature Communications* (NPG) **5**: 3698. arXiv:1401.5734. Bibcode:2014NatCo...5E3698H. doi:10.1038/ncomms4698. PMID 24739256. Retrieved April 2014.

[24] Miri, M.-A.; Heinrich, Matthias; Christodoulides, D. N. (2014). "SUSY-inspired one-dimensional transformation optics". *Optica* (OSA) **1** (2): 89. arXiv:1408.0832. doi:10.1364/OPTICA.1.000089. Retrieved August 2014.

[25] Krasnitz, Michael (2002). *Correlation functions in supersymmetric gauge theories from supergravity fluctuafluctuations hHKtions* (PDF). Princeton University Department of Physics: Princeton University Department of Physics. p. 91.

[26] Polchinski,J. *String theory. Vol. 2: Superstring theory and beyond*, Appendix B

[27] LEPSUSYWG, ALEPH, DELPHI, L3 and OPAL experiments, charginos, large m0 LEPSUSYWG/01-03.1

[28] The D0-Collaboration (2009). "Search for associated production of charginos and neutralinos in the trilepton final state using 2.3 fb^{-1} of data". arXiv:0901.0646. Bibcode:2009PhLB..680...34D. doi:10.1016/j.physletb.2009.08.011.

[29] The D0 Collaboration (2006). "Search for squarks and gluinos in events with jets and missing transverse energy using 2.1 fb-1 of pp$^-$ collision data at s=1.96 TeV". arXiv:0712.3805. Bibcode:2008PhLB..660..449D. doi:10.1016/j.physletb.2008.01.042.

[30] O. Buchmueller; et al. (2009). "Likelihood Functions for Supersymmetric Observables in Frequentist Analyses of the CMSSM and NUHM1". *The European Physical Journal C* **64** (3): 391–415. arXiv:0907.5568. Bibcode:2009EPJC...64..391B. doi:10.1140/epjc/s10052-009-1159-z.

[31] Roszkowski, Leszek; Sessolo, Enrico Maria; Williams, Andrew J. (11 August 2014). "What next for the CMSSM and the NUHM: improved prospects for superpartner and dark matter detection". *Journal of High Energy Physics* **2014** (8). doi:10.1007/JHEP08(2014)067.

[32] Marcela Carena and Howard E. Haber; Haber (1970). "Higgs Boson Theory and Phenomenology". *Progress in Particle and Nuclear Physics* **50**: 63–152. arXiv:hep-ph/0208209v3. Bibcode:2003PrPNP..50...63C. doi:10.1016/S0146-6410(02)00177-1.

[33] Patrick Draper; et al. (December 2011). "Implications of a 125 GeV Higgs for the MSSM and Low-Scale SUSY Breaking". *Physical Review D* **85** (9): 095007. arXiv:1112.3068. Bibcode:2012PhRvD..85i5007D. doi:10.1103/PhysRevD.85.095007.

3.9 Further reading

- Supersymmetry and Supergravity page in String Theory Wiki lists more books and reviews.

3.9.1 Theoretical introductions, free and online

- S. Martin (2011). "A Supersymmetry Primer". arXiv:hep-ph/9709356.

- Joseph D. Lykken (1996). "Introduction to Supersymmetry". arXiv:hep-th/9612114.

- Manuel Drees (1996). "An Introduction to Supersymmetry". arXiv:hep-ph/9611409.

- Adel Bilal (2001). "Introduction to Supersymmetry". arXiv:hep-th/0101055.

- An Introduction to Global Supersymmetry by Philip Arygres, 2001

3.9.2   Monographs

- Weak Scale Supersymmetry by Howard Baer and Xerxes Tata, 2006.

- Cooper, F.; Khare, A.; Sukhatme, U. (1995). "Supersymmetry and quantum mechanics". *Physics Reports* **251** (5–6): 267–385. doi:10.1016/0370-1573(94)00080-M. (arXiv:hep-th/9405029).

- Junker, G. (1996). "Supersymmetric Methods in Quantum and Statistical Physics". doi:10.1007/978-3-642-61194-0. ISBN 978-3-540-61591-0..

- Gordon L. Kane.*Supersymmetry: Unveiling the Ultimate Laws of Nature* Basic Books, New York (2001). ISBN 0-7382-0489-7.

- Gordon L. Kane and Shifman, M., eds. *The Supersymmetric World: The Beginnings of the Theory*, World Scientific, Singapore (2000). ISBN 981-02-4522-X.

- Weinberg, Steven, *The Quantum Theory of Fields, Volume 3: Supersymmetry*, Cambridge University Press, Cambridge, (1999). ISBN 0-521-66000-9.

- Wess, Julius, and Jonathan Bagger, *Supersymmetry and Supergravity*, Princeton University Press, Princeton, (1992). ISBN 0-691-02530-4.

- "Concise Encyclopedia of Supersymmetry". 2003. doi:10.1007/1-4020-4522-0. ISBN 978-1-4020-1338-6.

3.9.3   On experiments

- Bennett GW; Muon (g–2) Collaboration; Bousquet; Brown; Bunce; Carey; Cushman; Danby; Debevec; Deile; Deng; Dhawan; Druzhinin; Duong; Farley; Fedotovich; Gray; Grigoriev; Grosse-Perdekamp; Grossmann; Hare; Hertzog; Huang; Hughes; Iwasaki; Jungmann; Kawall; Khazin; Krienen; Kronkvist; et al. (2004). "Measurement of the negative muon anomalous magnetic moment to 0.7 ppm".

Physical Review Letters **92** (16): 161802. arXiv:hep-ex/0401008. Bibcode:2004PhRvL..92p1802B. doi:10.1103/PhysRevLett.92.161802. PMID 15169217.

- Brookhaven National Laboratory (Jan. 8, 2004). *New g−2 measurement deviates further from Standard Model.* Press Release.

- Fermi National Accelerator Laboratory (Sept 25, 2006). *Fermilab's CDF scientists have discovered the quick-change behavior of the B-sub-s meson.* Press Release.

3.10   External links

- Supersymmetry (physics) at *Encyclopædia Britannica*

- What do current LHC results (mid-August 2011) imply about supersymmetry? Matt Strassler

- ATLAS Experiment Supersymmetry search documents

- CMS Experiment Supersymmetry search documents

- "Particle wobble shakes up supersymmetry", *Cosmos* magazine, September 2006

- LHC results put supersymmetry theory 'on the spot' BBC news 27/8/2011

- SUSY running out of hiding places BBC news 12/11/2012

- Supersymmetry in optics? "Skulls in the Stars" blog 22/08/2013

Chapter 4

General relativity

For the book by Robert Wald, see General Relativity (book).

For a more accessible and less technical introduction to this topic, see Introduction to general relativity.

General relativity (GR, also known as the **general**

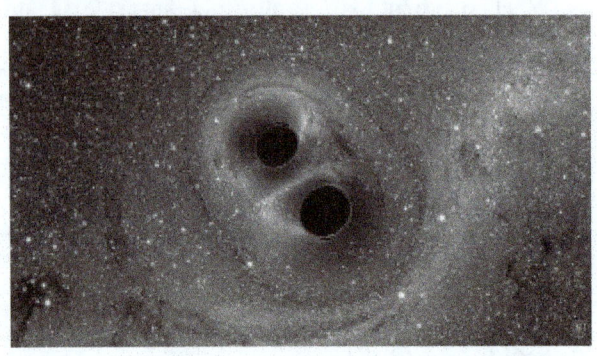

Slow motion computer simulation of the black hole binary system GW150914 as seen by a nearby observer, during 0.33 s of its final inspiral, merge, and ringdown. The star field behind the black holes is being heavily distorted and appears to rotate and move, due to extreme gravitational lensing, as space-time itself is distorted and dragged around by the rotating black holes.[1]

theory of relativity or **GTR**) is the geometric theory of gravitation published by Albert Einstein in 1915[2] and the current description of gravitation in modern physics. General relativity generalizes special relativity and Newton's law of universal gravitation, providing a unified description of gravity as a geometric property of space and time, or spacetime. In particular, the curvature of spacetime is directly related to the energy and momentum of whatever matter and radiation are present. The relation is specified by the Einstein field equations, a system of partial differential equations.

Some predictions of general relativity differ significantly from those of classical physics, especially concerning the passage of time, the geometry of space, the motion of bodies in free fall, and the propagation of light. Examples of such differences include gravitational time dilation, gravitational lensing, the gravitational redshift of light, and the gravitational time delay. The predictions of general relativity have been confirmed in all observations and experiments to date. Although general relativity is not the only relativistic theory of gravity, it is the simplest theory that is consistent with experimental data. However, unanswered questions remain, the most fundamental being how general relativity can be reconciled with the laws of quantum physics to produce a complete and self-consistent theory of quantum gravity.

Einstein's theory has important astrophysical implications. For example, it implies the existence of black holes—regions of space in which space and time are distorted in such a way that nothing, not even light, can escape—as an end-state for massive stars. There is ample evidence that the intense radiation emitted by certain kinds of astronomical objects is due to black holes; for example, microquasars and active galactic nuclei result from the presence of stellar black holes and black holes of a much more massive type, respectively. The bending of light by gravity can lead to the phenomenon of gravitational lensing, in which multiple images of the same distant astronomical object are visible in the sky. General relativity also predicts the existence of gravitational waves, which have since been observed directly by physics collaboration LIGO. In addition, general relativity is the basis of current cosmological models of a consistently expanding universe.

4.1 History

Main articles: History of general relativity and Classical theories of gravitation

Soon after publishing the special theory of relativity in 1905, Einstein started thinking about how to incorporate gravity into his new relativistic framework. In 1907, beginning with a simple thought experiment involving an observer in free fall, he embarked on what would be an eight-year search for a relativistic theory of gravity. After numerous detours and false starts, his work culminated in the presentation to the Prussian Academy of Science in Novem-

Albert Einstein developed the theories of special and general relativity. Picture from 1921.

ber 1915 of what are now known as the Einstein field equations. These equations specify how the geometry of space and time is influenced by whatever matter and radiation are present, and form the core of Einstein's general theory of relativity.[3]

The Einstein field equations are nonlinear and very difficult to solve. Einstein used approximation methods in working out initial predictions of the theory. But as early as 1916, the astrophysicist Karl Schwarzschild found the first non-trivial exact solution to the Einstein field equations, the so-called Schwarzschild metric. This solution laid the groundwork for the description of the final stages of gravitational collapse, and the objects known today as black holes. In the same year, the first steps towards generalizing Schwarzschild's solution to electrically charged objects were taken, which eventually resulted in the Reissner–Nordström solution, now associated with electrically charged black holes.[4] In 1917, Einstein applied his theory to the universe as a whole, initiating the field of relativistic cosmology. In line with contemporary thinking, he assumed a static universe, adding a new parameter to his original field equations—the cosmological constant—to match that observational presumption.[5] By 1929, however, the work of Hubble and others had shown that

our universe is expanding. This is readily described by the expanding cosmological solutions found by Friedmann in 1922, which do not require a cosmological constant. Lemaître used these solutions to formulate the earliest version of the Big Bang models, in which our universe has evolved from an extremely hot and dense earlier state.[6] Einstein later declared the cosmological constant the biggest blunder of his life.[7]

During that period, general relativity remained something of a curiosity among physical theories. It was clearly superior to Newtonian gravity, being consistent with special relativity and accounting for several effects unexplained by the Newtonian theory. Einstein himself had shown in 1915 how his theory explained the anomalous perihelion advance of the planet Mercury without any arbitrary parameters ("fudge factors").[8] Similarly, a 1919 expedition led by Eddington confirmed general relativity's prediction for the deflection of starlight by the Sun during the total solar eclipse of May 29, 1919,[9] making Einstein instantly famous.[10] Yet the theory entered the mainstream of theoretical physics and astrophysics only with the developments between approximately 1960 and 1975, now known as the golden age of general relativity.[11] Physicists began to understand the concept of a black hole, and to identify quasars as one of these objects' astrophysical manifestations.[12] Ever more precise solar system tests confirmed the theory's predictive power,[13] and relativistic cosmology, too, became amenable to direct observational tests.[14]

4.2 From classical mechanics to general relativity

General relativity can be understood by examining its similarities with and departures from classical physics. The first step is the realization that classical mechanics and Newton's law of gravity admit a geometric description. The combination of this description with the laws of special relativity results in a heuristic derivation of general relativity.[15]

4.2.1 Geometry of Newtonian gravity

At the base of classical mechanics is the notion that a body's motion can be described as a combination of free (or inertial) motion, and deviations from this free motion. Such deviations are caused by external forces acting on a body in accordance with Newton's second law of motion, which states that the net force acting on a body is equal to that body's (inertial) mass multiplied by its acceleration.[16] The preferred inertial motions are related to the geometry of space and time: in the standard reference frames of clas-

According to general relativity, objects in a gravitational field behave similarly to objects within an accelerating enclosure. For example, an observer will see a ball fall the same way in a rocket (left) as it does on Earth (right), provided that the acceleration of the rocket is equal to 9.8 m/s² (the acceleration due to gravity at the surface of the Earth).

sical mechanics, objects in free motion move along straight lines at constant speed. In modern parlance, their paths are geodesics, straight world lines in curved spacetime.[17]

Conversely, one might expect that inertial motions, once identified by observing the actual motions of bodies and making allowances for the external forces (such as electromagnetism or friction), can be used to define the geometry of space, as well as a time coordinate. However, there is an ambiguity once gravity comes into play. According to Newton's law of gravity, and independently verified by experiments such as that of Eötvös and its successors (see Eötvös experiment), there is a universality of free fall (also known as the weak equivalence principle, or the universal equality of inertial and passive-gravitational mass): the trajectory of a test body in free fall depends only on its position and initial speed, but not on any of its material properties.[18] A simplified version of this is embodied in Einstein's elevator experiment, illustrated in the figure on the right: for an observer in a small enclosed room, it is impossible to decide, by mapping the trajectory of bodies such as a dropped ball, whether the room is at rest in a gravitational field, or in free space aboard a rocket that is accelerating at a rate equal to that of the gravitational field.[19]

Given the universality of free fall, there is no observable distinction between inertial motion and motion under the influence of the gravitational force. This suggests the definition of a new class of inertial motion, namely that of objects in free fall under the influence of gravity. This new class of preferred motions, too, defines a geometry of space and time—in mathematical terms, it is the geodesic motion associated with a specific connection which depends on the gradient of the gravitational potential. Space, in this construction, still has the ordinary Euclidean geometry. How-

ever, space*time* as a whole is more complicated. As can be shown using simple thought experiments following the free-fall trajectories of different test particles, the result of transporting spacetime vectors that can denote a particle's velocity (time-like vectors) will vary with the particle's trajectory; mathematically speaking, the Newtonian connection is not integrable. From this, one can deduce that spacetime is curved. The resulting Newton–Cartan theory is a geometric formulation of Newtonian gravity using only covariant concepts, i.e. a description which is valid in any desired coordinate system.[20] In this geometric description, tidal effects—the relative acceleration of bodies in free fall—are related to the derivative of the connection, showing how the modified geometry is caused by the presence of mass.[21]

4.2.2 Relativistic generalization

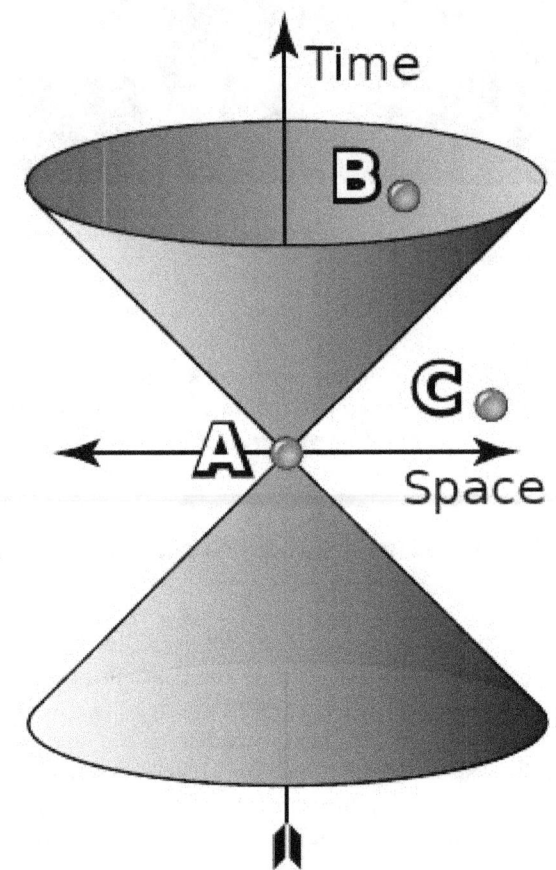

Light cone

As intriguing as geometric Newtonian gravity may be, its basis, classical mechanics, is merely a limiting case of (special) relativistic mechanics.[22] In the language of symmetry: where gravity can be neglected, physics is Lorentz invariant as in special relativity rather than Galilei invariant as in classical mechanics. (The defining symmetry of special relativity is the Poincaré group, which includes translations and rotations.) The differences between the two become significant when dealing with speeds approaching the speed of light, and with high-energy phenomena.[23]

With Lorentz symmetry, additional structures come into play. They are defined by the set of light cones (see image). The light-cones define a causal structure: for each event A, there is a set of events that can, in principle, either influence or be influenced by A via signals or interactions that do not need to travel faster than light (such as event B in the image), and a set of events for which such an influence is impossible (such as event C in the image). These sets are observer-independent.[24] In conjunction with the world-lines of freely falling particles, the light-cones can be used to reconstruct the space–time's semi-Riemannian metric, at least up to a positive scalar factor. In mathematical terms, this defines a Conformal structure[25] or conformal geometry.

Special relativity is defined in the absence of gravity, so for practical applications, it is a suitable model whenever gravity can be neglected. Bringing gravity into play, and assuming the universality of free fall, an analogous reasoning as in the previous section applies: there are no global inertial frames. Instead there are approximate inertial frames moving alongside freely falling particles. Translated into the language of spacetime: the straight time-like lines that define a gravity-free inertial frame are deformed to lines that are curved relative to each other, suggesting that the inclusion of gravity necessitates a change in spacetime geometry.[26]

A priori, it is not clear whether the new local frames in free fall coincide with the reference frames in which the laws of special relativity hold—that theory is based on the propagation of light, and thus on electromagnetism, which could have a different set of preferred frames. But using different assumptions about the special-relativistic frames (such as their being earth-fixed, or in free fall), one can derive different predictions for the gravitational redshift, that is, the way in which the frequency of light shifts as the light propagates through a gravitational field (cf. below). The actual measurements show that free-falling frames are the ones in which light propagates as it does in special relativity.[27] The generalization of this statement, namely that the laws of special relativity hold to good approximation in freely falling (and non-rotating) reference frames, is known as the Einstein equivalence principle, a crucial guiding principle for generalizing special-relativistic physics to include gravity.[28]

The same experimental data shows that time as measured by clocks in a gravitational field—proper time, to give the technical term—does not follow the rules of special relativ-

ity. In the language of spacetime geometry, it is not measured by the Minkowski metric. As in the Newtonian case, this is suggestive of a more general geometry. At small scales, all reference frames that are in free fall are equivalent, and approximately Minkowskian. Consequently, we are now dealing with a curved generalization of Minkowski space. The metric tensor that defines the geometry—in particular, how lengths and angles are measured—is not the Minkowski metric of special relativity, it is a generalization known as a semi- or pseudo-Riemannian metric. Furthermore, each Riemannian metric is naturally associated with one particular kind of connection, the Levi-Civita connection, and this is, in fact, the connection that satisfies the equivalence principle and makes space locally Minkowskian (that is, in suitable locally inertial coordinates, the metric is Minkowskian, and its first partial derivatives and the connection coefficients vanish).[29]

4.2.3   Einstein's equations

Main articles: Einstein field equations and Mathematics of general relativity

Having formulated the relativistic, geometric version of the effects of gravity, the question of gravity's source remains. In Newtonian gravity, the source is mass. In special relativity, mass turns out to be part of a more general quantity called the energy–momentum tensor, which includes both energy and momentum densities as well as stress (that is, pressure and shear).[30] Using the equivalence principle, this tensor is readily generalized to curved space-time. Drawing further upon the analogy with geometric Newtonian gravity, it is natural to assume that the field equation for gravity relates this tensor and the Ricci tensor, which describes a particular class of tidal effects: the change in volume for a small cloud of test particles that are initially at rest, and then fall freely. In special relativity, conservation of energy–momentum corresponds to the statement that the energy–momentum tensor is divergence-free. This formula, too, is readily generalized to curved spacetime by replacing partial derivatives with their curved-manifold counterparts, covariant derivatives studied in differential geometry. With this additional condition—the covariant divergence of the energy–momentum tensor, and hence of whatever is on the other side of the equation, is zero— the simplest set of equations are what are called Einstein's (field) equations:

On the left-hand side is the Einstein tensor, a specific divergence-free combination of the Ricci tensor $R_{\mu\nu}$ and the metric. Where $G_{\mu\nu}$ is symmetric. In particular,

$$R = g^{\mu\nu} R_{\mu\nu}$$

is the curvature scalar. The Ricci tensor itself is related to the more general Riemann curvature tensor as

$$R_{\mu\nu} = R^{\alpha}{}_{\mu\alpha\nu}.$$

On the right-hand side, $T_{\mu\nu}$ is the energy–momentum tensor. All tensors are written in abstract index notation.[31] Matching the theory's prediction to observational results for planetary orbits (or, equivalently, assuring that the weak-gravity, low-speed limit is Newtonian mechanics), the proportionality constant can be fixed as $\kappa = 8\pi G/c^4$, with G the gravitational constant and c the speed of light.[32] When there is no matter present, so that the energy–momentum tensor vanishes, the results are the vacuum Einstein equations,

$$R_{\mu\nu} = 0.$$

4.2.4   Alternatives to general relativity

Main article: Alternatives to general relativity

There are alternatives to general relativity built upon the same premises, which include additional rules and/or constraints, leading to different field equations. Examples are Brans–Dicke theory, teleparallelism, f(R) gravity and Einstein–Cartan theory.[33]

4.3   Definition and basic applications

See also: Mathematics of general relativity and Physical theories modified by general relativity

The derivation outlined in the previous section contains all the information needed to define general relativity, describe its key properties, and address a question of crucial importance in physics, namely how the theory can be used for model-building.

4.3.1   Definition and basic properties

General relativity is a metric theory of gravitation. At its core are Einstein's equations, which describe the relation between the geometry of a four-dimensional, pseudo-Riemannian manifold representing spacetime, and the energy–momentum contained in that spacetime.[34] Phenomena that in classical mechanics are ascribed to the action of the force of gravity (such as free-fall, orbital motion, and spacecraft trajectories), correspond to inertial motion within a curved geometry of spacetime in general relativity; there is no gravitational force deflecting objects from their natural, straight paths. Instead, gravity corresponds to changes in the properties of space and time, which in turn changes the straightest-possible paths that objects will naturally follow.[35] The curvature is, in turn, caused by the energy–momentum of matter. Paraphrasing the relativist John Archibald Wheeler, spacetime tells matter how to move; matter tells spacetime how to curve.[36]

While general relativity replaces the scalar gravitational potential of classical physics by a symmetric rank-two tensor, the latter reduces to the former in certain limiting cases. For weak gravitational fields and slow speed relative to the speed of light, the theory's predictions converge on those of Newton's law of universal gravitation.[37]

As it is constructed using tensors, general relativity exhibits general covariance: its laws—and further laws formulated within the general relativistic framework—take on the same form in all coordinate systems.[38] Furthermore, the theory does not contain any invariant geometric background structures, i.e. it is background independent. It thus satisfies a more stringent general principle of relativity, namely that the laws of physics are the same for all observers.[39] Locally, as expressed in the equivalence principle, spacetime is Minkowskian, and the laws of physics exhibit local Lorentz invariance.[40]

4.3.2   Model-building

The core concept of general-relativistic model-building is that of a solution of Einstein's equations. Given both Einstein's equations and suitable equations for the properties of matter, such a solution consists of a specific semi-Riemannian manifold (usually defined by giving the metric in specific coordinates), and specific matter fields defined on that manifold. Matter and geometry must satisfy Einstein's equations, so in particular, the matter's energy–momentum tensor must be divergence-free. The matter must, of course, also satisfy whatever additional equations were imposed on its properties. In short, such a solution is a model universe that satisfies the laws of general relativity, and possibly additional laws governing whatever matter might be present.[41]

Einstein's equations are nonlinear partial differential equations and, as such, difficult to solve exactly.[42] Nevertheless, a number of exact solutions are known, although only a few have direct physical applications.[43] The best-known exact solutions, and also those most interesting from a physics point of view, are the Schwarzschild solution, the Reissner–Nordström solution and the Kerr metric, each corresponding to a certain type of black hole in an otherwise empty universe,[44] and the Friedmann–Lemaître–Robertson–Walker and de Sitter universes, each describing an expanding cosmos.[45] Exact solutions of great theoretical interest include the Gödel universe (which opens up the intriguing possibility of time travel in curved spacetimes), the Taub-NUT solution (a model universe that is homogeneous, but anisotropic), and anti-de Sitter space (which has recently come to prominence in the context of what is called the Maldacena conjecture).[46]

Given the difficulty of finding exact solutions, Einstein's field equations are also solved frequently by numerical integration on a computer, or by considering small perturbations of exact solutions. In the field of numerical relativity, powerful computers are employed to simulate the geometry of spacetime and to solve Einstein's equations for interesting situations such as two colliding black holes.[47] In principle, such methods may be applied to any system, given sufficient computer resources, and may address fundamental questions such as naked singularities. Approximate solutions may also be found by perturbation theories such as linearized gravity[48] and its generalization, the post-Newtonian expansion, both of which were developed by Einstein. The latter provides a systematic approach to solving for the geometry of a spacetime that contains a distribution of matter that moves slowly compared with the speed of light. The expansion involves a series of terms; the first terms represent Newtonian gravity, whereas the later terms represent ever smaller corrections to Newton's theory due to general relativity.[49] An extension of this expansion is the parametrized post-Newtonian (PPN) formalism, which allows quantitative comparisons between the predictions of general relativity and alternative theories.[50]

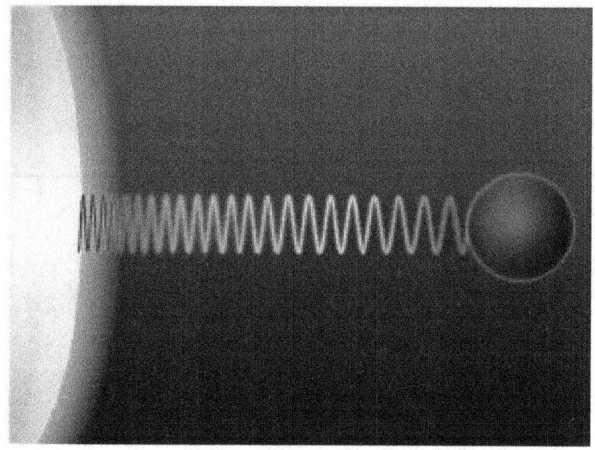

Schematic representation of the gravitational redshift of a light wave escaping from the surface of a massive body

redshifted; collectively, these two effects are known as the gravitational frequency shift. More generally, processes close to a massive body run more slowly when compared with processes taking place farther away; this effect is known as gravitational time dilation.[52]

Gravitational redshift has been measured in the laboratory[53] and using astronomical observations.[54] Gravitational time dilation in the Earth's gravitational field has been measured numerous times using atomic clocks,[55] while ongoing validation is provided as a side effect of the operation of the Global Positioning System (GPS).[56] Tests in stronger gravitational fields are provided by the observation of binary pulsars.[57] All results are in agreement with general relativity.[58] However, at the current level of accuracy, these observations cannot distinguish between general relativity and other theories in which the equivalence principle is valid.[59]

4.4 Consequences of Einstein's theory

General relativity has a number of physical consequences. Some follow directly from the theory's axioms, whereas others have become clear only in the course of many years of research that followed Einstein's initial publication.

4.4.1 Gravitational time dilation and frequency shift

Main article: Gravitational time dilation
Assuming that the equivalence principle holds,[51] gravity influences the passage of time. Light sent down into a gravity well is blueshifted, whereas light sent in the opposite direction (i.e., climbing out of the gravity well) is

4.4.2 Light deflection and gravitational time delay

Main articles: Kepler problem in general relativity, Gravitational lens and Shapiro delay
General relativity predicts that the path of light is bent in a gravitational field; light passing a massive body is deflected towards that body. This effect has been confirmed by observing the light of stars or distant quasars being deflected as it passes the Sun.[60]

This and related predictions follow from the fact that light follows what is called a light-like or null geodesic—a generalization of the straight lines along which light travels in classical physics. Such geodesics are the generalization of the invariance of lightspeed in special relativity.[61] As one examines suitable model spacetimes (either the exterior

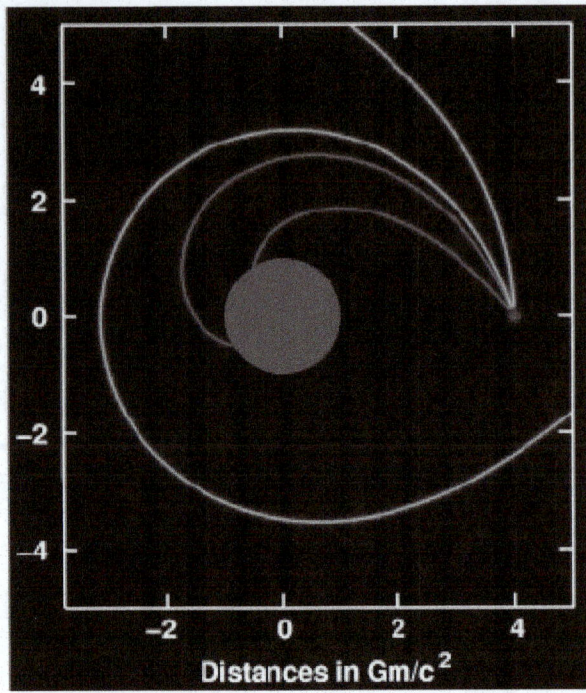

Deflection of light (sent out from the location shown in blue) near a compact body (shown in gray)

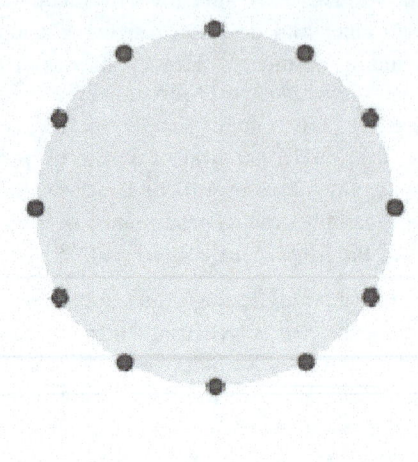

Ring of test particles influenced by gravitational wave

Schwarzschild solution or, for more than a single mass, the post-Newtonian expansion),[62] several effects of gravity on light propagation emerge. Although the bending of light can also be derived by extending the universality of free fall to light,[63] the angle of deflection resulting from such calculations is only half the value given by general relativity.[64]

Closely related to light deflection is the gravitational time delay (or Shapiro delay), the phenomenon that light signals take longer to move through a gravitational field than they would in the absence of that field. There have been numerous successful tests of this prediction.[65] In the parameterized post-Newtonian formalism (PPN), measurements of both the deflection of light and the gravitational time delay determine a parameter called γ, which encodes the influence of gravity on the geometry of space.[66]

4.4.3   Gravitational waves

Main article: Gravitational wave
 Predicted in 1916[67][68] by Albert Einstein, there are gravitational waves: ripples in the metric of spacetime that propagate at the speed of light. These are one of several analogies between weak-field gravity and electromagnetism in that, they are analogous to electromagnetic waves. On February 11, 2016, the Advanced LIGO team announced that they had directly detected gravitational waves from a pair of black holes merging.[69][70][71]

The simplest type of such a wave can be visualized by its action on a ring of freely floating particles. A sine wave propagating through such a ring towards the reader distorts the ring in a characteristic, rhythmic fashion (animated image to the right).[72] Since Einstein's equations are non-linear, arbitrarily strong gravitational waves do not obey linear superposition, making their description difficult. However, for weak fields, a linear approximation can be made. Such linearized gravitational waves are sufficiently accurate to describe the exceedingly weak waves that are expected to arrive here on Earth from far-off cosmic events, which typically result in relative distances increasing and decreasing by 10^{-21} or less. Data analysis methods routinely make use of the fact that these linearized waves can be Fourier decomposed.[73]

Some exact solutions describe gravitational waves without any approximation, e.g., a wave train traveling through empty space[74] or so-called Gowdy universes, varieties of an expanding cosmos filled with gravitational waves.[75] But for gravitational waves produced in astrophysically relevant situations, such as the merger of two black holes, numerical methods are presently the only way to construct appropriate models.[76]

4.4.4   Orbital effects and the relativity of direction

Main article: Kepler problem in general relativity

General relativity differs from classical mechanics in a

number of predictions concerning orbiting bodies. It predicts an overall rotation (precession) of planetary orbits, as well as orbital decay caused by the emission of gravitational waves and effects related to the relativity of direction.

Precession of apsides

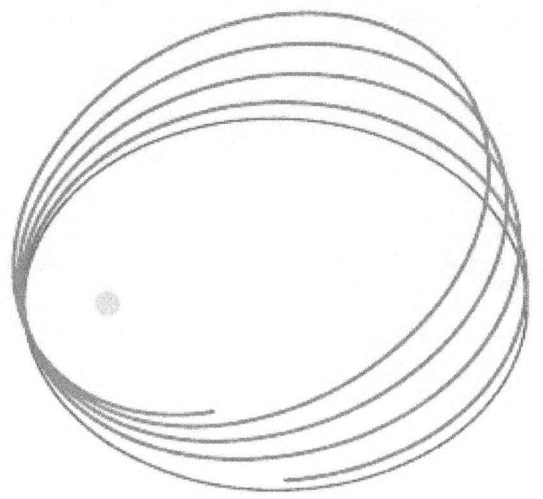

Newtonian (red) vs. Einsteinian orbit (blue) of a lone planet orbiting a star

In general relativity, the apsides of any orbit (the point of the orbiting body's closest approach to the system's center of mass) will precess—the orbit is not an ellipse, but akin to an ellipse that rotates on its focus, resulting in a rose curve-like shape (see image). Einstein first derived this result by using an approximate metric representing the Newtonian limit and treating the orbiting body as a test particle. For him, the fact that his theory gave a straightforward explanation of the anomalous perihelion shift of the planet Mercury, discovered earlier by Urbain Le Verrier in 1859, was important evidence that he had at last identified the correct form of the gravitational field equations.[77]

The effect can also be derived by using either the exact Schwarzschild metric (describing spacetime around a spherical mass)[78] or the much more general post-Newtonian formalism.[79] It is due to the influence of gravity on the geometry of space and to the contribution of self-energy to a body's gravity (encoded in the nonlinearity of Einstein's equations).[80] Relativistic precession has been observed for all planets that allow for accurate precession measurements (Mercury, Venus, and Earth),[81] as well as in binary pulsar systems, where it is larger by five orders of magnitude.[82]

In general relativity the perihelion shift σ, expressed in radians per revolution, is approximately given by:[83]

$$\sigma = \frac{24\pi^3 L^2}{T^2 c^2 (1 - e^2)} \, ,$$

where L is the semi-major axis, T is the orbital period, c is the speed of light, and e is the orbital eccentricity.

Orbital decay

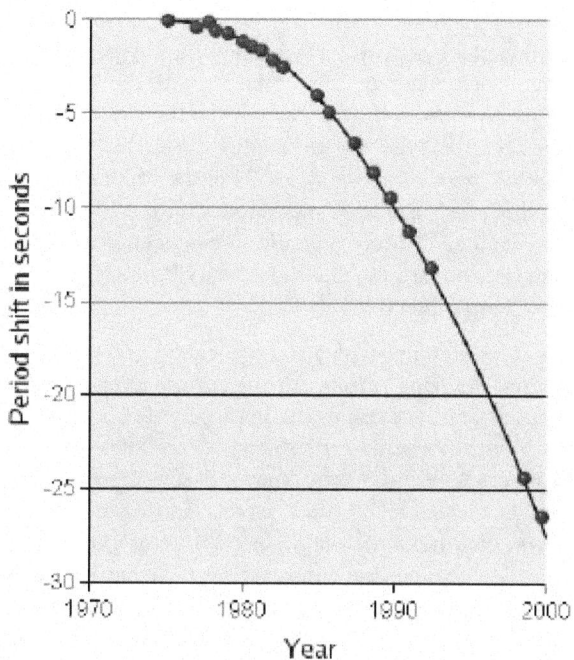

Orbital decay for PSR1913+16: time shift in seconds, tracked over three decades.[84]

According to general relativity, a binary system will emit gravitational waves, thereby losing energy. Due to this loss, the distance between the two orbiting bodies decreases, and so does their orbital period. Within the Solar System or for ordinary double stars, the effect is too small to be observable. This is not the case for a close binary pulsar, a system of two orbiting neutron stars, one of which is a pulsar: from the pulsar, observers on Earth receive a regular series of radio pulses that can serve as a highly accurate clock, which allows precise measurements of the orbital period. Because neutron stars are immensely compact, significant amounts of energy are emitted in the form of gravitational radiation.[85]

The first observation of a decrease in orbital period due to the emission of gravitational waves was made by Hulse and Taylor, using the binary pulsar PSR1913+16 they had discovered in 1974. This was the first detection of gravitational waves, albeit indirect, for which they were awarded the 1993 Nobel Prize in physics.[86] Since then, several other binary pulsars have been found, in particular the double pulsar PSR J0737-3039, in which both stars are pulsars.[87]

Geodetic precession and frame-dragging

Main articles: Geodetic precession and Frame dragging

Several relativistic effects are directly related to the relativity of direction.[88] One is geodetic precession: the axis direction of a gyroscope in free fall in curved spacetime will change when compared, for instance, with the direction of light received from distant stars—even though such a gyroscope represents the way of keeping a direction as stable as possible ("parallel transport").[89] For the Moon–Earth system, this effect has been measured with the help of lunar laser ranging.[90] More recently, it has been measured for test masses aboard the satellite Gravity Probe B to a precision of better than 0.3%.[91][92]

Near a rotating mass, there are so-called gravitomagnetic or frame-dragging effects. A distant observer will determine that objects close to the mass get "dragged around". This is most extreme for rotating black holes where, for any object entering a zone known as the ergosphere, rotation is inevitable.[93] Such effects can again be tested through their influence on the orientation of gyroscopes in free fall.[94] Somewhat controversial tests have been performed using the LAGEOS satellites, confirming the relativistic prediction.[95] Also the Mars Global Surveyor probe around Mars has been used.[96][97]

4.5   Astrophysical applications

4.5.1   Gravitational lensing

Main article: Gravitational lensing
The deflection of light by gravity is responsible for a new class of astronomical phenomena. If a massive object is situated between the astronomer and a distant target object with appropriate mass and relative distances, the astronomer will see multiple distorted images of the target. Such effects are known as gravitational lensing.[98] Depending on the configuration, scale, and mass distribution, there can be two or more images, a bright ring known as an Einstein ring, or partial rings called arcs.[99] The earliest example was discovered in 1979;[100] since then, more than a hundred gravitational lenses have been observed.[101] Even if the multiple images are too close to each other to be resolved, the effect can still be measured, e.g., as an overall brightening of the target object; a number of such

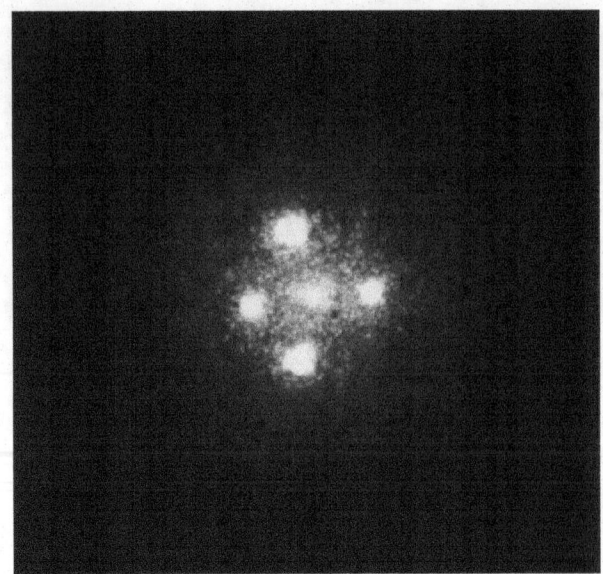

Einstein cross: four images of the same astronomical object, produced by a gravitational lens

"microlensing events" have been observed.[102]

Gravitational lensing has developed into a tool of observational astronomy. It is used to detect the presence and distribution of dark matter, provide a "natural telescope" for observing distant galaxies, and to obtain an independent estimate of the Hubble constant. Statistical evaluations of lensing data provide valuable insight into the structural evolution of galaxies.[103]

4.5.2   Gravitational wave astronomy

Main articles: Gravitational wave and Gravitational wave astronomy
Observations of binary pulsars provide strong indirect evidence for the existence of gravitational waves (see Orbital decay, above). Detection of these waves is a major goal of current relativity-related research.[104] Several land-based gravitational wave detectors are currently in operation, most notably the interferometric detectors GEO 600, LIGO (two detectors), TAMA 300 and VIRGO.[105] Various pulsar timing arrays are using millisecond pulsars to detect gravitational waves in the 10^{-9} to 10^{-6} Hertz frequency range, which originate from binary supermassive blackholes.[106] A European space-based detector, eLISA / NGO, is currently under development,[107] with a precursor mission (LISA Pathfinder) having launched in December 2015.[108]

Observations of gravitational waves promise to complement observations in the electromagnetic spectrum.[109] They are expected to yield information about black holes and other dense objects such as neutron stars and white dwarfs, about

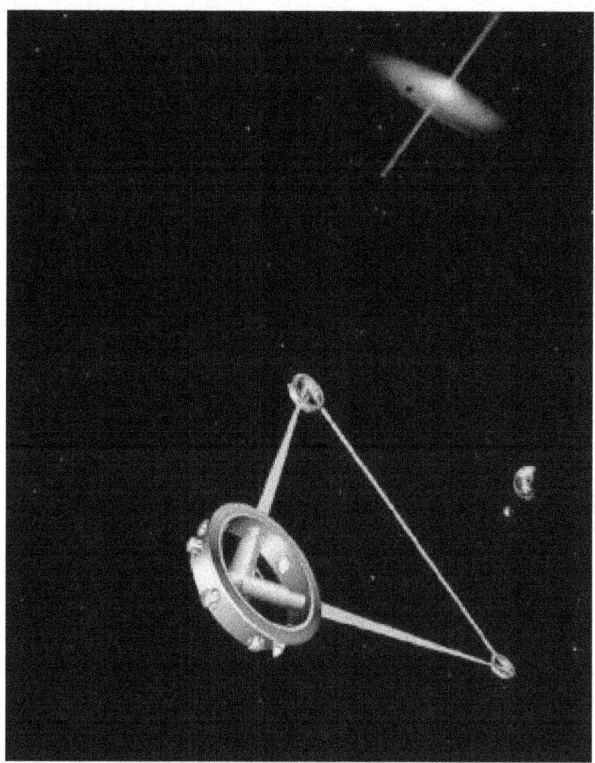

Artist's impression of the space-borne gravitational wave detector LISA

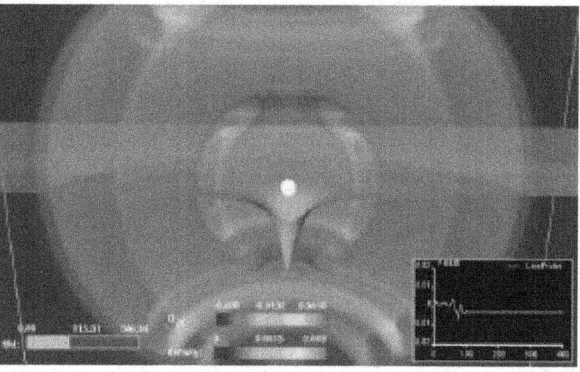

Simulation based on the equations of general relativity: a star collapsing to form a black hole while emitting gravitational waves

certain kinds of supernova implosions, and about processes in the very early universe, including the signature of certain types of hypothetical cosmic string.[110] In February 2016, the Advanced LIGO team announced that they had detected gravitational waves from a black hole merger.[69][70][111]

4.5.3 Black holes and other compact objects

Main article: Black hole

Whenever the ratio of an object's mass to its radius becomes sufficiently large, general relativity predicts the formation of a black hole, a region of space from which nothing, not even light, can escape. In the currently accepted models of stellar evolution, neutron stars of around 1.4 solar masses, and stellar black holes with a few to a few dozen solar masses, are thought to be the final state for the evolution of massive stars.[112] Usually a galaxy has one supermassive black hole with a few million to a few billion solar masses in its center,[113] and its presence is thought to have played an important role in the formation of the galaxy and larger cosmic structures.[114]

Astronomically, the most important property of compact objects is that they provide a supremely efficient mechanism for converting gravitational energy into electromag-

netic radiation.[115] Accretion, the falling of dust or gaseous matter onto stellar or supermassive black holes, is thought to be responsible for some spectacularly luminous astronomical objects, notably diverse kinds of active galactic nuclei on galactic scales and stellar-size objects such as microquasars.[116] In particular, accretion can lead to relativistic jets, focused beams of highly energetic particles that are being flung into space at almost light speed.[117] General relativity plays a central role in modelling all these phenomena,[118] and observations provide strong evidence for the existence of black holes with the properties predicted by the theory.[119]

Black holes are also sought-after targets in the search for gravitational waves (cf. Gravitational waves, above). Merging black hole binaries should lead to some of the strongest gravitational wave signals reaching detectors here on Earth, and the phase directly before the merger ("chirp") could be used as a "standard candle" to deduce the distance to the merger events–and hence serve as a probe of cosmic expansion at large distances.[120] The gravitational waves produced as a stellar black hole plunges into a supermassive one should provide direct information about the supermassive black hole's geometry.[121]

4.5.4 Cosmology

Main article: Physical cosmology

The current models of cosmology are based on Einstein's field equations, which include the cosmological constant Λ since it has important influence on the large-scale dynamics of the cosmos,

$$R_{\mu\nu} - \frac{1}{2} R\, g_{\mu\nu} + \Lambda\, g_{\mu\nu} = \frac{8\pi G}{c^4}\, T_{\mu\nu}$$

where $g_{\mu\nu}$ is the spacetime metric.[122] Isotropic and

This blue horseshoe is a distant galaxy that has been magnified and warped into a nearly complete ring by the strong gravitational pull of the massive foreground luminous red galaxy.

homogeneous solutions of these enhanced equations, the Friedmann–Lemaître–Robertson–Walker solutions,[123] allow physicists to model a universe that has evolved over the past 14 billion years from a hot, early Big Bang phase.[124] Once a small number of parameters (for example the universe's mean matter density) have been fixed by astronomical observation,[125] further observational data can be used to put the models to the test.[126] Predictions, all successful, include the initial abundance of chemical elements formed in a period of primordial nucleosynthesis,[127] the large-scale structure of the universe,[128] and the existence and properties of a "thermal echo" from the early cosmos, the cosmic background radiation.[129]

Astronomical observations of the cosmological expansion rate allow the total amount of matter in the universe to be estimated, although the nature of that matter remains mysterious in part. About 90% of all matter appears to be so-called dark matter, which has mass (or, equivalently, gravitational influence), but does not interact electromagnetically and, hence, cannot be observed directly.[130] There is no generally accepted description of this new kind of matter, within the framework of known particle physics[131] or otherwise.[132] Observational evidence from redshift surveys of distant supernovae and measurements of the cosmic background radiation also show that the evolution of our universe is significantly influenced by a cosmological constant resulting in an acceleration of cosmic expansion or, equivalently, by a form of energy with an unusual equation of state, known as dark energy, the nature of which remains unclear.[133]

A so-called inflationary phase,[134] an additional phase of strongly accelerated expansion at cosmic times of around 10^{-33} seconds, was hypothesized in 1980 to account for

several puzzling observations that were unexplained by classical cosmological models, such as the nearly perfect homogeneity of the cosmic background radiation.[135] Recent measurements of the cosmic background radiation have resulted in the first evidence for this scenario.[136] However, there is a bewildering variety of possible inflationary scenarios, which cannot be restricted by current observations.[137] An even larger question is the physics of the earliest universe, prior to the inflationary phase and close to where the classical models predict the big bang singularity. An authoritative answer would require a complete theory of quantum gravity, which has not yet been developed[138] (cf. the section on quantum gravity, below).

4.5.5   Time travel

Kurt Gödel showed[139] that solutions to Einstein's equations exist that contain closed timelike curves (CTCs), which allow for loops in time. The solutions require extreme physical conditions unlikely ever to occur in practice, and it remains an open question whether further laws of physics will eliminate them completely. Since then other—similarly impractical—GR solutions containing CTCs have been found, such as the Tipler cylinder and traversable wormholes.

4.6   Advanced concepts

4.6.1   Causal structure and global geometry

Main article: Causal structure
In general relativity, no material body can catch up with

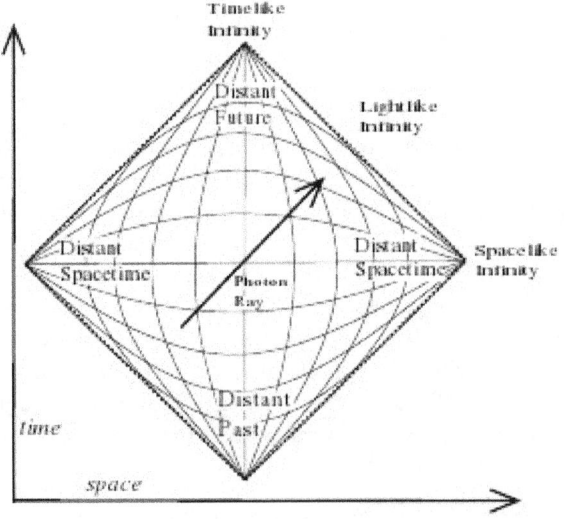

Penrose–Carter diagram of an infinite Minkowski universe

or overtake a light pulse. No influence from an event A can reach any other location X before light sent out at A to X. In consequence, an exploration of all light worldlines (null geodesics) yields key information about the spacetime's causal structure. This structure can be displayed using Penrose–Carter diagrams in which infinitely large regions of space and infinite time intervals are shrunk ("compactified") so as to fit onto a finite map, while light still travels along diagonals as in standard spacetime diagrams.[140]

Aware of the importance of causal structure, Roger Penrose and others developed what is known as global geometry. In global geometry, the object of study is not one particular solution (or family of solutions) to Einstein's equations. Rather, relations that hold true for all geodesics, such as the Raychaudhuri equation, and additional non-specific assumptions about the nature of matter (usually in the form of so-called energy conditions) are used to derive general results.[141]

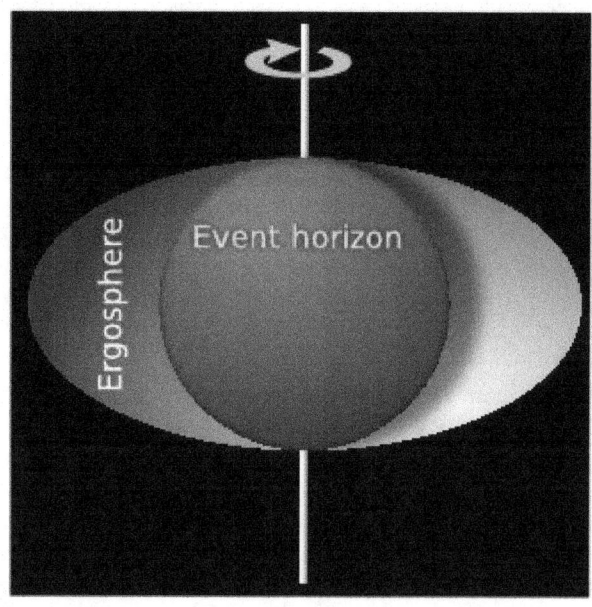

The ergosphere of a rotating black hole, which plays a key role when it comes to extracting energy from such a black hole

4.6.2 Horizons

Main articles: Horizon (general relativity), No hair theorem and Black hole mechanics

Using global geometry, some spacetimes can be shown to contain boundaries called horizons, which demarcate one region from the rest of spacetime. The best-known examples are black holes: if mass is compressed into a sufficiently compact region of space (as specified in the hoop conjecture, the relevant length scale is the Schwarzschild radius[142]), no light from inside can escape to the outside. Since no object can overtake a light pulse, all interior matter is imprisoned as well. Passage from the exterior to the interior is still possible, showing that the boundary, the black hole's *horizon*, is not a physical barrier.[143]

Early studies of black holes relied on explicit solutions of Einstein's equations, notably the spherically symmetric Schwarzschild solution (used to describe a static black hole) and the axisymmetric Kerr solution (used to describe a rotating, stationary black hole, and introducing interesting features such as the ergosphere). Using global geometry, later studies have revealed more general properties of black holes. In the long run, they are rather simple objects characterized by eleven parameters specifying energy, linear momentum, angular momentum, location at a specified time and electric charge. This is stated by the black hole uniqueness theorems: "black holes have no hair", that is, no distinguishing marks like the hairstyles of humans. Irrespective of the complexity of a gravitating object collapsing to form a black hole, the object that results (having emitted gravitational waves) is very simple.[144]

Even more remarkably, there is a general set of laws known as black hole mechanics, which is analogous to the laws of thermodynamics. For instance, by the second law of black hole mechanics, the area of the event horizon of a general black hole will never decrease with time, analogous to the entropy of a thermodynamic system. This limits the energy that can be extracted by classical means from a rotating black hole (e.g. by the Penrose process).[145] There is strong evidence that the laws of black hole mechanics are, in fact, a subset of the laws of thermodynamics, and that the black hole area is proportional to its entropy.[146] This leads to a modification of the original laws of black hole mechanics: for instance, as the second law of black hole mechanics becomes part of the second law of thermodynamics, it is possible for black hole area to decrease—as long as other processes ensure that, overall, entropy increases. As thermodynamical objects with non-zero temperature, black holes should emit thermal radiation. Semi-classical calculations indicate that indeed they do, with the surface gravity playing the role of temperature in Planck's law. This radiation is known as Hawking radiation (cf. the quantum theory section, below).[147]

There are other types of horizons. In an expanding universe, an observer may find that some regions of the past cannot be observed ("particle horizon"), and some regions of the future cannot be influenced (event horizon).[148] Even in flat Minkowski space, when described by an accelerated observer (Rindler space), there will be horizons associated with a semi-classical radiation known as Unruh radiation.[149]

4.6.3   Singularities

Main article: Spacetime singularity

Another general feature of general relativity is the appearance of spacetime boundaries known as singularities. Spacetime can be explored by following up on timelike and lightlike geodesics—all possible ways that light and particles in free fall can travel. But some solutions of Einstein's equations have "ragged edges"—regions known as spacetime singularities, where the paths of light and falling particles come to an abrupt end, and geometry becomes ill-defined. In the more interesting cases, these are "curvature singularities", where geometrical quantities characterizing spacetime curvature, such as the Ricci scalar, take on infinite values.[150] Well-known examples of spacetimes with future singularities—where worldlines end—are the Schwarzschild solution, which describes a singularity inside an eternal static black hole,[151] or the Kerr solution with its ring-shaped singularity inside an eternal rotating black hole.[152] The Friedmann–Lemaître–Robertson–Walker solutions and other spacetimes describing universes have past singularities on which worldlines begin, namely Big Bang singularities, and some have future singularities (Big Crunch) as well.[153]

Given that these examples are all highly symmetric—and thus simplified—it is tempting to conclude that the occurrence of singularities is an artifact of idealization.[154] The famous singularity theorems, proved using the methods of global geometry, say otherwise: singularities are a generic feature of general relativity, and unavoidable once the collapse of an object with realistic matter properties has proceeded beyond a certain stage[155] and also at the beginning of a wide class of expanding universes.[156] However, the theorems say little about the properties of singularities, and much of current research is devoted to characterizing these entities' generic structure (hypothesized e.g. by the so-called BKL conjecture).[157] The cosmic censorship hypothesis states that all realistic future singularities (no perfect symmetries, matter with realistic properties) are safely hidden away behind a horizon, and thus invisible to all distant observers. While no formal proof yet exists, numerical simulations offer supporting evidence of its validity.[158]

4.6.4   Evolution equations

Main article: Initial value formulation (general relativity)

Each solution of Einstein's equation encompasses the whole history of a universe — it is not just some snapshot of how things are, but a whole, possibly matter-filled, spacetime. It describes the state of matter and geometry everywhere and at every moment in that particular universe. Due to its general covariance, Einstein's theory is not sufficient by itself to determine the time evolution of the metric tensor. It must be combined with a coordinate condition, which is analogous to gauge fixing in other field theories.[159]

To understand Einstein's equations as partial differential equations, it is helpful to formulate them in a way that describes the evolution of the universe over time. This is done in so-called "3+1" formulations, where spacetime is split into three space dimensions and one time dimension. The best-known example is the ADM formalism.[160] These decompositions show that the spacetime evolution equations of general relativity are well-behaved: solutions always exist, and are uniquely defined, once suitable initial conditions have been specified.[161] Such formulations of Einstein's field equations are the basis of numerical relativity.[162]

4.6.5   Global and quasi-local quantities

Main article: Mass in general relativity

The notion of evolution equations is intimately tied in with another aspect of general relativistic physics. In Einstein's theory, it turns out to be impossible to find a general definition for a seemingly simple property such as a system's total mass (or energy). The main reason is that the gravitational field—like any physical field—must be ascribed a certain energy, but that it proves to be fundamentally impossible to localize that energy.[163]

Nevertheless, there are possibilities to define a system's total mass, either using a hypothetical "infinitely distant observer" (ADM mass)[164] or suitable symmetries (Komar mass).[165] If one excludes from the system's total mass the energy being carried away to infinity by gravitational waves, the result is the so-called Bondi mass at null infinity.[166] Just as in classical physics, it can be shown that these masses are positive.[167] Corresponding global definitions exist for momentum and angular momentum.[168] There have also been a number of attempts to define *quasi-local* quantities, such as the mass of an isolated system formulated using only quantities defined within a finite region of space containing that system. The hope is to obtain a quantity useful for general statements about isolated systems, such as a more precise formulation of the hoop conjecture.[169]

4.7 Relationship with quantum theory

If general relativity were considered to be one of the two pillars of modern physics, then quantum theory, the basis of understanding matter from elementary particles to solid state physics, would be the other.[170] However, how to reconcile quantum theory with general relativity is still an open question.

4.7.1 Quantum field theory in curved spacetime

Main article: Quantum field theory in curved spacetime

Ordinary quantum field theories, which form the basis of modern elementary particle physics, are defined in flat Minkowski space, which is an excellent approximation when it comes to describing the behavior of microscopic particles in weak gravitational fields like those found on Earth.[171] In order to describe situations in which gravity is strong enough to influence (quantum) matter, yet not strong enough to require quantization itself, physicists have formulated quantum field theories in curved spacetime. These theories rely on general relativity to describe a curved background spacetime, and define a generalized quantum field theory to describe the behavior of quantum matter within that spacetime.[172] Using this formalism, it can be shown that black holes emit a blackbody spectrum of particles known as Hawking radiation, leading to the possibility that they evaporate over time.[173] As briefly mentioned above, this radiation plays an important role for the thermodynamics of black holes.[174]

4.7.2 Quantum gravity

Main article: Quantum gravity
See also: String theory, Canonical general relativity, Loop quantum gravity, Causal Dynamical Triangulations, and Causal sets

The demand for consistency between a quantum description of matter and a geometric description of spacetime,[175] as well as the appearance of singularities (where curvature length scales become microscopic), indicate the need for a full theory of quantum gravity: for an adequate description of the interior of black holes, and of the very early universe, a theory is required in which gravity and the associated geometry of spacetime are described in the language of quantum physics.[176] Despite major efforts, no complete and consistent theory of quantum gravity is cur-

rently known, even though a number of promising candidates exist.[177][178]

Projection of a Calabi–Yau manifold, one of the ways of compactifying the extra dimensions posited by string theory

Attempts to generalize ordinary quantum field theories, used in elementary particle physics to describe fundamental interactions, so as to include gravity have led to serious problems.[179] Some have argued that at low energies, this approach proves successful, in that it results in an acceptable effective (quantum) field theory of gravity.[180] At very high energies, however, the perturbative results are badly divergent and lead to models devoid of predictive power ("perturbative non-renormalizability").[181]

One attempt to overcome these limitations is string theory, a quantum theory not of point particles, but of minute one-dimensional extended objects.[182] The theory promises to be a unified description of all particles and interactions, including gravity;[183] the price to pay is unusual features such as six extra dimensions of space in addition to the usual three.[184] In what is called the second superstring revolution, it was conjectured that both string theory and a unification of general relativity and supersymmetry known as supergravity[185] form part of a hypothesized eleven-dimensional model known as M-theory, which would constitute a uniquely defined and consistent theory of quantum gravity.[186]

Another approach starts with the canonical quantization procedures of quantum theory. Using the initial-value-formulation of general relativity (cf. evolution equations above), the result is the Wheeler–deWitt equation (an analogue of the Schrödinger equation) which, regrettably, turns out to be ill-defined without a proper ultraviolet (lattice)

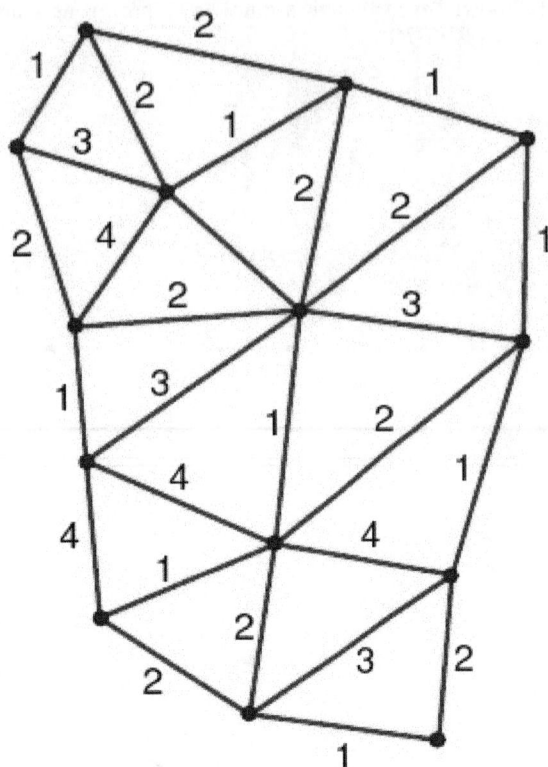

Simple spin network of the type used in loop quantum gravity

4.8 Current status

General relativity has emerged as a highly successful model of gravitation and cosmology, which has so far passed many unambiguous observational and experimental tests. However, there are strong indications the theory is incomplete.[196] The problem of quantum gravity and the question of the reality of spacetime singularities remain open.[197] Observational data that is taken as evidence for dark energy and dark matter could indicate the need for new physics.[198] Even taken as is, general relativity is rich with possibilities for further exploration. Mathematical relativists seek to understand the nature of singularities and the fundamental properties of Einstein's equations,[199] and increasingly powerful computer simulations (such as those describing merging black holes) are run.[200] In February 2016, it was announced that the existence of gravitational waves was directly detected by the Advanced LIGO team on September 14, 2015.[71][201][202] A century after its publication, general relativity remains a highly active area of research.[203]

4.9 See also

- Alcubierre drive (warp drive)
- Center of mass (relativistic)
- Contributors to general relativity
- Derivations of the Lorentz transformations
- Ehrenfest paradox
- Einstein–Hilbert action
- Introduction to mathematics of general relativity
- Relativity priority dispute
- Ricci calculus
- Tests of general relativity
- Timeline of gravitational physics and relativity
- Two-body problem in general relativity
- Weak Gravity Conjecture

cutoff.[187] However, with the introduction of what are now known as Ashtekar variables,[188] this leads to a promising model known as loop quantum gravity. Space is represented by a web-like structure called a spin network, evolving over time in discrete steps.[189]

Depending on which features of general relativity and quantum theory are accepted unchanged, and on what level changes are introduced,[190] there are numerous other attempts to arrive at a viable theory of quantum gravity, some examples being the lattice theory of gravity based on the Feynman Path Integral approach and Regge Calculus,[177] dynamical triangulations,[191] causal sets,[192] twistor models[193] or the path-integral based models of quantum cosmology.[194]

All candidate theories still have major formal and conceptual problems to overcome. They also face the common problem that, as yet, there is no way to put quantum gravity predictions to experimental tests (and thus to decide between the candidates where their predictions vary), although there is hope for this to change as future data from cosmological observations and particle physics experiments becomes available.[195]

4.10 Notes

[1] "GW150914: LIGO Detects Gravitational Waves". *Blackholes.org*. Retrieved 18 April 2016.

[2] O'Connor, J.J. and Robertson, E.F. (1996), *General relativity. Mathematical Physics index*, School of Mathematics and Statistics, University of St. Andrews, Scotland. Retrieved 2015-02-04.

[3] Pais 1982, ch. 9 to 15, Janssen 2005; an up-to-date collection of current research, including reprints of many of the original articles, is Renn 2007; an accessible overview can be found in Renn 2005, pp. 110ff. Einstein's original papers are found in Digital Einstein, volumes 4 and 6. An early key article is Einstein 1907, cf. Pais 1982, ch. 9. The publication featuring the field equations is Einstein 1915, cf. Pais 1982, ch. 11–15

[4] Schwarzschild 1916a, Schwarzschild 1916b and Reissner 1916 (later complemented in Nordström 1918)

[5] Einstein 1917, cf. Pais 1982, ch. 15e

[6] Hubble's original article is Hubble 1929; an accessible overview is given in Singh 2004, ch. 2–4

[7] As reported in Gamow 1970. Einstein's condemnation would prove to be premature, cf. the section Cosmology, below

[8] Pais 1982, pp. 253–254

[9] Kennefick 2005, Kennefick 2007

[10] Pais 1982, ch. 16

[11] Thorne, Kip (2003). "Warping spacetime". *The future of theoretical physics and cosmology: celebrating Stephen Hawking's 60th birthday*. Cambridge University Press. p. 74. ISBN 0-521-82081-2. Extract of page 74

[12] Israel 1987, ch. 7.8–7.10, Thorne 1994, ch. 3–9

[13] Sections Orbital effects and the relativity of direction, Gravitational time dilation and frequency shift and Light deflection and gravitational time delay, and references therein

[14] Section Cosmology and references therein; the historical development is in Overbye 1999

[15] The following exposition re-traces that of Ehlers 1973, sec. 1

[16] Arnold 1989, ch. 1

[17] Ehlers 1973, pp. 5f

[18] Will 1993, sec. 2.4, Will 2006, sec. 2

[19] Wheeler 1990, ch. 2

[20] Ehlers 1973, sec. 1.2, Havas 1964, Künzle 1972. The simple thought experiment in question was first described in Heckmann & Schücking 1959

[21] Ehlers 1973, pp. 10f

[22] Good introductions are, in order of increasing presupposed knowledge of mathematics, Giulini 2005, Mermin 2005, and Rindler 1991; for accounts of precision experiments, cf. part IV of Ehlers & Lämmerzahl 2006

[23] An in-depth comparison between the two symmetry groups can be found in Giulini 2006a

[24] Rindler 1991, sec. 22, Synge 1972, ch. 1 and 2

[25] Ehlers 1973, sec. 2.3

[26] Ehlers 1973, sec. 1.4, Schutz 1985, sec. 5.1

[27] Ehlers 1973, pp. 17ff; a derivation can be found in Mermin 2005, ch. 12. For the experimental evidence, cf. the section Gravitational time dilation and frequency shift, below

[28] Rindler 2001, sec. 1.13; for an elementary account, see Wheeler 1990, ch. 2; there are, however, some differences between the modern version and Einstein's original concept used in the historical derivation of general relativity, cf. Norton 1985

[29] Ehlers 1973, sec. 1.4 for the experimental evidence, see once more section Gravitational time dilation and frequency shift. Choosing a different connection with non-zero torsion leads to a modified theory known as Einstein–Cartan theory

[30] Ehlers 1973, p. 16, Kenyon 1990, sec. 7.2, Weinberg 1972, sec. 2.8

[31] Ehlers 1973, pp. 19–22; for similar derivations, see sections 1 and 2 of ch. 7 in Weinberg 1972. The Einstein tensor is the only divergence-free tensor that is a function of the metric coefficients, their first and second derivatives at most, and allows the spacetime of special relativity as a solution in the absence of sources of gravity, cf. Lovelock 1972. The tensors on both side are of second rank, that is, they can each be thought of as 4×4 matrices, each of which contains ten independent terms; hence, the above represents ten coupled equations. The fact that, as a consequence of geometric relations known as Bianchi identities, the Einstein tensor satisfies a further four identities reduces these to six independent equations, e.g. Schutz 1985, sec. 8.3

[32] Kenyon 1990, sec. 7.4

[33] Brans & Dicke 1961, Weinberg 1972, sec. 3 in ch. 7, Goenner 2004, sec. 7.2, and Trautman 2006, respectively

[34] Wald 1984, ch. 4, Weinberg 1972, ch. 7 or, in fact, any other textbook on general relativity

[35] At least approximately, cf. Poisson 2004

[36] Wheeler 1990, p. xi

[37] Wald 1984, sec. 4.4

[38] Wald 1984, sec. 4.1

[39] For the (conceptual and historical) difficulties in defining a general principle of relativity and separating it from the notion of general covariance, see Giulini 2006b

[40] section 5 in ch. 12 of Weinberg 1972

[41] Introductory chapters of Stephani et al. 2003

[42] A review showing Einstein's equation in the broader context of other PDEs with physical significance is Geroch 1996

[43] For background information and a list of solutions, cf. Stephani et al. 2003; a more recent review can be found in MacCallum 2006

[44] Chandrasekhar 1983, ch. 3,5,6

[45] Narlikar 1993, ch. 4, sec. 3.3

[46] Brief descriptions of these and further interesting solutions can be found in Hawking & Ellis 1973, ch. 5

[47] Lehner 2002

[48] For instance Wald 1984, sec. 4.4

[49] Will 1993, sec. 4.1 and 4.2

[50] Will 2006, sec. 3.2, Will 1993, ch. 4

[51] Rindler 2001, pp. 24–26 vs. pp. 236–237 and Ohanian & Ruffini 1994, pp. 164–172. Einstein derived these effects using the equivalence principle as early as 1907, cf. Einstein 1907 and the description in Pais 1982, pp. 196–198

[52] Rindler 2001, pp. 24–26; Misner, Thorne & Wheeler 1973, § 38.5

[53] Pound–Rebka experiment, see Pound & Rebka 1959, Pound & Rebka 1960; Pound & Snider 1964; a list of further experiments is given in Ohanian & Ruffini 1994, table 4.1 on p. 186

[54] Greenstein, Oke & Shipman 1971; the most recent and most accurate Sirius B measurements are published in Barstow, Bond et al. 2005.

[55] Starting with the Hafele–Keating experiment, Hafele & Keating 1972a and Hafele & Keating 1972b, and culminating in the Gravity Probe A experiment; an overview of experiments can be found in Ohanian & Ruffini 1994, table 4.1 on p. 186

[56] GPS is continually tested by comparing atomic clocks on the ground and aboard orbiting satellites; for an account of relativistic effects, see Ashby 2002 and Ashby 2003

[57] Stairs 2003 and Kramer 2004

[58] General overviews can be found in section 2.1. of Will 2006; Will 2003, pp. 32–36; Ohanian & Ruffini 1994, sec. 4.2

[59] Ohanian & Ruffini 1994, pp. 164–172

[60] Cf. Kennefick 2005 for the classic early measurements by the Eddington expeditions; for an overview of more recent measurements, see Ohanian & Ruffini 1994, ch. 4.3. For the most precise direct modern observations using quasars, cf. Shapiro et al. 2004

[61] This is not an independent axiom; it can be derived from Einstein's equations and the Maxwell Lagrangian using a WKB approximation, cf. Ehlers 1973, sec. 5

[62] Blanchet 2006, sec. 1.3

[63] Rindler 2001, sec. 1.16; for the historical examples, Israel 1987, pp. 202–204; in fact, Einstein published one such derivation as Einstein 1907. Such calculations tacitly assume that the geometry of space is Euclidean, cf. Ehlers & Rindler 1997

[64] From the standpoint of Einstein's theory, these derivations take into account the effect of gravity on time, but not its consequences for the warping of space, cf. Rindler 2001, sec. 11.11

[65] For the Sun's gravitational field using radar signals reflected from planets such as Venus and Mercury, cf. Shapiro 1964, Weinberg 1972, ch. 8, sec. 7; for signals actively sent back by space probes (transponder measurements), cf. Bertotti, Iess & Tortora 2003; for an overview, see Ohanian & Ruffini 1994, table 4.4 on p. 200; for more recent measurements using signals received from a pulsar that is part of a binary system, the gravitational field causing the time delay being that of the other pulsar, cf. Stairs 2003, sec. 4.4

[66] Will 1993, sec. 7.1 and 7.2

[67] Einstein, A (June 1916). "Näherungsweise Integration der Feldgleichungen der Gravitation". *Sitzungsberichte der Königlich Preussischen Akademie der Wissenschaften Berlin.* part 1: 688–696.

[68] Einstein, A (1918). "Über Gravitationswellen". *Sitzungsberichte der Königlich Preussischen Akademie der Wissenschaften Berlin.* part 1: 154–167.

[69] Castelvecchi, Davide; Witze, Witze (February 11, 2016). "Einstein's gravitational waves found at last". *Nature News.* doi:10.1038/nature.2016.19361. Retrieved 2016-02-11.

[70] B. P. Abbott et al. (LIGO Scientific Collaboration and Virgo Collaboration) (2016). "Observation of Gravitational Waves from a Binary Black Hole Merger". *Physical Review Letters* **116** (6). doi:10.1103/PhysRevLett.116.061102.

[71] "Gravitational waves detected 100 years after Einstein's prediction I NSF - National Science Foundation". *www.nsf.gov.* Retrieved 2016-02-11.

[72] Most advanced textbooks on general relativity contain a description of these properties, e.g. Schutz 1985, ch. 9

[73] For example Jaranowski & Królak 2005

[74] Rindler 2001, ch. 13

[75] Gowdy 1971, Gowdy 1974

[76] See Lehner 2002 for a brief introduction to the methods of numerical relativity, and Seidel 1998 for the connection with gravitational wave astronomy

[77] Schutz 2003, pp. 48–49, Pais 1982, pp. 253–254

[78] Rindler 2001, sec. 11.9

[79] Will 1993, pp. 177–181

[80] In consequence, in the parameterized post-Newtonian formalism (PPN), measurements of this effect determine a linear combination of the terms β and γ, cf. Will 2006, sec. 3.5 and Will 1993, sec. 7.3

[81] The most precise measurements are VLBI measurements of planetary positions; see Will 1993, ch. 5, Will 2006, sec. 3.5, Anderson et al. 1992; for an overview, Ohanian & Ruffini 1994, pp. 406–407

[82] Kramer et al. 2006

[83] Dediu, Adrian-Horia; Magdalena, Luis; Martín-Vide, Carlos (2015). *Theory and Practice of Natural Computing: Fourth International Conference, TPNC 2015, Mieres, Spain, December 15-16, 2015. Proceedings* (illustrated ed.). Springer. p. 141. ISBN 978-3-319-26841-5. Extract of page 141

[84] A figure that includes error bars is fig. 7 in Will 2006, sec. 5.1

[85] Stairs 2003, Schutz 2003, pp. 317–321, Bartusiak 2000, pp. 70–86

[86] Weisberg & Taylor 2003; for the pulsar discovery, see Hulse & Taylor 1975; for the initial evidence for gravitational radiation, see Taylor 1994

[87] Kramer 2004

[88] Penrose 2004, §14.5, Misner, Thorne & Wheeler 1973, §11.4

[89] Weinberg 1972, sec. 9.6, Ohanian & Ruffini 1994, sec. 7.8

[90] Bertotti, Ciufolini & Bender 1987, Nordtvedt 2003

[91] Kahn 2007

[92] A mission description can be found in Everitt et al. 2001; a first post-flight evaluation is given in Everitt, Parkinson & Kahn 2007; further updates will be available on the mission website Kahn 1996–2012.

[93] Townsend 1997, sec. 4.2.1, Ohanian & Ruffini 1994, pp. 469–471

[94] Ohanian & Ruffini 1994, sec. 4.7, Weinberg 1972, sec. 9.7; for a more recent review, see Schäfer 2004

[95] Ciufolini & Pavlis 2004, Ciufolini, Pavlis & Peron 2006, Iorio 2009

[96] Iorio L. (August 2006), "COMMENTS, REPLIES AND NOTES: A note on the evidence of the gravitomagnetic field of Mars", *Classical Quantum Gravity* 23 (17): 5451–5454, arXiv:gr-qc/0606092, Bibcode:2006CQGra..23.5451I, doi:10.1088/0264-9381/23/17/N01

[97] Iorio L. (June 2010), "On the Lense–Thirring test with the Mars Global Surveyor in the gravitational field of Mars", *Central European Journal of Physics* 8 (3): 509–513, arXiv:gr-qc/0701146, Bibcode:2010CEJPh...8..509I, doi:10.2478/s11534-009-0117-6

[98] For overviews of gravitational lensing and its applications, see Ehlers, Falco & Schneider 1992 and Wambsganss 1998

[99] For a simple derivation, see Schutz 2003, ch. 23; cf. Narayan & Bartelmann 1997, sec. 3

[100] Walsh, Carswell & Weymann 1979

[101] Images of all the known lenses can be found on the pages of the CASTLES project, Kochanek et al. 2007

[102] Roulet & Mollerach 1997

[103] Narayan & Bartelmann 1997, sec. 3.7

[104] Barish 2005, Bartusiak 2000, Blair & McNamara 1997

[105] Hough & Rowan 2000

[106] Hobbs, George; Archibald, A.; Arzoumanian, Z.; Backer, D.; Bailes, M.; Bhat, N. D. R.; Burgay, M.; Burke-Spolaor, S.; et al. (2010), "The international pulsar timing array project: using pulsars as a gravitational wave detector", *Classical and Quantum Gravity* 27 (8): 084013, arXiv:0911.5206, Bibcode:2010CQGra..27h4013H, doi:10.1088/0264-9381/27/8/084013

[107] Danzmann & Rüdiger 2003

[108] "LISA pathfinder overview". ESA. Retrieved 2012-04-23.

[109] Thorne 1995

[110] Cutler & Thorne 2002

[111] "Gravitational waves detected 100 years after Einstein's prediction | NSF - National Science Foundation". *www.nsf.gov*. Retrieved 2016-02-11.

[112] Miller 2002, lectures 19 and 21

[113] Celotti, Miller & Sciama 1999, sec. 3

[114] Springel et al. 2005 and the accompanying summary Gnedin 2005

[115] Blandford 1987, sec. 8.2.4

[116] For the basic mechanism, see Carroll & Ostlie 1996, sec. 17.2; for more about the different types of astronomical objects associated with this, cf. Robson 1996

[117] For a review, see Begelman, Blandford & Rees 1984. To a distant observer, some of these jets even appear to move faster than light; this, however, can be explained as an optical illusion that does not violate the tenets of relativity, see Rees 1966

[118] For stellar end states, cf. Oppenheimer & Snyder 1939 or, for more recent numerical work, Font 2003, sec. 4.1; for supernovae, there are still major problems to be solved, cf. Buras et al. 2003; for simulating accretion and the formation of jets, cf. Font 2003, sec. 4.2. Also, relativistic lensing effects are thought to play a role for the signals received from X-ray pulsars, cf. Kraus 1998

[119] The evidence includes limits on compactness from the observation of accretion-driven phenomena ("Eddington luminosity"), see Celotti, Miller & Sciama 1999, observations of stellar dynamics in the center of our own Milky Way galaxy, cf. Schödel et al. 2003, and indications that at least some of the compact objects in question appear to have no solid surface, which can be deduced from the examination of X-ray bursts for which the central compact object is either a neutron star or a black hole; cf. Remillard et al. 2006 for an overview, Narayan 2006, sec. 5. Observations of the "shadow" of the Milky Way galaxy's central black hole horizon are eagerly sought for, cf. Falcke, Melia & Agol 2000

[120] Dalal et al. 2006

[121] Barack & Cutler 2004

[122] Originally Einstein 1917; cf. Pais 1982, pp. 285–288

[123] Carroll 2001, ch. 2

[124] Bergström & Goobar 2003, ch. 9–11; use of these models is justified by the fact that, at large scales of around hundred million light-years and more, our own universe indeed appears to be isotropic and homogeneous, cf. Peebles et al. 1991

[125] E.g. with WMAP data, see Spergel et al. 2003

[126] These tests involve the separate observations detailed further on, see, e.g., fig. 2 in Bridle et al. 2003

[127] Peebles 1966; for a recent account of predictions, see Coc, Vangioni-Flam et al. 2004; an accessible account can be found in Weiss 2006; compare with the observations in Olive & Skillman 2004, Bania, Rood & Balser 2002, O'Meara et al. 2001, and Charbonnel & Primas 2005

[128] Lahav & Suto 2004, Bertschinger 1998, Springel et al. 2005

[129] Alpher & Herman 1948, for a pedagogical introduction, see Bergström & Goobar 2003, ch. 11; for the initial detection, see Penzias & Wilson 1965 and, for precision measurements by satellite observatories, Mather et al. 1994 (COBE) and Bennett et al. 2003 (WMAP). Future measurements could also reveal evidence about gravitational waves in the early universe; this additional information is contained in the background radiation's polarization, cf. Kamionkowski, Kosowsky & Stebbins 1997 and Seljak & Zaldarriaga 1997

[130] Evidence for this comes from the determination of cosmological parameters and additional observations involving the dynamics of galaxies and galaxy clusters cf. Peebles 1993, ch. 18, evidence from gravitational lensing, cf. Peacock 1999, sec. 4.6, and simulations of large-scale structure formation, see Springel et al. 2005

[131] Peacock 1999, ch. 12, Peskin 2007; in particular, observations indicate that all but a negligible portion of that matter is not in the form of the usual elementary particles ("non-baryonic matter"), cf. Peacock 1999, ch. 12

[132] Namely, some physicists have questioned whether or not the evidence for dark matter is, in fact, evidence for deviations from the Einsteinian (and the Newtonian) description of gravity cf. the overview in Mannheim 2006, sec. 9

[133] Carroll 2001; an accessible overview is given in Caldwell 2004. Here, too, scientists have argued that the evidence indicates not a new form of energy, but the need for modifications in our cosmological models, cf. Mannheim 2006, sec. 10; aforementioned modifications need not be modifications of general relativity, they could, for example, be modifications in the way we treat the inhomogeneities in the universe, cf. Buchert 2007

[134] A good introduction is Linde 1990; for a more recent review, see Linde 2005

[135] More precisely, these are the flatness problem, the horizon problem, and the monopole problem; a pedagogical introduction can be found in Narlikar 1993, sec. 6.4, see also Börner 1993, sec. 9.1

[136] Spergel et al. 2007, sec. 5,6

[137] More concretely, the potential function that is crucial to determining the dynamics of the inflaton is simply postulated, but not derived from an underlying physical theory

[138] Brandenberger 2007, sec. 2

[139] Gödel 1949

[140] Frauendiener 2004, Wald 1984, sec. 11.1, Hawking & Ellis 1973, sec. 6.8, 6.9

[141] Wald 1984, sec. 9.2–9.4 and Hawking & Ellis 1973, ch. 6

[142] Thorne 1972; for more recent numerical studies, see Berger 2002, sec. 2.1

[143] Israel 1987. A more exact mathematical description distinguishes several kinds of horizon, notably event horizons and apparent horizons cf. Hawking & Ellis 1973, pp. 312–320 or Wald 1984, sec. 12.2; there are also more intuitive definitions for isolated systems that do not require knowledge of spacetime properties at infinity, cf. Ashtekar & Krishnan 2004

[144] For first steps, cf. Israel 1971; see Hawking & Ellis 1973, sec. 9.3 or Heusler 1996, ch. 9 and 10 for a derivation, and Heusler 1998 as well as Beig & Chruściel 2006 as overviews of more recent results

[145] The laws of black hole mechanics were first described in Bardeen, Carter & Hawking 1973; a more pedagogical presentation can be found in Carter 1979; for a more recent review, see Wald 2001, ch. 2. A thorough, book-length introduction including an introduction to the necessary mathematics Poisson 2004. For the Penrose process, see Penrose 1969

[146] Bekenstein 1973, Bekenstein 1974

[147] The fact that black holes radiate, quantum mechanically, was first derived in Hawking 1975; a more thorough derivation can be found in Wald 1975. A review is given in Wald 2001, ch. 3

[148] Narlikar 1993, sec. 4.4.4, 4.4.5

[149] Horizons: cf. Rindler 2001, sec. 12.4. Unruh effect: Unruh 1976, cf. Wald 2001, ch. 3

[150] Hawking & Ellis 1973, sec. 8.1, Wald 1984, sec. 9.1

[151] Townsend 1997, ch. 2; a more extensive treatment of this solution can be found in Chandrasekhar 1983, ch. 3

[152] Townsend 1997, ch. 4; for a more extensive treatment, cf. Chandrasekhar 1983, ch. 6

[153] Ellis & Van Elst 1999; a closer look at the singularity itself is taken in Börner 1993, sec. 1.2

[154] Here one should remind to the well-known fact that the important "quasi-optical" singularities of the so-called eikonal approximations of many wave-equations, namely the "caustics", are resolved into finite peaks beyond that approximation.

[155] Namely when there are trapped null surfaces, cf. Penrose 1965

[156] Hawking 1966

[157] The conjecture was made in Belinskii, Khalatnikov & Lifschitz 1971; for a more recent review, see Berger 2002. An accessible exposition is given by Garfinkle 2007

[158] The restriction to future singularities naturally excludes initial singularities such as the big bang singularity, which in principle be visible to observers at later cosmic time. The cosmic censorship conjecture was first presented in Penrose 1969; a textbook-level account is given in Wald 1984, pp. 302–305. For numerical results, see the review Berger 2002, sec. 2.1

[159] Hawking & Ellis 1973, sec. 7.1

[160] Arnowitt, Deser & Misner 1962; for a pedagogical introduction, see Misner, Thorne & Wheeler 1973, §21.4–§21.7

[161] Fourès-Bruhat 1952 and Bruhat 1962; for a pedagogical introduction, see Wald 1984, ch. 10; an online review can be found in Reula 1998

[162] Gourgoulhon 2007; for a review of the basics of numerical relativity, including the problems arising from the peculiarities of Einstein's equations, see Lehner 2001

[163] Misner, Thorne & Wheeler 1973, §20.4

[164] Arnowitt, Deser & Misner 1962

[165] Komar 1959; for a pedagogical introduction, see Wald 1984, sec. 11.2; although defined in a totally different way, it can be shown to be equivalent to the ADM mass for stationary spacetimes, cf. Ashtekar & Magnon-Ashtekar 1979

[166] For a pedagogical introduction, see Wald 1984, sec. 11.2

[167] Wald 1984, p. 295 and refs therein; this is important for questions of stability—if there were negative mass states, then flat, empty Minkowski space, which has mass zero, could evolve into these states

[168] Townsend 1997, ch. 5

[169] Such quasi-local mass–energy definitions are the Hawking energy, Geroch energy, or Penrose's quasi-local energy-momentum based on twistor methods; cf. the review article Szabados 2004

[170] An overview of quantum theory can be found in standard textbooks such as Messiah 1999; a more elementary account is given in Hey & Walters 2003

[171] Ramond 1990, Weinberg 1995, Peskin & Schroeder 1995; a more accessible overview is Auyang 1995

[172] Wald 1994, Birrell & Davies 1984

[173] For Hawking radiation Hawking 1975, Wald 1975; an accessible introduction to black hole evaporation can be found in Traschen 2000

[174] Wald 2001, ch. 3

[175] Put simply, matter is the source of spacetime curvature, and once matter has quantum properties, we can expect spacetime to have them as well. Cf. Carlip 2001, sec. 2

[176] Schutz 2003, p. 407

[177] Hamber 2009

[178] A timeline and overview can be found in Rovelli 2000

[179] t'Hooft 1974

[180] Donoghue 1995

[181] In particular, a perturbative technique known as renormalization, an integral part of deriving predictions which take into account higher-energy contributions, cf. Weinberg 1996, ch. 17, 18, fails in this case; cf. Veltman 1975, Goroff & Sagnotti 1985; for a recent comprehensive review of the failure of perturbative renormalizability for quantum gravity see Hamber 2009

[182] An accessible introduction at the undergraduate level can be found in Zwiebach 2004; more complete overviews can be found in Polchinski 1998a and Polchinski 1998b

[183] At the energies reached in current experiments, these strings are indistinguishable from point-like particles, but, crucially, different modes of oscillation of one and the same type of fundamental string appear as particles with different (electric and other) charges, e.g. Ibanez 2000. The theory is successful in that one mode will always correspond to a graviton, the messenger particle of gravity, e.g. Green, Schwarz & Witten 1987, sec. 2.3, 5.3

[184] Green, Schwarz & Witten 1987, sec. 4.2

[185] Weinberg 2000, ch. 31

[186] Townsend 1996, Duff 1996

[187] Kuchař 1973, sec. 3

[188] These variables represent geometric gravity using mathematical analogues of electric and magnetic fields; cf. Ashtekar 1986, Ashtekar 1987

[189] For a review, see Thiemann 2006; more extensive accounts can be found in Rovelli 1998, Ashtekar & Lewandowski 2004 as well as in the lecture notes Thiemann 2003

[190] Isham 1994, Sorkin 1997

[191] Loll 1998

[192] Sorkin 2005

[193] Penrose 2004, ch. 33 and refs therein

[194] Hawking 1987

[195] Ashtekar 2007, Schwarz 2007

[196] Maddox 1998, pp. 52–59, 98–122; Penrose 2004, sec. 34.1, ch. 30

[197] section Quantum gravity, above

[198] section Cosmology, above

[199] Friedrich 2005

[200] A review of the various problems and the techniques being developed to overcome them, see Lehner 2002

[201] See Bartusiak 2000 for an account up to that year; up-to-date news can be found on the websites of major detector collaborations such as GEO 600 and LIGO

[202] For the most recent papers on gravitational wave polarizations of inspiralling compact binaries, see Blanchet et al. 2008, and Arun et al. 2007; for a review of work on compact binaries, see Blanchet 2006 and Futamase & Itoh 2006; for a general review of experimental tests of general relativity, see Will 2006

[203] See, e.g., the electronic review journal Living Reviews in Relativity

4.11   References

- Alpher, R. A.; Herman, R. C. (1948), "Evolution of the universe", *Nature* **162** (4124): 774–775, Bibcode:1948Natur.162..774A, doi:10.1038/162774b0

- Anderson, J. D.; Campbell, J. K.; Jurgens, R. F.; Lau, E. L. (1992), "Recent developments in solar-system tests of general relativity", in Sato, H.; Nakamura, T., *Proceedings of the Sixth Marcel Großmann Meeting on General Relativity*, World Scientific, pp. 353–355, ISBN 981-02-0950-9

- Arnold, V. I. (1989), *Mathematical Methods of Classical Mechanics*, Springer, ISBN 3-540-96890-3

- Arnowitt, Richard; Deser, Stanley; Misner, Charles W. (1962), "The dynamics of general relativity", in Witten, Louis, *Gravitation: An Introduction to Current Research*, Wiley, pp. 227–265

- Arun, K.G.; Blanchet, L.; Iyer, B. R.; Qusailah, M. S. S. (2007), "Inspiralling compact binaries in quasi-elliptical orbits: The complete 3PN energy flux", *Physical Review D* **77** (6), arXiv:0711.0302, Bibcode:2008PhRvD..77f4035A, doi:10.1103/PhysRevD.77.064035

- Ashby, Neil (2002), "Relativity and the Global Positioning System" (PDF), *Physics Today* **55** (5): 41–47, Bibcode:2002PhT....55e..41A, doi:10.1063/1.1485583

- Ashby, Neil (2003), "Relativity in the Global Positioning System", *Living Reviews in Relativity* **6**, doi:10.12942/lrr-2003-1, retrieved 2007-07-06

- Ashtekar, Abhay (1986), "New variables for classical and quantum gravity", *Phys. Rev. Lett.* **57** (18): 2244–2247, Bibcode:1986PhRvL..57.2244A, doi:10.1103/PhysRevLett.57.2244, PMID 10033673

- Ashtekar, Abhay (1987), "New Hamiltonian formulation of general relativity", *Phys. Rev.* **D36** (6): 1587–1602, Bibcode:1987PhRvD..36.1587A, doi:10.1103/PhysRevD.36.1587

- Ashtekar, Abhay (2007), "LOOP QUANTUM GRAVITY: FOUR RECENT ADVANCES AND A DOZEN FREQUENTLY ASKED QUESTIONS", *The Eleventh Marcel Grossmann Meeting - on Recent Developments in Theoretical and Experimental General Relativity, Gravitation and Relativistic Field Theories - Proceedings of the MG11 Meeting on General Relativity*, p. 126, arXiv:0705.2222, Bibcode:2008mgm..conf..126A, doi:10.1142/9789812834300_0008, ISBN 9789812834263

- Ashtekar, Abhay; Krishnan, Badri (2004), "Isolated and Dynamical Horizons and Their Applications", *Living Reviews in Relativity* 7, arXiv:gr-qc/0407042, Bibcode:2004LRR.....7...10A, doi:10.12942/lrr-2004-10, retrieved 2007-08-28

- Ashtekar, Abhay; Lewandowski, Jerzy (2004), "Background Independent Quantum Gravity: A Status Report", *Class. Quant. Grav.* 21 (15): R53–R152, arXiv:gr-qc/0404018, Bibcode:2004CQGra..21R..53A, doi:10.1088/0264-9381/21/15/R01

- Ashtekar, Abhay; Magnon-Ashtekar, Anne (1979), "On conserved quantities in general relativity", *Journal of Mathematical Physics* 20 (5): 793–800, Bibcode:1979JMP....20..793A, doi:10.1063/1.524151

- Auyang, Sunny Y. (1995), *How is Quantum Field Theory Possible?*, Oxford University Press, ISBN 0-19-509345-3

- Bania, T. M.; Rood, R. T.; Balser, D. S. (2002), "The cosmological density of baryons from observations of 3He+ in the Milky Way", *Nature* 415 (6867): 54–57, Bibcode:2002Natur.415...54B, doi:10.1038/415054a, PMID 11780112

- Barack, Leor; Cutler, Curt (2004), "LISA Capture Sources: Approximate Waveforms, Signal-to-Noise Ratios, and Parameter Estimation Accuracy", *Phys. Rev.* D69 (8): 082005, arXiv:gr-qc/0310125, Bibcode:2004PhRvD..69h2005B, doi:10.1103/PhysRevD.69.082005

- Bardeen, J. M.; Carter, B.; Hawking, S. W. (1973), "The Four Laws of Black Hole Mechanics", *Comm. Math. Phys.* 31 (2): 161–170, Bibcode:1973CMaPh..31..161B, doi:10.1007/BF01645742

- Barish, Barry (2005), "Towards detection of gravitational waves", in Florides, P.; Nolan, B.; Ottewil, A., *General Relativity and Gravitation. Proceedings of the 17th International Conference*, World Scientific, pp. 24–34, ISBN 981-256-424-1

- Barstow, M; Bond, Howard E.; Holberg, J. B.; Burleigh, M. R.; Hubeny, I.; Koester, D. (2005), "Hubble Space Telescope Spectroscopy of the Balmer lines in Sirius B", *Mon. Not. Roy. Astron. Soc.* 362 (4): 1134–1142, arXiv:astro-ph/0506600, Bibcode:2005MNRAS.362.1134B, doi:10.1111/j.1365-2966.2005.09359.x

- Bartusiak, Marcia (2000), *Einstein's Unfinished Symphony: Listening to the Sounds of Space-Time*, Berkley, ISBN 978-0-425-18620-6

- Begelman, Mitchell C.; Blandford, Roger D.; Rees, Martin J. (1984), "Theory of extragalactic radio sources", *Rev. Mod. Phys.* 56 (2): 255–351, Bibcode:1984RvMP...56..255B, doi:10.1103/RevModPhys.56.255

- Beig, Robert; Chruściel, Piotr T. (2006), "Stationary black holes", in Françoise, J.-P.; Naber, G.; Tsou, T.S., *Encyclopedia of Mathematical Physics, Volume 2*, Elsevier, p. 2041, arXiv:gr-qc/0502041, Bibcode:2005gr.qc.....2041B, ISBN 0-12-512660-3

- Bekenstein, Jacob D. (1973), "Black Holes and Entropy", *Phys. Rev.* D7 (8): 2333–2346, Bibcode:1973PhRvD...7.2333B, doi:10.1103/PhysRevD.7.2333

- Bekenstein, Jacob D. (1974), "Generalized Second Law of Thermodynamics in Black-Hole Physics", *Phys. Rev.* D9 (12): 3292–3300, Bibcode:1974PhRvD...9.3292B, doi:10.1103/PhysRevD.9.3292

- Belinskii, V. A.; Khalatnikov, I. M.; Lifschitz, E. M. (1971), "Oscillatory approach to the singular point in relativistic cosmology", *Advances in Physics* 19 (80): 525–573, Bibcode:1970AdPhy..19..525B, doi:10.1080/00018737000101171; original paper in Russian: Belinsky, V. A.; Lifshits, I. M.; Khalatnikov, E. M. (1970), "Колебательный Режим Приближения К Особой Точке В Релятивистской Космологии", *Uspekhi Fizicheskikh Nauk (Успехи Физических Наук)*, 102(3) (11): 463–500, Bibcode:1970UsFiN.102..463B

- Bennett, C. L.; Halpern, M.; Hinshaw, G.; Jarosik, N.; Kogut, A.; Limon, M.; Meyer, S. S.; Page, L.; et al. (2003), "First Year Wilkinson Microwave Anisotropy Probe (WMAP) Observations: Preliminary Maps and Basic Results", *Astrophys. J. Suppl.* 148 (1): 1–27, arXiv:astro-ph/0302207, Bibcode:2003ApJS..148....1B, doi:10.1086/377253

- Berger, Beverly K. (2002), "Numerical Approaches to Spacetime Singularities", *Living Reviews in Relativity* 5, arXiv:gr-qc/0201056, Bibcode:2002LRR.....5....1B, doi:10.12942/lrr-2002-1, retrieved 2007-08-04

- Bergström, Lars; Goobar, Ariel (2003), *Cosmology and Particle Astrophysics* (2nd ed.), Wiley & Sons, ISBN 3-540-43128-4

- Bertotti, Bruno; Ciufolini, Ignazio; Bender, Peter L. (1987), "New test of general relativity: Measurement of de Sitter geodetic precession rate for lunar perigee", *Physical Review Letters* **58** (11): 1062–1065, Bibcode:1987PhRvL..58.1062B, doi:10.1103/PhysRevLett.58.1062, PMID 10034329

- Bertotti, Bruno; Iess, L.; Tortora, P. (2003), "A test of general relativity using radio links with the Cassini spacecraft", *Nature* **425** (6956): 374–376, Bibcode:2003Natur.425..374B, doi:10.1038/nature01997, PMID 14508481

- Bertschinger, Edmund (1998), "Simulations of structure formation in the universe", *Annu. Rev. Astron. Astrophys.* **36** (1): 599–654, Bibcode:1998ARA&A..36..599B, doi:10.1146/annurev.astro.36.1.599

- Birrell, N. D.; Davies, P. C. (1984), *Quantum Fields in Curved Space*, Cambridge University Press, ISBN 0-521-27858-9

- Blair, David; McNamara, Geoff (1997), *Ripples on a Cosmic Sea. The Search for Gravitational Waves*, Perseus, ISBN 0-7382-0137-5

- Blanchet, L.; Faye, G.; Iyer, B. R.; Sinha, S. (2008), "The third post-Newtonian gravitational wave polarisations and associated spherical harmonic modes for inspiralling compact binaries in quasi-circular orbits", *Classical and Quantum Gravity* **25** (16): 165003, arXiv:0802.1249, Bibcode:2008CQGra..25p5003B, doi:10.1088/0264-9381/25/16/165003

- Blanchet, Luc (2006), "Gravitational Radiation from Post-Newtonian Sources and Inspiralling Compact Binaries", *Living Reviews in Relativity* **9**, Bibcode:2006LRR.....9....4B, doi:10.12942/lrr-2006-4, retrieved 2007-08-07

- Blandford, R. D. (1987), "Astrophysical Black Holes", in Hawking, Stephen W.; Israel, Werner, *300 Years of Gravitation*, Cambridge University Press, pp. 277–329, ISBN 0-521-37976-8

- Börner, Gerhard (1993), *The Early Universe. Facts and Fiction*, Springer, ISBN 0-387-56729-1

- Brandenberger, Robert H. (2007), "Conceptual Problems of Inflationary Cosmology and a New Approach to Cosmological Structure Formation", *Inflationary Cosmology*, Lecture Notes in Physics **738**, p. 393, arXiv:hep-th/0701111, Bibcode:2008LNP...738..393B, doi:10.1007/978-3-540-74353-8_11, ISBN 978-3-540-74352-1

- Brans, C. H.; Dicke, R. H. (1961), "Mach's Principle and a Relativistic Theory of Gravitation", *Physical Review* **124** (3): 925–935, Bibcode:1961PhRv..124..925B, doi:10.1103/PhysRev.124.925

- Bridle, Sarah L.; Lahav, Ofer; Ostriker, Jeremiah P.; Steinhardt, Paul J. (2003), "Precision Cosmology? Not Just Yet", *Science* **299** (5612): 1532–1533, arXiv:astro-ph/0303180, Bibcode:2003Sci...299.1532B, doi:10.1126/science.1082158, PMID 12624255

- Bruhat, Yvonne (1962), "The Cauchy Problem", in Witten, Louis, *Gravitation: An Introduction to Current Research*, Wiley, p. 130, ISBN 978-1-114-29166-9

- Buchert, Thomas (2007), "Dark Energy from Structure—A Status Report", *General Relativity and Gravitation* **40** (2–3): 467–527, arXiv:0707.2153, Bibcode:2008GReGr..40..467B, doi:10.1007/s10714-007-0554-8

- Buras, R.; Rampp, M.; Janka, H.-Th.; Kifonidis, K. (2003), "Improved Models of Stellar Core Collapse and Still no Explosions: What is Missing?", *Phys. Rev. Lett.* **90** (24): 241101, arXiv:astro-ph/0303171, Bibcode:2003PhRvL..90x1101B, doi:10.1103/PhysRevLett.90.241101, PMID 12857181

- Caldwell, Robert R. (2004), "Dark Energy", *Physics World* **17** (5): 37–42

- Carlip, Steven (2001), "Quantum Gravity: a Progress Report", *Rept. Prog. Phys.* **64** (8): 885–942, arXiv:gr-qc/0108040, Bibcode:2001RPPh...64..885C, doi:10.1088/0034-4885/64/8/301

- Carroll, Bradley W.; Ostlie, Dale A. (1996), *An Introduction to Modern Astrophysics*, Addison-Wesley, ISBN 0-201-54730-9

- Carroll, Sean M. (2001), "The Cosmological Constant", *Living Reviews in Relativity* **4**, arXiv:astro-ph/0004075, Bibcode:2001LRR.....4....1C, doi:10.12942/lrr-2001-1, retrieved 2007-07-21

- Carter, Brandon (1979), "The general theory of the mechanical, electromagnetic and thermodynamic properties of black holes", in Hawking, S. W.; Israel, W., *General Relativity, an Einstein Centenary Survey*, Cambridge University Press, pp. 294–369 and 860–863, ISBN 0-521-29928-4

- Celotti, Annalisa; Miller, John C.; Sciama, Dennis W. (1999), "Astrophysical evidence for the existence of black holes", *Class. Quant. Grav.*

16 (12A): A3–A21, arXiv:astro-ph/9912186, doi:10.1088/0264-9381/16/12A/301

- Chandrasekhar, Subrahmanyan (1983), *The Mathematical Theory of Black Holes*, Oxford University Press, ISBN 0-19-850370-9

- Charbonnel, C.; Primas, F. (2005), "The Lithium Content of the Galactic Halo Stars", *Astronomy & Astrophysics* **442** (3): 961–992, arXiv:astro-ph/0505247, Bibcode:2005A&A...442..961C, doi:10.1051/0004-6361:20042491

- Ciufolini, Ignazio; Pavlis, Erricos C. (2004), "A confirmation of the general relativistic prediction of the Lense-Thirring effect", *Nature* **431** (7011): 958–960, Bibcode:2004Natur.431..958C, doi:10.1038/nature03007, PMID 15496915

- Ciufolini, Ignazio; Pavlis, Erricos C.; Peron, R. (2006), "Determination of frame-dragging using Earth gravity models from CHAMP and GRACE", *New Astron.* **11** (8): 527–550, Bibcode:2006NewA...11..527C, doi:10.1016/j.newast.2006.02.001

- Coc, A.; Vangioni-Flam, Elisabeth; Descouvemont, Pierre; Adahchour, Abderrahim; Angulo, Carmen (2004), "Updated Big Bang Nucleosynthesis confronted to WMAP observations and to the Abundance of Light Elements", *Astrophysical Journal* **600** (2): 544–552, arXiv:astro-ph/0309480, Bibcode:2004ApJ...600..544C, doi:10.1086/380121

- Cutler, Curt; Thorne, Kip S. (2002), "An overview of gravitational wave sources", in Bishop, Nigel; Maharaj, Sunil D., *Proceedings of 16th International Conference on General Relativity and Gravitation (GR16)*, World Scientific, p. 4090, arXiv:gr-qc/0204090, Bibcode:2002gr.qc.....4090C, ISBN 981-238-171-6

- Dalal, Neal; Holz, Daniel E.; Hughes, Scott A.; Jain, Bhuvnesh (2006), "Short GRB and binary black hole standard sirens as a probe of dark energy", *Phys.Rev.* **D74** (6): 063006, arXiv:astro-ph/0601275, Bibcode:2006PhRvD..74f3006D, doi:10.1103/PhysRevD.74.063006

- Danzmann, Karsten; Rüdiger, Albrecht (2003), "LISA Technology—Concepts, Status, Prospects" (PDF), *Class. Quant. Grav.* **20** (10): S1–S9, Bibcode:2003CQGra..20S...1D, doi:10.1088/0264-9381/20/10/301

- Dirac, Paul (1996), *General Theory of Relativity*, Princeton University Press, ISBN 0-691-01146-X

- Donoghue, John F. (1995), "Introduction to the Effective Field Theory Description of Gravity", in Cornet, Fernando, *Effective Theories: Proceedings of the Advanced School, Almunecar, Spain, 26 June–1 July 1995*, Singapore: World Scientific, p. 12024, arXiv:gr-qc/9512024, Bibcode:1995gr.qc....12024D, ISBN 981-02-2908-9

- Duff, Michael (1996), "M-Theory (the Theory Formerly Known as Strings)", *Int. J. Mod. Phys.* **A11** (32): 5623–5641, arXiv:hep-th/9608117, Bibcode:1996IJMPA..11.5623D, doi:10.1142/S0217751X96002583

- Ehlers, Jürgen (1973), "Survey of general relativity theory", in Israel, Werner, *Relativity, Astrophysics and Cosmology*, D. Reidel, pp. 1–125, ISBN 90-277-0369-8

- Ehlers, Jürgen; Falco, Emilio E.; Schneider, Peter (1992), *Gravitational lenses*, Springer, ISBN 3-540-66506-4

- Ehlers, Jürgen; Lämmerzahl, Claus, eds. (2006), *Special Relativity—Will it Survive the Next 101 Years?*, Springer, ISBN 3-540-34522-1

- Ehlers, Jürgen; Rindler, Wolfgang (1997), "Local and Global Light Bending in Einstein's and other Gravitational Theories", *General Relativity and Gravitation* **29** (4): 519–529, Bibcode:1997GReGr..29..519E, doi:10.1023/A:1018843001842

- Einstein, Albert (1907), "Über das Relativitätsprinzip und die aus demselben gezogene Folgerungen" (PDF), *Jahrbuch der Radioaktivität und Elektronik* **4**: 411, retrieved 2008-05-05

- Einstein, Albert (1915), "Die Feldgleichungen der Gravitation", *Sitzungsberichte der Preussischen Akademie der Wissenschaften zu Berlin*: 844–847, retrieved 2006-09-12

- Einstein, Albert (1916), "Die Grundlage der allgemeinen Relativitätstheorie", *Annalen der Physik* **49**: 769–822, Bibcode:1916AnP...354..769E, doi:10.1002/andp.19163540702, archived from the original (PDF) on 2006-08-29, retrieved 2016-02-14

- Einstein, Albert (1917), "Kosmologische Betrachtungen zur allgemeinen Relativitätstheorie", *Sitzungsberichte der Preußischen Akademie der Wissenschaften*: 142

- Ellis, George F R; Van Elst, Henk (1999), Lachièze-Rey, Marc, ed., "Theoretical and Observational Cosmology: Cosmological models (Cargèse lectures

1998)", *Theoretical and observational cosmology : proceedings of the NATO Advanced Study Institute on Theoretical and Observational Cosmology* (Kluwer): 1–116, arXiv:gr-qc/9812046, Bibcode:1999toc..conf....1E, doi:10.1007/978-94-011-4455-1_1, ISBN 978-0-7923-5946-3

- Everitt, C. W. F.; Buchman, S.; DeBra, D. B.; Keiser, G. M. (2001), "Gravity Probe B: Countdown to launch", in Lämmerzahl, C.; Everitt, C. W. F.; Hehl, F. W., *Gyros, Clocks, and Interferometers: Testing Relativistic Gravity in Space (Lecture Notes in Physics 562)*, Springer, pp. 52–82, ISBN 3-540-41236-0

- Everitt, C. W. F.; Parkinson, Bradford; Kahn, Bob (2007), *The Gravity Probe B experiment. Post Flight Analysis—Final Report (Preface and Executive Summary)* (PDF), Project Report: NASA, Stanford University and Lockheed Martin, retrieved 2007-08-05

- Falcke, Heino; Melia, Fulvio; Agol, Eric (2000), "Viewing the Shadow of the Black Hole at the Galactic Center", *Astrophysical Journal* **528** (1): L13–L16, arXiv:astro-ph/9912263, Bibcode:2000ApJ...528L..13F, doi:10.1086/312423, PMID 10587484

- Flanagan, Éanna É.; Hughes, Scott A. (2005), "The basics of gravitational wave theory", *New J.Phys.* **7**: 204, arXiv:gr-qc/0501041, Bibcode:2005NJPh....7..204F, doi:10.1088/1367-2630/7/1/204

- Font, José A. (2003), "Numerical Hydrodynamics in General Relativity", *Living Reviews in Relativity* **6**, doi:10.12942/lrr-2003-4, retrieved 2007-08-19

- Fourès-Bruhat, Yvonne (1952), "Théoréme d'existence pour certains systémes d'équations aux derivées partielles non linéaires", *Acta Mathematica* **88** (1): 141–225, Bibcode:1952AcM....88..141F, doi:10.1007/BF02392131

- Frauendiener, Jörg (2004), "Conformal Infinity", *Living Reviews in Relativity* **7**, Bibcode:2004LRR.....7....1F, doi:10.12942/lrr-2004-1, retrieved 2007-07-21

- Friedrich, Helmut (2005), "Is general relativity 'essentially understood'?", *Annalen der Physik* **15** (1–2): 84–108, arXiv:gr-qc/0508016, Bibcode:2006AnP...518...84F, doi:10.1002/andp.200510173

- Futamase, T.; Itoh, Y. (2006), "The Post-Newtonian Approximation for Relativistic Compact Binaries", *Living Reviews in Relativity* **10**, doi:10.12942/lrr-2007-2, retrieved 2008-02-29

- Gamow, George (1970), *My World Line*, Viking Press, ISBN 0-670-50376-2

- Garfinkle, David (2007), "Of singularities and breadmaking", *Einstein Online*, retrieved 2007-08-03

- Geroch, Robert (1996). "Partial Differential Equations of Physics". arXiv:gr-qc/9602055 [gr-qc].

- Giulini, Domenico (2005), *Special Relativity: A First Encounter*, Oxford University Press, ISBN 0-19-856746-4

- Giulini, Domenico (2006a), "Algebraic and Geometric Structures in Special Relativity", in Ehlers, Jürgen; Lämmerzahl, Claus, *Special Relativity—Will it Survive the Next 101 Years?*, Springer, pp. 45–111, arXiv:math-ph/0602018, Bibcode:2006math.ph...2018G, ISBN 3-540-34522-1

- Giulini, Domenico (2006b), Stamatescu, I. O., ed., "An assessment of current paradigms in the physics of fundamental interactions: Some remarks on the notions of general covariance and background independence", *Approaches to Fundamental Physics*, Lecture Notes in Physics (Springer) **721**: 105, arXiv:gr-qc/0603087, Bibcode:2007LNP...721..105G, doi:10.1007/978-3-540-71117-9_6, ISBN 978-3-540-71115-5

- Gnedin, Nickolay Y. (2005), "Digitizing the Universe", *Nature* **435** (7042): 572–573, Bibcode:2005Natur.435..572G, doi:10.1038/435572a, PMID 15931201

- Goenner, Hubert F. M. (2004), "On the History of Unified Field Theories", *Living Reviews in Relativity* **7**, Bibcode:2004LRR.....7....2G, doi:10.12942/lrr-2004-2, retrieved 2008-02-28

- Goroff, Marc H.; Sagnotti, Augusto (1985), "Quantum gravity at two loops", *Phys. Lett.* **160B** (1–3): 81–86, Bibcode:1985PhLB..160...81G, doi:10.1016/0370-2693(85)91470-4

- Gourgoulhon, Eric (2007). "3+1 Formalism and Bases of Numerical Relativity". arXiv:gr-qc/0703035 [gr-qc].

- Gowdy, Robert H. (1971), "Gravitational Waves in Closed Universes", *Phys. Rev. Lett.* **27** (12): 826–829, Bibcode:1971PhRvL..27..826G, doi:10.1103/PhysRevLett.27.826

- Gowdy, Robert H. (1974), "Vacuum spacetimes with two-parameter spacelike isometry groups and compact invariant hypersurfaces: Topologies and boundary conditions", *Annals of Physics* **83** (1): 203–241, Bibcode:1974AnPhy..83..203G, doi:10.1016/0003-4916(74)90384-4

- Green, M. B.; Schwarz, J. H.; Witten, E. (1987), *Superstring theory. Volume 1: Introduction*, Cambridge University Press, ISBN 0-521-35752-7

- Greenstein, J. L.; Oke, J. B.; Shipman, H. L. (1971), "Effective Temperature, Radius, and Gravitational Redshift of Sirius B", *Astrophysical Journal* **169**: 563, Bibcode:1971ApJ...169..563G, doi:10.1086/151174

- Hamber, Herbert W. (2009), *Quantum Gravitation - The Feynman Path Integral Approach*, Springer Publishing, doi:10.1007/978-3-540-85293-3, ISBN 978-3-540-85292-6

- Gödel, Kurt (1949). "An Example of a New Type of Cosmological Solution of Einstein's Field Equations of Gravitation". *Rev. Mod. Phys.* **21** (3): 447. Bibcode:1949RvMP...21..447G. doi:10.1103/RevModPhys.21.447.

- Hafele, J. C.; Keating, R. E. (July 14, 1972). "Around-the-World Atomic Clocks: Predicted Relativistic Time Gains". *Science* **177** (4044): 166–168. Bibcode:1972Sci...177..166H. doi:10.1126/science.177.4044.166. PMID 17779917.

- Hafele, J. C.; Keating, R. E. (July 14, 1972). "Around-the-World Atomic Clocks: Observed Relativistic Time Gains". *Science* **177** (4044): 168–170. Bibcode:1972Sci...177..168H. doi:10.1126/science.177.4044.168. PMID 17779918.

- Havas, P. (1964), "Four-Dimensional Formulation of Newtonian Mechanics and Their Relation to the Special and the General Theory of Relativity", *Rev. Mod. Phys.* **36** (4): 938–965, Bibcode:1964RvMP...36..938H, doi:10.1103/RevModPhys.36.938

- Hawking, Stephen W. (1966), "The occurrence of singularities in cosmology", *Proceedings of the Royal Society* **A294** (1439): 511–521, Bibcode:1966RSPSA.294..511H, doi:10.1098/rspa.1966.0221

- Hawking, S. W. (1975), "Particle Creation by Black Holes", *Communications in Mathematical Physics* **43** (3): 199–220, Bibcode:1975CMaPh..43..199H, doi:10.1007/BF02345020

- Hawking, Stephen W. (1987), "Quantum cosmology", in Hawking, Stephen W.; Israel, Werner, *300 Years of Gravitation*, Cambridge University Press, pp. 631–651, ISBN 0-521-37976-8

- Hawking, Stephen W.; Ellis, George F. R. (1973), *The large scale structure of space-time*, Cambridge University Press, ISBN 0-521-09906-4

- Heckmann, O. H. L.; Schücking, E. (1959), "Newtonsche und Einsteinsche Kosmologie", in Flügge, S., *Encyclopedia of Physics* **53**, p. 489

- Heusler, Markus (1998), "Stationary Black Holes: Uniqueness and Beyond", *Living Reviews in Relativity* **1**, doi:10.12942/lrr-1998-6, retrieved 2007-08-04

- Heusler, Markus (1996), *Black Hole Uniqueness Theorems*, Cambridge University Press, ISBN 0-521-56735-1

- Hey, Tony; Walters, Patrick (2003), *The new quantum universe*, Cambridge University Press, ISBN 0-521-56457-3

- Hough, Jim; Rowan, Sheila (2000), "Gravitational Wave Detection by Interferometry (Ground and Space)", *Living Reviews in Relativity* **3**, retrieved 2007-07-21

- Hubble, Edwin (1929), "A Relation between Distance and Radial Velocity among Extra-Galactic Nebulae" (PDF), *Proc. Nat. Acad. Sci.* **15** (3): 168–173, Bibcode:1929PNAS...15..168H, doi:10.1073/pnas.15.3.168, PMC 522427, PMID 16577160

- Hulse, Russell A.; Taylor, Joseph H. (1975), "Discovery of a pulsar in a binary system", *Astrophys. J.* **195**: L51–L55, Bibcode:1975ApJ...195L..51H, doi:10.1086/181708

- Ibanez, L. E. (2000), "The second string (phenomenology) revolution", *Class. Quant. Grav.* **17** (5): 1117–1128, arXiv:hep-ph/9911499, Bibcode:2000CQGra..17.1117I, doi:10.1088/0264-9381/17/5/321

- Iorio, L. (2009), "An Assessment of the Systematic Uncertainty in Present and Future Tests of the Lense-Thirring Effect with Satellite Laser Ranging", *Space Sci. Rev.* **148** (1–4): 363, arXiv:0809.1373, Bibcode:2009SSRv..148..363I, doi:10.1007/s11214-008-9478-1

- Isham, Christopher J. (1994), "Prima facie questions in quantum gravity", in Ehlers, Jürgen; Friedrich, Helmut, *Canonical Gravity: From Classical to Quantum*, Springer, ISBN 3-540-58339-4

- Israel, Werner (1971), "Event Horizons and Gravitational Collapse", *General Relativity and Gravitation* **2** (1): 53–59, Bibcode:1971GReGr...2...53I, doi:10.1007/BF02450518

- Israel, Werner (1987), "Dark stars: the evolution of an idea", in Hawking, Stephen W.; Israel, Werner, *300 Years of Gravitation*, Cambridge University Press, pp. 199–276, ISBN 0-521-37976-8

- Janssen, Michel (2005), "Of pots and holes: Einstein's bumpy road to general relativity" (PDF), *Annalen der Physik* **14** (S1): 58–85, Bibcode:2005AnP...517S..58J, doi:10.1002/andp.200410130

- Jaranowski, Piotr; Królak, Andrzej (2005), "Gravitational-Wave Data Analysis. Formalism and Sample Applications: The Gaussian Case", *Living Reviews in Relativity* **8**, doi:10.12942/lrr-2005-3, retrieved 2007-07-30

- Kahn, Bob (1996–2012), *Gravity Probe B Website*, Stanford University, retrieved 2012-04-20

- Kahn, Bob (April 14, 2007), *Was Einstein right? Scientists provide first public peek at Gravity Probe B results (Stanford University Press Release)* (PDF), Stanford University News Service

- Kamionkowski, Marc; Kosowsky, Arthur; Stebbins, Albert (1997), "Statistics of Cosmic Microwave Background Polarization", *Phys. Rev.* **D55** (12): 7368–7388, arXiv:astro-ph/9611125, Bibcode:1997PhRvD..55.7368K, doi:10.1103/PhysRevD.55.7368

- Kennefick, Daniel (2005), "Astronomers Test General Relativity: Light-bending and the Solar Redshift", in Renn, Jürgen, *One hundred authors for Einstein*, Wiley-VCH, pp. 178–181, ISBN 3-527-40574-7

- Kennefick, Daniel (2007), "Not Only Because of Theory: Dyson, Eddington and the Competing Myths of the 1919 Eclipse Expedition", *Proceedings of the 7th Conference on the History of General Relativity, Tenerife, 2005* **0709**, p. 685, arXiv:0709.0685, Bibcode:2007arXiv0709.0685K

- Kenyon, I. R. (1990), *General Relativity*, Oxford University Press, ISBN 0-19-851996-6

- Kochanek, C.S.; Falco, E.E.; Impey, C.; Lehar, J. (2007), *CASTLES Survey Website*, Harvard-Smithsonian Center for Astrophysics, retrieved 2007-08-21

- Komar, Arthur (1959), "Covariant Conservation Laws in General Relativity", *Phys. Rev.* **113** (3): 934–936, Bibcode:1959PhRv..113..934K, doi:10.1103/PhysRev.113.934

- Kramer, Michael (2004), Karshenboim, S. G.; Peik, E., eds., "Astrophysics, Clocks and Fundamental Constants: Millisecond Pulsars as Tools of Fundamental Physics", *Lecture Notes in Physics* (Springer) **648**: 33–54, arXiv:astro-ph/0405178, Bibcode:2004LNP...648...33K, doi:10.1007/978-3-540-40991-5_3, ISBN 978-3-540-21967-5

- Kramer, M.; Stairs, I. H.; Manchester, R. N.; McLaughlin, M. A.; Lyne, A. G.; Ferdman, R. D.; Burgay, M.; Lorimer, D. R.; et al. (2006), "Tests of general relativity from timing the double pulsar", *Science* **314** (5796): 97–102, arXiv:astro-ph/0609417, Bibcode:2006Sci...314...97K, doi:10.1126/science.1132305, PMID 16973838

- Kraus, Ute (1998), "Light Deflection Near Neutron Stars", *Relativistic Astrophysics*, Vieweg, pp. 66–81, ISBN 3-528-06909-0

- Kuchař, Karel (1973), "Canonical Quantization of Gravity", in Israel, Werner, *Relativity, Astrophysics and Cosmology*, D. Reidel, pp. 237–288, ISBN 90-277-0369-8

- Künzle, H. P. (1972), "Galilei and Lorentz Structures on spacetime: comparison of the corresponding geometry and physics", *Annales de l'Institut Henri Poincaré A* **17**: 337–362

- Lahav, Ofer; Suto, Yasushi (2004), "Measuring our Universe from Galaxy Redshift Surveys", *Living Reviews in Relativity* **7**, arXiv:astro-ph/0310642, Bibcode:2004LRR.....7....8L, doi:10.12942/lrr-2004-8, retrieved 2007-08-19

- Landgraf, M.; Hechler, M.; Kemble, S. (2005), "Mission design for LISA Pathfinder", *Class. Quant. Grav.* **22** (10): S487–S492, arXiv:gr-qc/0411071, Bibcode:2005CQGra..22S.487L, doi:10.1088/0264-9381/22/10/048

- Lehner, Luis (2001), "Numerical Relativity: A review", *Class. Quant. Grav.* **18** (17): R25–R86, arXiv:gr-qc/0106072, Bibcode:2001CQGra..18R..25L, doi:10.1088/0264-9381/18/17/202

- Lehner, Luis (2002), "NUMERICAL RELATIVITY: STATUS AND PROSPECTS", *General Relativity and Gravitation - Proceedings of the 16th International Conference*, p. 210, arXiv:gr-qc/0202055, Bibcode:2002grg..conf..210L,

doi:10.1142/9789812776556_0010,   ISBN
9789812381712

- Linde, Andrei (1990), *Particle Physics and Infla-tionary Cosmology*, Harwood, p. 3203, arXiv:hep-th/0503203, Bibcode:2005hep.th....3203L, ISBN 3-7186-0489-2

- Linde, Andrei (2005), "Towards inflation in string theory", *J. Phys. Conf. Ser.* **24**: 151–160, arXiv:hep-th/0503195,   Bibcode:2005JPhCS..24..151L, doi:10.1088/1742-6596/24/1/018

- Loll, Renate (1998), "Discrete Approaches to Quantum Gravity in Four Dimensions", *Living Reviews in Relativity* **1**, arXiv:gr-qc/9805049, Bibcode:1998LRR.....1...13L, doi:10.12942/lrr-1998-13, retrieved 2008-03-09

- Lovelock, David (1972), "The Four-Dimensionality of Space and the Einstein Tensor", *J. Math. Phys.* **13** (6): 874–876, Bibcode:1972JMP....13..874L, doi:10.1063/1.1666069

- Ludyk, Günter (2013). *Einstein in Matrix Form* (1st ed.). Berlin: Springer. ISBN 9783642357978.

- MacCallum, M. (2006), "Finding and using exact solutions of the Einstein equations", in Mornas, L.; Alonso, J. D., *A Century of Relativity Physics (ERE05, the XXVIII Spanish Relativity Meeting)* **841**, American Institute of Physics, p. 129, arXiv:gr-qc/0601102,   Bibcode:2006AIPC..841..129M, doi:10.1063/1.2218172

- Maddox, John (1998), *What Remains To Be Discovered*, Macmillan, ISBN 0-684-82292-X

- Mannheim, Philip D. (2006), "Alternatives to Dark Matter and Dark Energy", *Prog. Part. Nucl. Phys.* **56** (2): 340–445, arXiv:astro-ph/0505266,   Bibcode:2006PrPNP..56..340M, doi:10.1016/j.ppnp.2005.08.001

- Mather, J. C.; Cheng, E. S.; Cottingham, D. A.; Eplee, R. E.; Fixsen, D. J.; Hewagama, T.; Isaacman, R. B.; Jensen, K. A.; et al. (1994), "Measurement of the cosmic microwave spectrum by the COBE FIRAS instrument", *Astrophysical Journal* **420**: 439–444, Bibcode:1994ApJ...420..439M, doi:10.1086/173574

- Mermin, N. David (2005), *It's About Time. Understanding Einstein's Relativity*, Princeton University Press, ISBN 0-691-12201-6

- Messiah, Albert (1999), *Quantum Mechanics*, Dover Publications, ISBN 0-486-40924-4

- Miller, Cole (2002), *Stellar Structure and Evolution (Lecture notes for Astronomy 606)*, University of Maryland, retrieved 2007-07-25

- Misner, Charles W.; Thorne, Kip. S.; Wheeler, John A. (1973), *Gravitation*, W. H. Freeman, ISBN 0-7167-0344-0

- Møller, Christian (1952), *The Theory of Relativity* (3rd ed.), Oxford University Press

- Narayan, Ramesh (2006), "Black holes in astrophysics", *New Journal of Physics* **7**: 199, arXiv:gr-qc/0506078, Bibcode:2005NJPh....7..199N, doi:10.1088/1367-2630/7/1/199

- Narayan, Ramesh; Bartelmann, Matthias (1997). "Lectures on Gravitational Lensing". arXiv:astro-ph/9606001 [astro-ph].

- Narlikar, Jayant V. (1993), *Introduction to Cosmology*, Cambridge University Press, ISBN 0-521-41250-1

- Nieto, Michael Martin (2006), "The quest to understand the Pioneer anomaly" (PDF), *Europhysics-sNews* **37** (6): 30–34, Bibcode:2006ENews..37...30N, doi:10.1051/epn:2006604

- Nordström, Gunnar (1918), "On the Energy of the Gravitational Field in Einstein's Theory", *Verhandl. Koninkl. Ned. Akad. Wetenschap.* **26**: 1238–1245

- Nordtvedt, Kenneth (2003). "Lunar Laser Ranging—a comprehensive probe of post-Newtonian gravity". arXiv:gr-qc/0301024 [gr-qc].

- Norton, John D. (1985), "What was Einstein's principle of equivalence?" (PDF), *Studies in History and Philosophy of Science* **16** (3): 203–246, doi:10.1016/0039-3681(85)90002-0, retrieved 2007-06-11

- Ohanian, Hans C.; Ruffini, Remo (1994), *Gravitation and Spacetime*, W. W. Norton & Company, ISBN 0-393-96501-5

- Olive, K. A.; Skillman, E. A. (2004), "A Realistic Determination of the Error on the Primordial Helium Abundance", *Astrophysical Journal* **617** (1): 29–49, arXiv:astro-ph/0405588, Bibcode:2004ApJ...617...29O, doi:10.1086/425170

- O'Meara, John M.; Tytler, David; Kirkman, David; Suzuki, Nao; Prochaska, Jason X.; Lubin, Dan; Wolfe, Arthur M. (2001), "The Deuterium to Hydrogen Abundance Ratio Towards a Fourth QSO: HS0105+1619", *Astrophysical Journal* **552** (2): 718–730, arXiv:astro-ph/0011179, Bibcode:2001ApJ...552..718O, doi:10.1086/320579

- Oppenheimer, J. Robert; Snyder, H. (1939), "On continued gravitational contraction", *Physical Review* **56** (5): 455–459, Bibcode:1939PhRv...56..455O, doi:10.1103/PhysRev.56.455

- Overbye, Dennis (1999), *Lonely Hearts of the Cosmos: the story of the scientific quest for the secret of the Universe*, Back Bay, ISBN 0-316-64896-5

- Pais, Abraham (1982), *'Subtle is the Lord...' The Science and life of Albert Einstein*, Oxford University Press, ISBN 0-19-853907-X

- Peacock, John A. (1999), *Cosmological Physics*, Cambridge University Press, ISBN 0-521-41072-X

- Peebles, P. J. E. (1966), "Primordial Helium abundance and primordial fireball II", *Astrophysical Journal* **146**: 542–552, Bibcode:1966ApJ...146..542P, doi:10.1086/148918

- Peebles, P. J. E. (1993), *Principles of physical cosmology*, Princeton University Press, ISBN 0-691-01933-9

- Peebles, P.J.E.; Schramm, D.N.; Turner, E.L.; Kron, R.G. (1991), "The case for the relativistic hot Big Bang cosmology", *Nature* **352** (6338): 769–776, Bibcode:1991Natur.352..769P, doi:10.1038/352769a0

- Penrose, Roger (1965), "Gravitational collapse and spacetime singularities", *Physical Review Letters* **14** (3): 57–59, Bibcode:1965PhRvL..14...57P, doi:10.1103/PhysRevLett.14.57

- Penrose, Roger (1969), "Gravitational collapse: the role of general relativity", *Rivista del Nuovo Cimento* **1**: 252–276, Bibcode:1969NCimR...1..252P

- Penrose, Roger (2004), *The Road to Reality*, A. A. Knopf, ISBN 0-679-45443-8

- Penzias, A. A.; Wilson, R. W. (1965), "A measurement of excess antenna temperature at 4080 Mc/s", *Astrophysical Journal* **142**: 419–421, Bibcode:1965ApJ...142..419P, doi:10.1086/148307

- Peskin, Michael E.; Schroeder, Daniel V. (1995), *An Introduction to Quantum Field Theory*, Addison-Wesley, ISBN 0-201-50397-2

- Peskin, Michael E. (2007), "Dark Matter and Particle Physics", *Journal of the Physical Society of Japan* **76** (11): 111017, arXiv:0707.1536, Bibcode:2007JPSJ...76k1017P, doi:10.1143/JPSJ.76.111017

- Poisson, Eric (2004), "The Motion of Point Particles in Curved Spacetime", *Living Reviews in Relativity* **7**, doi:10.12942/lrr-2004-6, retrieved 2007-06-13

- Poisson, Eric (2004), *A Relativist's Toolkit. The Mathematics of Black-Hole Mechanics*, Cambridge University Press, ISBN 0-521-83091-5

- Polchinski, Joseph (1998a), *String Theory Vol. I: An Introduction to the Bosonic String*, Cambridge University Press, ISBN 0-521-63303-6

- Polchinski, Joseph (1998b), *String Theory Vol. II: Superstring Theory and Beyond*, Cambridge University Press, ISBN 0-521-63304-4

- Pound, R. V.; Rebka, G. A. (1959), "Gravitational Red-Shift in Nuclear Resonance", *Physical Review Letters* **3** (9): 439–441, Bibcode:1959PhRvL...3..439P, doi:10.1103/PhysRevLett.3.439

- Pound, R. V.; Rebka, G. A. (1960), "Apparent weight of photons", *Phys. Rev. Lett.* **4** (7): 337–341, Bibcode:1960PhRvL...4..337P, doi:10.1103/PhysRevLett.4.337

- Pound, R. V.; Snider, J. L. (1964), "Effect of Gravity on Nuclear Resonance", *Phys. Rev. Lett.* **13** (18): 539–540, Bibcode:1964PhRvL..13..539P, doi:10.1103/PhysRevLett.13.539

- Ramond, Pierre (1990), *Field Theory: A Modern Primer*, Addison-Wesley, ISBN 0-201-54611-6

- Rees, Martin (1966), "Appearance of Relativistically Expanding Radio Sources", *Nature* **211** (5048): 468–470, Bibcode:1966Natur.211..468R, doi:10.1038/211468a0

- Reissner, H. (1916), "Über die Eigengravitation des elektrischen Feldes nach der Einsteinschen Theorie", *Annalen der Physik* **355** (9): 106–120, Bibcode:1916AnP...355..106R, doi:10.1002/andp.19163550905

- Remillard, Ronald A.; Lin, Dacheng; Cooper, Randall L.; Narayan, Ramesh (2006), "The Rates of Type I X-Ray Bursts from Transients Observed with RXTE: Evidence for Black Hole Event Horizons", *Astrophysical Journal* **646** (1): 407–419, arXiv:astro-ph/0509758, Bibcode:2006ApJ...646..407R, doi:10.1086/504862

- Renn, Jürgen, ed. (2007), *The Genesis of General Relativity (4 Volumes)*, Dordrecht: Springer, ISBN 1-4020-3999-9

- Renn, Jürgen, ed. (2005), *Albert Einstein—Chief Engineer of the Universe: Einstein's Life and Work in Context*, Berlin: Wiley-VCH, ISBN 3-527-40571-2

- Reula, Oscar A. (1998), "Hyperbolic Methods for Einstein's Equations", *Living Reviews in Relativity* **1**, Bibcode:1998LRR.....1....3R, doi:10.12942/lrr-1998-3, retrieved 2007-08-29

- Rindler, Wolfgang (2001), *Relativity. Special, General and Cosmological*, Oxford University Press, ISBN 0-19-850836-0

- Rindler, Wolfgang (1991), *Introduction to Special Relativity*, Clarendon Press, Oxford, ISBN 0-19-853952-5

- Robson, Ian (1996), *Active galactic nuclei*, John Wiley, ISBN 0-471-95853-0

- Roulet, E.; Mollerach, S. (1997), "Microlensing", *Physics Reports* **279** (2): 67–118, arXiv:astro-ph/9603119, Bibcode:1997PhR...279...67R, doi:10.1016/S0370-1573(96)00020-8

- Rovelli, Carlo (2000). "Notes for a brief history of quantum gravity". arXiv:gr-qc/0006061 [gr-qc].

- Rovelli, Carlo (1998), "Loop Quantum Gravity", *Living Reviews in Relativity* **1**, doi:10.12942/lrr-1998-1, retrieved 2008-03-13

- Schäfer, Gerhard (2004), "Gravitomagnetic Effects", *General Relativity and Gravitation* **36** (10): 2223–2235, arXiv:gr-qc/0407116, Bibcode:2004GReGr..36.2223S, doi:10.1023/B:GERG.0000046180.97877.32

- Schödel, R.; Ott, T.; Genzel, R.; Eckart, A.; Mouawad, N.; Alexander, T. (2003), "Stellar Dynamics in the Central Arcsecond of Our Galaxy", *Astrophysical Journal* **596** (2): 1015–1034, arXiv:astro-ph/0306214, Bibcode:2003ApJ...596.1015S, doi:10.1086/378122

- Schutz, Bernard F. (1985), *A first course in general relativity*, Cambridge University Press, ISBN 0-521-27703-5

- Schutz, Bernard F. (2001), "Gravitational radiation", in Murdin, Paul, *Encyclopedia of Astronomy and Astrophysics*, Grove's Dictionaries, ISBN 1-56159-268-4

- Schutz, Bernard F. (2003), *Gravity from the ground up*, Cambridge University Press, ISBN 0-521-45506-5

- Schwarz, John H. (2007), "String Theory: Progress and Problems", *Progress of Theoretical Physics Supplement* **170**: 214, arXiv:hep-th/0702219, Bibcode:2007PThPS.170..214S, doi:10.1143/PTPS.170.214

- Schwarzschild, Karl (1916a), "Über das Gravitationsfeld eines Massenpunktes nach der Einsteinschen Theorie", *Sitzungsber. Preuss. Akad. D. Wiss.*: 189–196

- Schwarzschild, Karl (1916b), "Über das Gravitationsfeld eines Kugel aus inkompressibler Flüssigkeit nach der Einsteinschen Theorie", *Sitzungsber. Preuss. Akad. D. Wiss.*: 424–434

- Seidel, Edward (1998), "Numerical Relativity: Towards Simulations of 3D Black Hole Coalescence", in Narlikar, J. V.; Dadhich, N., *Gravitation and Relativity: At the turn of the millennium (Proceedings of the GR-15 Conference, held at IUCAA, Pune, India, December 16–21, 1997)*, IUCAA, p. 6088, arXiv:gr-qc/9806088, Bibcode:1998gr.qc.....6088S, ISBN 81-900378-3-8

- Seljak, Uroš; Zaldarriaga, Matias (1997), "Signature of Gravity Waves in the Polarization of the Microwave Background", *Phys. Rev. Lett.* **78** (11): 2054–2057, arXiv:astro-ph/9609169, Bibcode:1997PhRvL..78.2054S, doi:10.1103/PhysRevLett.78.2054

- Shapiro, S. S.; Davis, J. L.; Lebach, D. E.; Gregory, J. S. (2004), "Measurement of the solar gravitational deflection of radio waves using geodetic very-long-baseline interferometry data, 1979–1999", *Phys. Rev. Lett.* **92** (12): 121101, Bibcode:2004PhRvL..92l1101S, doi:10.1103/PhysRevLett.92.121101, PMID 15089661

- Shapiro, Irwin I. (1964), "Fourth test of general relativity", *Phys. Rev. Lett.* **13** (26): 789–791, Bibcode:1964PhRvL..13..789S, doi:10.1103/PhysRevLett.13.789

- Shapiro, I. I.; Pettengill, Gordon; Ash, Michael; Stone, Melvin; Smith, William; Ingalls, Richard; Brockelman, Richard (1968), "Fourth test of general relativity: preliminary results", *Phys. Rev. Lett.* **20** (22): 1265–1269, Bibcode:1968PhRvL..20.1265S, doi:10.1103/PhysRevLett.20.1265

- Singh, Simon (2004), *Big Bang: The Origin of the Universe*, Fourth Estate, ISBN 0-00-715251-5

- Sorkin, Rafael D. (2005), "Causal Sets: Discrete Gravity", in Gomberoff, Andres; Marolf, Donald,

Lectures on Quantum Gravity, Springer, p. 9009, arXiv:gr-qc/0309009, Bibcode:2003gr.qc.....9009S, ISBN 0-387-23995-2

- Sorkin, Rafael D. (1997), "Forks in the Road, on the Way to Quantum Gravity", *Int. J. Theor. Phys.* **36** (12): 2759–2781, arXiv:gr-qc/9706002,   Bibcode:1997IJTP...36.2759S, doi:10.1007/BF02435709

- Spergel, D. N.; Verde, L.; Peiris, H. V.; Komatsu, E.; Nolta, M. R.; Bennett, C. L.; Halpern, M.; Hinshaw, G.; et al. (2003), "First Year Wilkinson Microwave Anisotropy Probe (WMAP) Observations: Determination of Cosmological Parameters", *Astrophys. J. Suppl.* **148** (1): 175–194, arXiv:astro-ph/0302209, Bibcode:2003ApJS..148..175S, doi:10.1086/377226

- Spergel, D. N.; Bean, R.; Doré, O.; Nolta, M. R.; Bennett, C. L.; Dunkley, J.; Hinshaw, G.; Jarosik, N.; et al. (2007), "Wilkinson Microwave Anisotropy Probe (WMAP) Three Year Results: Implications for Cosmology", *Astrophysical Journal Supplement* **170** (2): 377–408, arXiv:astro-ph/0603449, Bibcode:2007ApJS..170..377S, doi:10.1086/513700

- Springel, Volker; White, Simon D. M.; Jenkins, Adrian; Frenk, Carlos S.; Yoshida, Naoki; Gao, Liang; Navarro, Julio; Thacker, Robert; et al. (2005), "Simulations of the formation, evolution and clustering of galaxies and quasars", *Nature* **435** (7042): 629–636, arXiv:astro-ph/0504097,   Bibcode:2005Natur.435..629S, doi:10.1038/nature03597, PMID 15931216

- Stairs, Ingrid H. (2003), "Testing General Relativity with Pulsar Timing", *Living Reviews in Relativity* **6**, arXiv:astro-ph/0307536, Bibcode:2003LRR.....6....5S, doi:10.12942/lrr-2003-5, retrieved 2007-07-21

- Stephani, H.; Kramer, D.; MacCallum, M.; Hoenselaers, C.; Herlt, E. (2003), *Exact Solutions of Einstein's Field Equations* (2 ed.), Cambridge University Press, ISBN 0-521-46136-7

- Synge, J. L. (1972), *Relativity: The Special Theory*, North-Holland Publishing Company, ISBN 0-7204-0064-3

- Szabados, László B. (2004), "Quasi-Local Energy-Momentum and Angular Momentum in GR", *Living Reviews in Relativity* **7**, doi:10.12942/lrr-2004-4, retrieved 2007-08-23

- Taylor, Joseph H. (1994), "Binary pulsars and relativistic gravity", *Rev. Mod. Phys.* **66**

(3):   711–719,   Bibcode:1994RvMP...66..711T, doi:10.1103/RevModPhys.66.711

- Thiemann, Thomas (2006), "Approaches to Fundamental Physics: Loop Quantum Gravity: An Inside View", *Lecture Notes in Physics* **721**: 185–263, arXiv:hep-th/0608210, Bibcode:2007LNP...721..185T, doi:10.1007/978-3-540-71117-9_10, ISBN 978-3-540-71115-5

- Thiemann, Thomas (2003), "Lectures on Loop Quantum Gravity", *Lecture Notes in Physics* **631**: 41–135, arXiv:gr-qc/0210094, doi:10.1007/978-3-540-45230-0_3, ISBN 978-3-540-40810-9

- 't Hooft, Gerard; Veltman, Martinus (1974), "One Loop Divergencies in the Theory of Gravitation", *Ann. Inst. Poincare* **20**: 69

- Thorne, Kip S. (1972), "Nonspherical Gravitational Collapse—A Short Review", in Klauder, J., *Magic without Magic*, W. H. Freeman, pp. 231–258

- Thorne, Kip S. (1994), *Black Holes and Time Warps: Einstein's Outrageous Legacy*, W W Norton & Company, ISBN 0-393-31276-3

- Thorne, Kip S. (1995), "Gravitational radiation", *Particle and Nuclear Astrophysics and Cosmology in the Next Millenium*: 160, arXiv:gr-qc/9506086, Bibcode:1995pnac.conf..160T, ISBN 0-521-36853-7

- Townsend, Paul K. (1997). "Black Holes (Lecture notes)". arXiv:gr-qc/9707012 [gr-qc].

- Townsend, Paul K. (1996). "Four Lectures on M-Theory". arXiv:hep-th/9612121 [hep-th].

- Traschen, Jenny (2000), Bytsenko, A.; Williams, F., eds., "An Introduction to Black Hole Evaporation", *Mathematical Methods of Physics (Proceedings of the 1999 Londrina Winter School)* (World Scientific): 180, arXiv:gr-qc/0010055, Bibcode:2000mmp..conf..180T

- Trautman, Andrzej (2006), "Einstein–Cartan theory", in Françoise, J.-P.; Naber, G. L.; Tsou, S. T., *Encyclopedia of Mathematical Physics, Vol. 2*, Elsevier, pp. 189–195, arXiv:gr-qc/0606062, Bibcode:2006gr.qc.....6062T

- Unruh, W. G. (1976), "Notes on Black Hole Evaporation", *Phys. Rev.* **D 14** (4): 870–892,   Bibcode:1976PhRvD..14..870U, doi:10.1103/PhysRevD.14.870

- Valtonen, M. J.; Lehto, H. J.; Nilsson, K.; Heidt, J.; Takalo, L. O.; Sillanpää, A.; Villforth, C.;

Kidger, M.; et al. (2008), "A massive binary black-hole system in OJ 287 and a test of general relativity", *Nature* **452** (7189): 851–853, arXiv:0809.1280, Bibcode:2008Natur.452..851V, doi:10.1038/nature06896, PMID 18421348

- Veltman, Martinus (1975), "Quantum Theory of Gravitation", in Balian, Roger; Zinn-Justin, Jean, *Methods in Field Theory - Les Houches Summer School in Theoretical Physics.* **77**, North Holland

- Wald, Robert M. (1975), "On Particle Creation by Black Holes", *Commun. Math. Phys.* **45** (3): 9–34, Bibcode:1975CMaPh..45....9W, doi:10.1007/BF01609863

- Wald, Robert M. (1984), *General Relativity*, University of Chicago Press, ISBN 0-226-87033-2

- Wald, Robert M. (1994), *Quantum field theory in curved spacetime and black hole thermodynamics*, University of Chicago Press, ISBN 0-226-87027-8

- Wald, Robert M. (2001), "The Thermodynamics of Black Holes", *Living Reviews in Relativity* **4**, Bibcode:2001LRR.....4....6W, doi:10.12942/lrr-2001-6, retrieved 2007-08-08

- Walsh, D.; Carswell, R. F.; Weymann, R. J. (1979), "0957 + 561 A, B: twin quasistellar objects or gravitational lens?", *Nature* **279** (5712): 381–4, Bibcode:1979Natur.279..381W, doi:10.1038/279381a0, PMID 16068158

- Wambsganss, Joachim (1998), "Gravitational Lensing in Astronomy", *Living Reviews in Relativity* **1**, arXiv:astro-ph/9812021, Bibcode:1998LRR.....1...12W, doi:10.12942/lrr-1998-12, retrieved 2007-07-20

- Weinberg, Steven (1972), *Gravitation and Cosmology*, John Wiley, ISBN 0-471-92567-5

- Weinberg, Steven (1995), *The Quantum Theory of Fields I: Foundations*, Cambridge University Press, ISBN 0-521-55001-7

- Weinberg, Steven (1996), *The Quantum Theory of Fields II: Modern Applications*, Cambridge University Press, ISBN 0-521-55002-5

- Weinberg, Steven (2000), *The Quantum Theory of Fields III: Supersymmetry*, Cambridge University Press, ISBN 0-521-66000-9

- Weisberg, Joel M.; Taylor, Joseph H. (2003), "The Relativistic Binary Pulsar B1913+16"", in Bailes, M.; Nice, D. J.; Thorsett, S. E., *Proceedings of "Radio Pulsars," Chania, Crete, August, 2002*, ASP Conference Series

- Weiss, Achim (2006), "Elements of the past: Big Bang Nucleosynthesis and observation", *Einstein Online* (Max Planck Institute for Gravitational Physics), retrieved 2007-02-24

- Wheeler, John A. (1990), *A Journey Into Gravity and Spacetime*, Scientific American Library, San Francisco: W. H. Freeman, ISBN 0-7167-6034-7

- Will, Clifford M. (1993), *Theory and experiment in gravitational physics*, Cambridge University Press, ISBN 0-521-43973-6

- Will, Clifford M. (2006), "The Confrontation between General Relativity and Experiment", *Living Reviews in Relativity* **9**, arXiv:gr-qc/0510072, Bibcode:2006LRR.....9....3W, doi:10.12942/lrr-2006-3, retrieved 2007-06-12

- Zwiebach, Barton (2004), *A First Course in String Theory*, Cambridge University Press, ISBN 0-521-83143-1

4.12 Further reading

Popular books

- Geroch, R (1981), *General Relativity from A to B*, Chicago: University of Chicago Press, ISBN 0-226-28864-1

- Lieber, Lillian (2008), *The Einstein Theory of Relativity: A Trip to the Fourth Dimension*, Philadelphia: Paul Dry Books, Inc., ISBN 978-1-58988-044-3

- Wald, Robert M. (1992), *Space, Time, and Gravity: the Theory of the Big Bang and Black Holes*, Chicago: University of Chicago Press, ISBN 0-226-87029-4

- Wheeler, John; Ford, Kenneth (1998), *Geons, Black Holes, & Quantum Foam: a life in physics*, New York: W. W. Norton, ISBN 0-393-31991-1

Beginning undergraduate textbooks

- Callahan, James J. (2000), *The Geometry of Spacetime: an Introduction to Special and General Relativity*, New York: Springer, ISBN 0-387-98641-3

- Taylor, Edwin F.; Wheeler, John Archibald (2000), *Exploring Black Holes: Introduction to General Relativity*, Addison Wesley, ISBN 0-201-38423-X

Advanced undergraduate textbooks

- B. F. Schutz (2009), *A First Course in General Relativity (Second Edition)*, Cambridge University Press, ISBN 978-0-521-88705-2

- Cheng, Ta-Pei (2005), *Relativity, Gravitation and Cosmology: a Basic Introduction*, Oxford and New York: Oxford University Press, ISBN 0-19-852957-0

- Gron, O.; Hervik, S. (2007), *Einstein's General theory of Relativity*, Springer, ISBN 978-0-387-69199-2

- Hartle, James B. (2003), *Gravity: an Introduction to Einstein's General Relativity*, San Francisco: Addison-Wesley, ISBN 0-8053-8662-9

- Hughston, L. P. & Tod, K. P. (1991), *Introduction to General Relativity*, Cambridge: Cambridge University Press, ISBN 0-521-33943-X

- d'Inverno, Ray (1992), *Introducing Einstein's Relativity*, Oxford: Oxford University Press, ISBN 0-19-859686-3

- Ludyk, Günter (2013). *Einstein in Matrix Form* (1st ed.). Berlin: Springer. ISBN 9783642357978.

Graduate-level textbooks

- Carroll, Sean M. (2004), *Spacetime and Geometry: An Introduction to General Relativity*, San Francisco: Addison-Wesley, ISBN 0-8053-8732-3

- Grøn, Øyvind; Hervik, Sigbjørn (2007), *Einstein's General Theory of Relativity*, New York: Springer, ISBN 978-0-387-69199-2

- Landau, Lev D.; Lifshitz, Evgeny F. (1980), *The Classical Theory of Fields (4th ed.)*, London: Butterworth-Heinemann, ISBN 0-7506-2768-9

- Misner, Charles W.; Thorne, Kip. S.; Wheeler, John A. (1973), *Gravitation*, W. H. Freeman, ISBN 0-7167-0344-0

- Stephani, Hans (1990), *General Relativity: An Introduction to the Theory of the Gravitational Field*, Cambridge: Cambridge University Press, ISBN 0-521-37941-5

- Wald, Robert M. (1984), *General Relativity*, University of Chicago Press, ISBN 0-226-87033-2

4.13 External links

- Einstein Online – Articles on a variety of aspects of relativistic physics for a general audience; hosted by the Max Planck Institute for Gravitational Physics

- NCSA Spacetime Wrinkles – produced by the numerical relativity group at the NCSA, with an elementary introduction to general relativity

- **Courses**

- **Lectures**

- **Tutorials**

- Einstein's General Theory of Relativity on YouTube (lecture by Leonard Susskind recorded September 22, 2008 at Stanford University).

- Series of lectures on General Relativity given in 2006 at the Institut Henri Poincaré (introductory/advanced).

- General Relativity Tutorials by John Baez.

- Brown, Kevin. "Reflections on relativity". *Mathpages.com*. Retrieved May 29, 2005.

- Carroll, Sean M. "Lecture Notes on General Relativity". Retrieved January 5, 2014.

- Moor, Rafi. "Understanding General Relativity". Retrieved July 11, 2006.

- Waner, Stefan. "Introduction to Differential Geometry and General Relativity" (PDF). Retrieved 2015-04-05.

Chapter 5

Local symmetry

In physics, a **local symmetry** is symmetry of some physical quantity, which smoothly depends on the point of the base manifold. Such quantities can be for example an observable, a tensor or the Lagrangian of a theory. If a symmetry is local in this sense, then one can apply a local transformation (resp. local gauge transformation), which means that the representation of the symmetry group is a function of the manifold and can thus be taken to act differently on different points of spacetime.

The diffeomorphism group is a local symmetry and thus every geometrical or generally covariant theory (i.e. a theory whose equations are tensor equations, for example general relativity) has local symmetries.

Often the term local symmetry is specifically associated with local gauge symmetries in Yang–Mills theory (see also standard model) where the Lagrangian is locally symmetric under some compact Lie group. Local gauge symmetries always come together with some bosonic gauge fields, like the photon or gluon field, which induce a force in addition to requiring conservation laws.[1]

5.1 Examples

- General relativity has a local symmetry (general covariance, diffeomorphisms) which can be seen as generating the gravitational force.[2] Special relativity only has a global symmetry (Lorentz symmetry or more generally Poincaré symmetry)

- There are many global symmetries (such as SU(2) of isospin symmetry) and local symmetries (like SU(2) of weak interactions) in particle physics. The standard model of particle physics consists of Yang-Mills Theories

- The symmetry group of Supergravity is a local symmetry, whereas supersymmetry is a global symmetry.

5.2 See also

- Field (physics)
- Global spacetime structure
- Local spacetime structure
- Gauge theory
- Gravitation (book)

5.3 References

[1] Kaku, Michio (1993). *Quantum Field Theory: A Modern Introduction*. New York: Oxford University Press. ISBN 0-19-507652-4.

[2] Misner, Charles W.; Thorne, Kip S.; Wheeler, John Archibald (1973-09-15). "Gravitation". San Francisco: W. H. Freeman. ISBN 978-0-7167-0344-0.

Chapter 6

Minimal Supersymmetric Standard Model

The **Minimal Supersymmetric Standard Model (MSSM)** is an extension to the Standard Model that realizes supersymmetry. MSSM is the minimal supersymmetrical model as it considers only "the [minimum] number of new particle states and new interactions consistent with phenomenology".[1] Supersymmetry pairs bosons with fermions; therefore every Standard Model particle has a partner that has yet to be discovered. If the superparticles are found, it may be analogous to discovering dark matter [2] and depending on the details of what might be found, it could provide evidence for grand unification and might even, in principle, provide hints as to whether string theory describes nature. The failure to find evidence for supersymmetry using the Large Hadron Collider since 2010 has led to suggestions that the theory should be abandoned.[3]

6.1 Background

The MSSM was originally proposed in 1981 to stabilize the weak scale, solving the hierarchy problem.[4] The Higgs boson mass of the Standard Model is unstable to quantum corrections and the theory predicts that weak scale should be much weaker than what is observed to be. In the MSSM, the Higgs boson has a fermionic superpartner, the Higgsino, that has the same mass as it would if supersymmetry were an exact symmetry. Because fermion masses are radiatively stable, the Higgs mass inherits this stability. However, in MSSM there is a need for more than one Higgs field, as described below.

The only unambiguous way to claim discovery of supersymmetry is to produce superparticles in the laboratory. Because superparticles are expected to be 100 to 1000 times heavier than the proton, it requires a huge amount of energy to make these particles that can only be achieved at particle accelerators. The Tevatron was actively looking for evidence of the production of supersymmetric particles before it was shut down on 30 September 2011. Most physicists believe that supersymmetry must be discovered at the LHC

if it is responsible for stabilizing the weak scale. There are five classes of particle that superpartners of the Standard Model fall into: squarks, gluinos, charginos, neutralinos, and sleptons. These superparticles have their interactions and subsequent decays described by the MSSM and each has characteristic signatures.

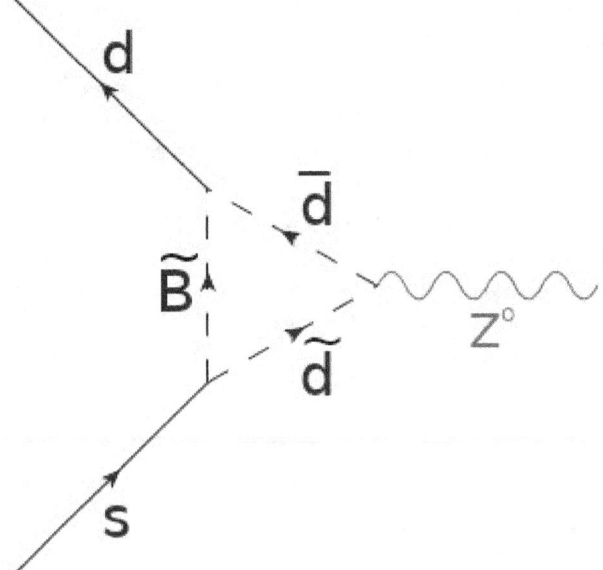

An example of a flavor changing neutral current process in MSSM. A strange quark emits a bino, turning into a sdown-type quark, which then emits a Z boson and reabsorbs the bino, turning into a down quark. If the MSSM squark masses are flavor violating, such a process can occur.

The MSSM imposes R-parity to explain the stability of the proton. It adds supersymmetry breaking by introducing explicit soft supersymmetry breaking operators into the Lagrangian that is communicated to it by some unknown (and unspecified) dynamics. This means that there are 120 new parameters in the MSSM. Most of these parameters lead to unacceptable phenomenology such as large flavor changing neutral currents or large electric dipole moments for the neutron and electron. To avoid these problems, the MSSM takes all of the soft supersymmetry breaking to be diagonal in flavor space and for all of the new CP violating phases to vanish.

6.2 Theoretical motivations

There are three principal motivations for the MSSM over other theoretical extensions of the Standard Model, namely:

- Naturalness
- Gauge coupling unification
- Dark Matter

These motivations come out without much effort and they are the primary reasons why the MSSM is the leading candidate for a new theory to be discovered at collider experiments such as the Tevatron or the LHC.

6.2.1 Naturalness

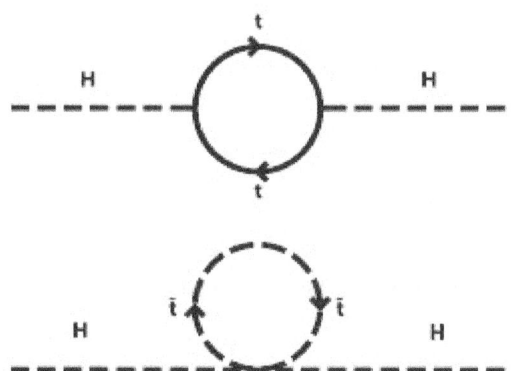

Cancellation of the Higgs boson quadratic mass renormalization between fermionic top quark loop and scalar top squark Feynman diagrams in a supersymmetric extension of the Standard Model

The original motivation for proposing the MSSM was to stabilize the Higgs mass to radiative corrections that are quadratically divergent in the Standard Model (hierarchy problem). In supersymmetric models, scalars are related to fermions and have the same mass. Since fermion masses are logarithmically divergent, scalar masses inherit the same radiative stability. The Higgs vacuum expectation value is related to the negative scalar mass in the Lagrangian. In order for the radiative corrections to the Higgs mass to not be dramatically larger than the actual value, the mass of the

superpartners of the Standard Model should not be significantly heavier than the Higgs VEV—roughly 100 GeV. In 2012, the Higgs particle was discovered at the LHC, and its mass was found to be 125-126 GeV.

6.2.2 Gauge-coupling unification

If the superpartners of the Standard Model are near the TeV scale, then measured gauge couplings of the three gauge groups unify at high energies.[5] [6] [7] The beta-functions for the MSSM gauge couplings are given by

where α_1^{-1} is measured in SU(5) normalization—a factor of $\frac{3}{5}$ different than the Standard Model's normalization and predicted by Georgi–Glashow SU(5) .

The condition for gauge coupling unification at one loop is whether the following expression is satisfied $\frac{\alpha_3^{-1}-\alpha_2^{-1}}{\alpha_2^{-1}-\alpha_1^{-1}} = \frac{b_{0\,3}-b_{0\,2}}{b_{0\,2}-b_{0\,1}}$.

Remarkably, this is precisely satisfied to experimental errors in the values of $\alpha^{-1}(M_{Z^0})$. There are two loop corrections and both TeV-scale and GUT-scale threshold corrections that alter this condition on gauge coupling unification, and the results of more extensive calculations reveal that gauge coupling unification occurs to an accuracy of 1%, though this is about 3 standard deviations from the theoretical expectations.

This prediction is generally considered as indirect evidence for both the MSSM and SUSY GUTs.[8] It should be noted that gauge coupling unification does not necessarily imply grand unification and there exist other mechanisms to reproduce gauge coupling unification. However, if superpartners are found in the near future, the apparent success of gauge coupling unification would suggest that a supersymmetric grand unified theory is a promising candidate for high scale physics.

6.2.3 Dark matter

If R-parity is preserved, then the lightest superparticle (LSP) of the MSSM is stable and is a Weakly interacting massive particle (WIMP) — i.e. it does not have electromagnetic or strong interactions. This makes the LSP a good dark matter candidate and falls into the category of cold dark matter (CDM) particle.

6.3 Predictions of the MSSM regarding hadron colliders

The Tevatron and LHC have active experimental programs searching for supersymmetric particles. Since both of these machines are hadron colliders — proton antiproton for the Tevatron and proton proton for the LHC — they search best for strongly interacting particles. Therefore, most experimental signature involve production of squarks or gluinos. Since the MSSM has R-parity, the lightest supersymmetric particle is stable and after the squarks and gluinos decay each decay chain will contain one LSP that will leave the detector unseen. This leads to the generic prediction that the MSSM will produce a 'missing energy' signal from these particles leaving the detector.

6.3.1 Neutralinos

There are four neutralinos that are fermions and are electrically neutral, the lightest of which is typically stable. They are typically labeled N0

1, N0

2, N0

3, N0

4 (although sometimes $\tilde{\chi}_1^0, \ldots, \tilde{\chi}_4^0$ is used instead). These four states are mixtures of the Bino and the neutral Wino (which are the neutral electroweak Gauginos), and the neutral Higgsinos. As the neutralinos are Majorana fermions, each of them is identical with its antiparticle. Because these particles only interact with the weak vector bosons, they are not directly produced at hadron colliders in copious numbers. They primarily appear as particles in cascade decays of heavier particles usually originating from colored supersymmetric particles such as squarks or gluinos.

In R-parity conserving models, the lightest neutralino is stable and all supersymmetric cascades decays end up decaying into this particle which leaves the detector unseen and its existence can only be inferred by looking for unbalanced momentum in a detector.

The heavier neutralinos typically decay through a Z0 to a lighter neutralino or through a W± to chargino. Thus a typical decay is

The mass splittings between the different Neutralinos will dictate which patterns of decays are allowed.

6.3.2 Charginos

There are two Charginos that are fermions and are electrically charged. They are typically labeled C $\tilde{\chi}\pm$
1 and C $\tilde{\chi}\pm$

2 (although sometimes $\tilde{\chi}_1^\pm$ and $\tilde{\chi}_2^\pm$ is used instead). The heavier chargino can decay through Z0 to the lighter chargino. Both can decay through a W± to neutralino.

6.3.3 Squarks

The squarks are the scalar superpartners of the quarks and there is one version for each Standard Model quark. Due to phenomenological constraints from flavor changing neutral currents, typically the lighter two generations of squarks have to be nearly the same in mass and therefore are not given distinct names. The superpartners of the top and bottom quark can be split from the lighter squarks and are called *stop* and *sbottom*.

On the other way, there may be a remarkable left-right mixing of the stops $\tilde{t}$ and of the sbottoms $\tilde{b}$ because of the high masses of the partner quarks top and bottom: [9]

- $\tilde{t}_1 = e^{+i\phi}\cos(\theta)\tilde{t_L} + \sin(\theta)\tilde{t_R}$

- $\tilde{t}_2 = e^{-i\phi}\cos(\theta)\tilde{t_R} - \sin(\theta)\tilde{t_L}$

Same holds for bottom $\tilde{b}$ with its own parameters ϕ and θ .

Squarks can be produced through strong interactions and therefore are easily produced at hadron colliders. They decay to quarks and neutralinos or charginos which further decay. In R-parity conserving scenarios, squarks are pair produced and therefore a typical signal is

- $\tilde{q}\tilde{q} \to q\tilde{N}_1^0\bar{q}\tilde{N}_1^0 \to 2$ jets + missing energy

- $\tilde{q}\tilde{q} \to q\tilde{N}_2^0\bar{q}\tilde{N}_1^0 \to q\tilde{N}_1^0\ell\bar{\ell}\bar{q}\tilde{N}_1^0 \to 2$ jets + 2 leptons + missing energy

6.3.4 Gluinos

Gluinos are Majorana fermionic partners of the gluon which means that they are their own antiparticles. They interact strongly and therefore can be produced significantly at the LHC. They can only decay to a quark and a squark and thus a typical gluino signal is

- $\tilde{g}\tilde{g} \to (q\bar{\tilde{q}})(\bar{q}\tilde{q}) \to (q\bar{q}\tilde{N}_1^0)(\bar{q}q\tilde{N}_1^0) \to 4$ jets + Missing energy

Because gluinos are Majorana, gluinos can decay to either a quark+anti-squark or an anti-quark+squark with equal probability. Therefore, pairs of gluinos can decay to

- $\tilde{g}\tilde{g} \to (\bar{q}\tilde{q})(\bar{q}\tilde{q}) \to (q\bar{q}\tilde{C}_1^+)(q\bar{q}\tilde{C}_1^+) \to (q\bar{q}W^+)(q\bar{q}W^+) \to 4$ jets+ $\ell^+\ell^+$ + Missing energy

This is a distinctive signature because it has same-sign di-leptons and has very little background in the Standard Model.

6.3.5 Sleptons

Sleptons are the scalar partners of the leptons of the Standard Model. They are not strongly interacting and therefore are not produced very often at hadron colliders unless they are very light.

Because of the high mass of the tau lepton there will be left-right mixing of the stau similar to that of stop and sbottom (see above).

Sfermions will typically be found in decays of a charginos and neutralinos if they are light enough to be a decay product

- $\tilde{C}^+ \to \tilde{\ell}^+ \nu$

- $\tilde{N}^0 \to \tilde{\ell}^+ \ell^-$

6.4 MSSM fields

Fermions have bosonic superpartners (called sfermions), and bosons have fermionic superpartners (called bosinos). For most of the Standard Model particles, doubling is very straightforward. However, for the Higgs boson, it is more complicated.

A single Higgsino (the fermionic superpartner of the Higgs boson) would lead to a gauge anomaly and would cause the theory to be inconsistent. However, if two Higgsinos are added, there is no gauge anomaly. The simplest theory is one with two Higgsinos and therefore two scalar Higgs doublets. Another reason for having two scalar Higgs doublets rather than one is in order to have Yukawa couplings between the Higgs and both down-type quarks and up-type quarks; these are the terms responsible for the quarks' masses. In the Standard Model the down-type quarks couple to the Higgs field (which has Y=−1/2) and the up-type quarks to its complex conjugate (which has Y=+1/2). However, in a supersymmetric theory this is not allowed, so two types of Higgs fields are needed.

6.4.1 MSSM superfields

In supersymmetric theories, every field and its superpartner can be written together as a superfield. The superfield formulation of supersymmetry is very convenient to write down manifestly supersymmetric theories (i.e. one does not have to tediously check that the theory is supersymmetric term by term in the Lagrangian). The MSSM contains vector superfields associated with the Standard Model gauge groups which contain the vector bosons and associated gauginos. It also contains chiral superfields for the Standard Model fermions and Higgs bosons (and their respective superpartners).

6.4.2 MSSM Higgs Mass

The MSSM Higgs Mass is a prediction of the Minimal Supersymmetric Standard Model. The mass of the lightest Higgs boson is set by the Higgs *quartic coupling*. Quartic couplings are not soft supersymmetry-breaking parameters since they lead to a quadratic divergence of the Higgs mass. Furthermore, there are no supersymmetric parameters to make the Higgs mass a free parameter in the MSSM (though not in non-minimal extensions). This means that Higgs mass is a prediction of the MSSM. The LEP II and the IV experiments placed a lower limit on the Higgs mass of 114.4 GeV. This lower limit is significantly above where the MSSM would typically predict it to be, and while it does not rule out the MSSM, the discovery of the Higgs with a mass of 125 GeV makes proponents of the MSSM nervous.[10][11]

Formulas

The only susy-preserving operator that creates a quartic coupling for the Higgs in the MSSM arise for the D-terms of the SU(2) and U(1) gauge sector and the magnitude of the quartic coupling is set by the size of the gauge couplings.

This leads to the prediction that the Standard Model-like Higgs mass (the scalar that couples approximately to the vev) is limited to be less than the Z mass

$$m_{h^0}^2 \leq m_{Z^0}^2 \cos^2 2\beta .$$

Since supersymmetry is broken, there are radiative corrections to the quartic coupling that can increase the Higgs mass. These dominantly arise from the 'top sector'

$$m_{h^0}^2 \leq m_{Z^0}^2 \cos^2 2\beta + \frac{3}{\pi^2} \frac{m_t^4 \sin^4 \beta}{v^2} \log \frac{m_{\tilde{t}}}{m_t}$$

where m_t is the top mass and $m_{\tilde{t}}$ is the mass of the top squark. This result can be interpreted as the RG running of the Higgs quartic coupling from the scale of supersymmetry to the top mass—however since the top squark mass should be relatively close to the top mass, this is usually a fairly modest contribution and increases the Higgs mass to roughly the LEP II bound of 114 GeV before the top squark becomes too heavy.

Finally there is a contribution from the top squark A-terms

$$\mathcal{L} = y_t\, m_{\tilde{t}}\, a\, h_u \tilde{q}_3 \tilde{u}_3^c$$

where a is a dimensionless number. This contributes an additional term to the Higgs mass at loop level, but is not logarithmically enhanced

$$m_{h^0}^2 \leq m_{Z^0}^2 \cos^2 2\beta + \frac{3}{\pi^2} \frac{m_t^4 \sin^4\beta}{v^2} \left(\log\frac{m_{\tilde{t}}}{m_t} + a^2(1 - a^2/12) \right)$$

by pushing $a \to \sqrt{6}$ (known as 'maximal mixing') it is possible to push the Higgs mass to 125 GeV without decoupling the top squark or adding new dynamics to the MSSM.

As the Higgs was found at around 125 GeV (along with no other superparticles) at the LHC, this strongly hints at new dynamics beyond the MSSM, such as the 'Next to Minimal Supersymmetric Standard Model' (NMSSM); and suggests some correlation to the little hierarchy problem.

6.5 The MSSM Lagrangian

The Lagrangian for the MSSM contains several pieces.

- The first is the Kähler potential for the matter and Higgs fields which produces the kinetic terms for the fields.

- The second piece is the gauge field superpotential that produces the kinetic terms for the gauge bosons and gauginos.

- The next term is the superpotential for the matter and Higgs fields. These produce the Yukawa couplings for the Standard Model fermions and also the mass term for the Higgsinos. After imposing R-parity, the renormalizable, gauge invariant operators in the superpotential are

$$W = \mu H_u H_d + y_u H_u Q U^c + y_d H_d Q D^c + y_l H_d L E^c$$

The constant term is unphysical in global supersymmetry (as opposed to supergravity).

6.5.1 Soft Susy breaking

Main article: Soft SUSY breaking

The last piece of the MSSM Lagrangian is the soft supersymmetry breaking Lagrangian. The vast majority of the parameters of the MSSM are in the susy breaking Lagrangian. The soft susy breaking are divided into roughly three pieces.

- The first are the gaugino masses

$$\mathcal{L} \supset m_{\frac{1}{2}} \tilde{\lambda}\tilde{\lambda} + \text{h.c.}$$

Where $\tilde{\lambda}$ are the gauginos and $m_{\frac{1}{2}}$ is different for the wino, bino and gluino.

- The next are the soft masses for the scalar fields

$$\mathcal{L} \supset m_0^2 \phi^\dagger \phi$$

where ϕ are any of the scalars in the MSSM and m_0 are 3×3 Hermitian matrices for the squarks and sleptons of a given set of gauge quantum numbers. The eigenvalues of these matrices are actually the masses squared, rather than the masses.

- There are the A and B terms which are given by

$$\mathcal{L} \supset B_\mu h_u h_d + A h_u \tilde{q}\tilde{u^c} + A h_d \tilde{q}\tilde{d^c} + A h_d \tilde{l}\tilde{e^c} + \text{h.c.}$$

The A terms are 3×3 complex matrices much as the scalar masses are.

- Although not often mentioned with regard to soft terms, to be consistent with observation, one must also include Gravitino and Goldstino soft masses given by

$$\mathcal{L} \supset m_{3/2} \Psi_\mu^\alpha (\sigma^{\mu\nu})_\alpha^\beta \Psi_\beta + m_{3/2} G^\alpha G_\alpha + \text{h.c.}$$

The reason these soft terms are not often mentioned are that they arise through local supersymmetry and not global supersymmetry, although they are required otherwise if the Goldstino were massless it would contradict observation. The Goldstino mode is eaten by the Gravitino to become massive, through a gauge shift, which also absorbs the would-be "mass" term of the Goldstino.

6.6 Problems with the MSSM

There are several problems with the MSSM — most of them falling into understanding the parameters.

- The mu problem: The Higgsino mass parameter μ appears as the following term in the superpotential: μHᵤHd. It should have the same order of magnitude as the electroweak scale, many orders of magnitude smaller than that of the Planck scale, which is the natural cutoff scale. The soft supersymmetry breaking terms should also be of the same order of magnitude as the electroweak scale. This brings about a problem of naturalness: why are these scales so much smaller than the cutoff scale yet happen to fall so close to each other?

- Flavor universality of soft masses and A-terms: since no flavor mixing additional to that predicted by the standard model has been discovered so far, the coefficients of the additional terms in the MSSM Lagrangian must be, at least approximately, flavor invariant (i.e. the same for all flavors).

- Smallness of CP violating phases: since no CP violation additional to that predicted by the standard model has been discovered so far, the additional terms in the MSSM Lagrangian must be, at least approximately, CP invariant, so that their CP violating phases are small.

6.7 Theories of supersymmetry breaking

A large amount of theoretical effort has been spent trying to understand the mechanism for soft supersymmetry breaking that produces the desired properties in the superpartner masses and interactions. The three most extensively studied mechanisms are:

6.7.1 Gravity-mediated supersymmetry breaking

Gravity-mediated supersymmetry breaking is a method of communicating supersymmetry breaking to the supersymmetric Standard Model through gravitational interactions. It was the first method proposed to communicate supersymmetry breaking. In gravity-mediated supersymmetry-breaking models, there is a part of the theory that only interacts with the MSSM through gravitational interaction. This hidden sector of the theory breaks supersymmetry. Through the supersymmetric version of the Higgs mechanism, the gravitino, the supersymmetric version of the graviton, acquires a mass. After the gravitino has a mass, gravitational radiative corrections to soft masses are incompletely cancelled beneath the gravitino's mass.

It is currently believed that it is not generic to have a sector completely decoupled from the MSSM and there should be higher dimension operators that couple different sectors together with the higher dimension operators suppressed by the Planck scale. These operators give as large of a contribution to the soft supersymmetry breaking masses as the gravitational loops; therefore, today people usually consider gravity mediation to be gravitational sized direct interactions between the hidden sector and the MSSM.

mSUGRA stands for minimal supergravity. The construction of a realistic model of interactions within $N = 1$ supergravity framework where supersymmetry breaking is communicated through the supergravity interactions was carried out by Ali Chamseddine, Richard Arnowitt, and Pran Nath in 1982.[12] mSUGRA is one of the most widely investigated models of particle physics due to its predictive power requiring only 4 input parameters and a sign, to determine the low energy phenomenology from the scale of Grand Unification. The most widely used set of parameters is:

Gravity-Mediated Supersymmetry Breaking was assumed to be flavor universal because of the universality of gravity; however, in 1986 Hall, Kostelecky, and Raby [13] showed that Planck-scale physics that are necessary to generate the Standard-Model Yukawa couplings spoil the universality of the supersymmetry breaking.

6.7.2 Gauge-mediated supersymmetry breaking (GMSB)

Gauge-mediated supersymmetry breaking is method of communicating supersymmetry breaking to the supersymmetric Standard Model through the Standard Model's gauge interactions. Typically a hidden sector breaks supersymmetry and communicates it to massive messenger fields that are charged under the Standard Model. These messenger fields induce a gaugino mass at one loop and then this is transmitted on to the scalar superpartners at two loops. Requiring stop squarks below 2 TeV, the maximum Higgs boson mass predicted is just 121.5GeV.[14] With the Higgs being discovered at 125GeV - this model requires stops above 2 TeV.

6.7.3 Anomaly-mediated supersymmetry breaking (AMSB)

Anomaly-mediated supersymmetry breaking is a special type of gravity mediated supersymmetry breaking that results in supersymmetry breaking being communicated to the supersymmetric Standard Model through the conformal anomaly.[15][16] Requiring stop squarks below 2 TeV, the maximum Higgs boson mass predicted is just 121.0GeV.[14] With the Higgs being discovered at 125GeV - this scenario requires stops heavier than 2 TeV.

6.8 Phenomenological MSSM (pMSSM)

The unconstrained MSSM has more than 100 parameters in addition to the Standard Model parameters. This makes any phenomenological analysis (e.g. finding regions in parameter space consistent with observed data) impractical. Under the following three assumptions:

- no new source of CP-violation

- no Flavour Changing Neutral Currents

- first and second generation universality

one can reduce the number of additional parameters to the following 19 quantities of the phenomenological MSSM (pMSSM):[17] The large parameter space of pMSSM makes searches in pMSSM extremely challenging and makes pMSSM difficult to exclude.

6.9   See also

- Desert (particle physics)

6.10   References

[1] Howard Baer; Xerxes Tata (2006). "8 - The Minimal Supersymmetric Standard Model". *Weak Scale Supersymmetry From Superfields to Scattering Events.* Cambridge: Cambridge University Press. p. 127. ISBN 9780511617270. It is minimal in the sense that it contains the smallest number of new particle states and new interactions consistent with phenomenology.

[2] Murayama, Hitoshi (2000). "Supersymmetry phenomenology". arXiv:hep-ph/0002232.

[3] Wolchover, Natalie (November 29, 2012). "Supersymmetry Fails Test, Forcing Physics to Seek New Ideas". *Scientific American.*

[4] S. Dimopoulos, H. Georgi; Georgi (1981). "Softly Broken Supersymmetry and SU(5)". *Nuclear Physics B* **193**: 150. Bibcode:1981NuPhB.193..150D. doi:10.1016/0550-3213(81)90522-8.

[5] S. Dimopoulos, S. Raby and F. Wilczek; Raby; Wilczek (1981). "Supersymmetry and the Scale of Unification". *Physical Review D* **24** (6): 1681–1683. Bibcode:1981PhRvD..24.1681D. doi:10.1103/PhysRevD.24.1681.

[6] L.E. Ibanez and G.G. Ross; Ross (1981). "Low-energy predictions in supersymmetric grand unified theories". *Physics Letters B* **105** (6): 439. Bibcode:1981PhLB..105..439I. doi:10.1016/0370-2693(81)91200-4.

[7] W.J. Marciano and G. Senjanovic; Senjanović (1982). "Predictions of supersymmetric grand unified theories". *Physical Review D* **25** (11): 3092. Bibcode:1982PhRvD..25.3092M. doi:10.1103/PhysRevD.25.3092.

[8] Gordon Kane, "The Dawn of Physics Beyond the Standard Model", *Scientific American*, June 2003, page 60 and *The frontiers of physics*, special edition, Vol 15, #3, page 8 "Indirect evidence for supersymmetry comes from the extrapolation of interactions to high energies."

[9] Bartl, A.; Hesselbach, S.; Hidaka, K.; Kernreiter, T.; Porod, W. (2003). "Impact of SUSY CP Phases on Stop and Sbottom Decays in the MSSM". arXiv:hep-ph/0306281 [hep-ph].

[10] Heinemeyer, S.; Stål, O.; Weiglein, G. (2012). "Interpreting the LHC Higgs search results in the MSSM". *Physics Letters B* **710**: 201. arXiv:1112.3026v3. Bibcode:2012PhLB..710..201H. doi:10.1016/j.physletb.2012.02.084.

[11] Carena, M.; Heinemeyer, S.; Wagner, C. E. M.; Weiglein, G. (2006). "MSSM Higgs boson searches at the evatron and the LHC: Impact of different benchmark scenarios" (PDF). *The European Physical Journal C* **45** (3): 797. arXiv:hep-ph/0511023. Bibcode:2006EPJC...45..797C. doi:10.1140/epjc/s2005-02470-y.

[12] A. Chamseddine, R. Arnowitt, P. Nath; Arnowitt; Nath (1982). "Locally Supersymmetric Grand Unification". *Physical Review Letters* **49** (14): 970–974. Bibcode:1982PhRvL..49..970C. doi:10.1103/PhysRevLett.49.970.

[13] L.J. Hall, V.A. Kostelecky, S. Raby; Kostelecky; Raby (1986). "New Flavor Violations in Supergravity Models". *Nuclear Physics B* **267** (2): 415. Bibcode:1986NuPhB.267..415H. doi:10.1016/0550-3213(86)90397-4.

[14] Arbey, A.; Battaglia, M.; Djouadi, A.; Mahmoudi, F.; Quevillon, J. (2011). "Implications of a 125 GeV Higgs for supersymmetric models". *Physics Letters B.* 3 **708** (2012): 162–169. arXiv:1112.3028. Bibcode:2012PhLB..708..162A. doi:10.1016/j.physletb.2012.01.053.

[15] L. Randall, R. Sundrum; Sundrum (1999). "Out of this world supersymmetry breaking". *Nuclear Physics B* **557**: 79–118. arXiv:hep-th/9810155. Bibcode:1999NuPhB.557...79R. doi:10.1016/S0550-3213(99)00359-4.

[16] G. Giudice, M. Luty, H. Murayama, R. Rattazzi; Rattazzi; Luty; Murayama (1998). "Gaugino mass without singlets". *Journal of High Energy Physics* **9812** (12): 027. arXiv:hep-ph/9810442. Bibcode:1998JHEP...12..027G. doi:10.1088/1126-6708/1998/12/027.

[17] Djouadi, A.; Rosier-Lees, S.; Bezouh, M.; Bizouard, M. A.; Boehm, C.; Borzumati, F.; Briot, C.; Carr, J.; Causse, M. B.; Charles, F.; Chereau, X.; Colas, P.; Duflot, L.; Dupperin, A.; Ealet, A.; El-Mamouni, H.; Ghodbane, N.; Gieres, F.; Gonzalez-Pineiro, B.; Gourmelen, S.; Grenier, G.; Gris, Ph.;

Grivaz, J. -F.; Hebrard, C.; Ille, B.; Kneur, J. -L.; Kostantinidis, N.; Layssac, J.; Lebrun, P.; et al. (1999). "The Minimal Supersymmetric Standard Model: Group Summary Report". arXiv:hep-ph/9901246.

6.11 External links

- MSSM on arxiv.org

- Stephen P. Martin (1997). "A Supersymmetry Primer". arXiv:hep-ph/9709356.

- Particle Data Group review of MSSM and search for MSSM predicted particles

- Ian J R Aitchison (2005). "Supersymmetry and the MSSM: An Elementary Introduction". arXiv:hep-ph/0505105.

Chapter 7

Poincaré group

Henri Poincaré

7.1 Overview

A Minkowski spacetime isometry has the property that the interval between events is left invariant. For example, if everything was postponed by two hours, including the two events and the path you took to go from one to the other, then the time interval between the events recorded by a stop-watch you carried with you would be the same. Or if everything was shifted five miles to the west, or turned 60 degrees to the right, you would also see no change in the interval. It turns out that the proper length of an object is also unaffected by such a shift. A time or space reversal (a reflection) is also an isometry of this group.

In Minkowski space (i.e. ignoring the effects of gravity), there are ten degrees of freedom of the isometries, which may be thought of as translation through time or space (four degrees, one per dimension); reflection through a plane (three degrees, the freedom in orientation of this plane); or a "boost" in any of the three spatial directions (three degrees). Composition of transformations is the operator of the Poincaré group, with proper rotations being produced as the composition of an even number of reflections.

In classical physics, the Galilean group is a comparable ten-parameter group that acts on absolute time and space. Instead of boosts, it features shear mappings to relate comoving frames of reference.

7.2 Details

For the Poincaré group (fundamental group) of a topological space, see Fundamental group.

The **Poincaré group**, named after Henri Poincaré (1906),[1] was first defined by Minkowski (1908) being the group of Minkowski spacetime isometries.[2][3] It is a ten-generator non-abelian Lie group of fundamental importance in physics.

The Poincaré group is the group of Minkowski spacetime isometries. It is a ten-dimensional noncompact Lie group. The abelian group of translations is a normal subgroup, while the Lorentz group is also a subgroup, the stabilizer of the origin. The Poincaré group itself is the minimal subgroup of the affine group which includes all translations and Lorentz transformations. More precisely, it is a semidirect product of the translations and the Lorentz group,

$\mathbf{R}^{1,3} \rtimes SO(1,3)$.

Another way of putting this is that the Poincaré group is a group extension of the Lorentz group by a vector representation of it; it is sometimes dubbed, informally, as the *"inhomogeneous Lorentz group"*. In turn, it can also be obtained as a group contraction of the de Sitter group $SO(4,1) \sim Sp(2,2)$, as the de Sitter radius goes to infinity.

Its positive energy unitary irreducible representations are indexed by mass (nonnegative number) and spin (integer or half integer) and are associated with particles in quantum mechanics (see Wigner's classification).

In accordance with the Erlangen program, the geometry of Minkowski space is defined by the Poincaré group: Minkowski space is considered as a homogeneous space for the group.

The **Poincaré algebra** is the Lie algebra of the Poincaré group. It is a Lie algebra extension of the Lie algebra of the Lorentz group. More specifically, the proper (detΛ=1), orthochronous ($\Lambda^0{}_0 \geq 1$) part of the Lorentz subgroup (its identity component), $SO^+(1, 3)$, is connected to the identity and is thus provided by the exponentiation $\exp(ia_\mu P^\mu) \exp(i\omega_{\mu\nu} M^{\mu\nu}/2)$ of this Lie algebra. In component form, the Poincaré algebra is given by the commutation relations:[4][5]

where P is the generator of translations, M is the generator of Lorentz transformations, and η is the (+,−,−,−) Minkowski metric (see Sign convention).

The bottom commutation relation is the ("homogeneous") Lorentz group, consisting of rotations, $J_i = -\epsilon_{imn}M^{mn}/2$, and boosts, $K_i = M_{i0}$. In this notation, the entire Poincaré algebra is expressible in noncovariant (but more practical) language as

$$[J_m, P_n] = i\epsilon_{mnk}P_k ,$$

$$[J_i, P_0] = 0 ,$$

$$[K_i, P_k] = i\eta_{ik}P_0 ,$$

$$[K_i, P_0] = -iP_i ,$$

$$[J_m, J_n] = i\epsilon_{mnk}J_k ,$$

$$[J_m, K_n] = i\epsilon_{mnk}K_k ,$$

$$[K_m, K_n] = -i\epsilon_{mnk}J_k ,$$

where the bottom line commutator of two boosts is often referred to as a "Wigner rotation". Note the important simplification $[J_m + i K_m , J_n - i K_n] = 0$, which permits reduction of the Lorentz subalgebra to $\mathbf{su(2)} \oplus \mathbf{su(2)}$ and efficient treatment of its associated representations.

The Casimir invariants of this algebra are $P_\mu P^\mu$ and $W_\mu W^\mu$ where W_μ is the Pauli–Lubanski pseudovector; they serve as labels for the representations of the group.

The Poincaré group is the full symmetry group of any relativistic field theory. As a result, all elementary particles fall in representations of this group. These are usually specified by the *four-momentum* squared of each particle (i.e. its mass squared) and the intrinsic quantum numbers J^{PC}, where J is the spin quantum number, P is the parity and C is the charge-conjugation quantum number. In practice, charge conjugation and parity are violated by many quantum field theories; where this occurs, P and C are forfeited. Since CPT symmetry is invariant in quantum field theory, a time-reversal quantum number may be constructed from those given.

As a topological space, the group has four connected components: the component of the identity; the time reversed component; the spatial inversion component; and the component which is both time-reversed and spatially inverted.

7.3 Poincaré symmetry

Poincaré symmetry is the full symmetry of special relativity. It includes:

- *translations* (displacements) in time and space (**P**), forming the abelian Lie group of translations on spacetime;

- *rotations* in space, forming the non-Abelian Lie group of three-dimensional rotations (**J**);

- *boosts*, transformations connecting two uniformly moving bodies (**K**).

The last two symmetries, **J** and **K**, together make the Lorentz group (see also Lorentz invariance); the semi-direct product of the translations group and the Lorentz group then produce the Poincaré group. Objects which are invariant under this group are then said to possess **Poincaré invariance** or **relativistic invariance**.

7.4 See also

- Euclidean group

- Representation theory of the Poincaré group

- Wigner's classification

- Symmetry in quantum mechanics

- Center of mass (relativistic)

- Pauli–Lubanski pseudovector

- Particle physics and representation theory

7.5   Notes

[1] Poincaré, Henri, "Sur la dynamique de l'électron", *Rendiconti del Circolo matematico di Palermo* **21**: 129–176, doi:10.1007/bf03013466 (Wikisource translation: On the Dynamics of the Electron). The group defined in this paper would now be described as the homogeneous Lorentz group with scalar multipliers.

[2] Minkowski, Hermann, "Die Grundgleichungen für die elektromagnetischen Vorgänge in bewegten Körpern", *Nachrichten von der Gesellschaft der Wissenschaften zu Göttingen, Mathematisch-Physikalische Klasse*: 53–111 (Wikisource translation: The Fundamental Equations for Electromagnetic Processes in Moving Bodies).

[3] Minkowski, Hermann, "Raum und Zeit", *Physikalische Zeitschrift* **10**: 75–88

[4] N.N. Bogolubov (1989). *General Principles of Quantum Field Theory* (2nd ed.). Springer. p. 272. ISBN 0-7923-0540-X.

[5] T. Ohlsson (2011). *Relativistic Quantum Physics: From Advanced Quantum Mechanics to Introductory Quantum Field Theory*. Cambridge University Press. p. 10. ISBN 1-13950-4320.

7.6   References

- Wu-Ki Tung (1985). *Group Theory in Physics*. World Scientific Publishing. ISBN 9971-966-57-3.

- Weinberg, Steven (1995). *The Quantum Theory of Fields* **1**. Cambridge: Cambridge University press. ISBN 978-0-521-55001-7.

- L.H. Ryder (1996). *Quantum Field Theory* (2nd ed.). Cambridge University Press. p. 62. ISBN 0-52147-8146.

Chapter 8

Super-Poincaré algebra

In theoretical physics, a **super-Poincaré algebra** is an extension of the Poincaré algebra to incorporate supersymmetry, a relation between bosons and fermions. They are examples of supersymmetry algebras (without central charges or internal symmetries), and are Lie superalgebras. Thus a super-Poincaré algebra is a Z_2-graded vector space with a graded Lie bracket such that the even part is a Lie algebra containing the Poincaré algebra, and the odd part is built from spinors on which there is an anticommutation relation with values in the even part.

The simplest supersymmetric extension of the Poincaré algebra contains two Weyl spinors with the following anti-commutation relation:

$$\{Q_\alpha, \bar{Q}_\beta\} = 2\sigma^\mu{}_{\alpha\beta} P_\mu$$

and all other anti-commutation relations between the Qs and Ps vanish.[1] In the above expression P_μ are the generators of translation and σ^μ are the Pauli matrices.

To do this in full form it is easy to introduce the General Relativity metric. The Pauli and Dirac matrices should then depend on the metric $g^{\mu\nu}$ as:

$$\{\gamma^\mu, \gamma^\nu\} = 2g^{\mu\nu}$$

and

$$\sigma^{\mu\nu} = \frac{i}{2}[\gamma^\mu, \gamma^\nu]$$

This then gives the full algebra[2]

$$[M^{\mu\nu}, Q_\alpha] = \frac{1}{2}(\sigma^{\mu\nu})^\beta_\alpha Q_\beta$$

$$[Q_\alpha, P^\mu] = 0$$

$$\{Q_\alpha, \bar{Q}_\beta\} = 2(\sigma^\mu)_{\alpha\dot\beta} P_\mu$$

with the addition of the normal Poincaré algebra. It is a closed algebra since all Jacobi identities are satisfied and can have since explicit matrix representations. Following this line of reasoning will lead to Supergravity.

8.1 SUSY in 3 + 1 Minkowski space-time

In 3+1 Minkowski spacetime, the Haag-Lopuszanski-Sohnius theorem states that the SUSY algebra with N spinor generators is as follows.

The even part of the star Lie superalgebra is the direct sum of the Poincaré algebra and a reductive Lie algebra B (such that its self-adjoint part is the tangent space of a real compact Lie group). The odd part of the algebra would be

$$\left(\frac{1}{2}, 0\right) \otimes V \oplus \left(0, \frac{1}{2}\right) \otimes V^*$$

where $(1/2, 0)$ and $(0, 1/2)$ are specific representations of the Poincaré algebra. Both components are conjugate to each other under the * conjugation. V is an N-dimensional complex representation of B and V^* is its dual representation. The Lie bracket for the odd part is given by a symmetric equivariant pairing $\{.,.\}$ on the odd part with values in the even part. In particular, its reduced intertwiner from $[(\frac{1}{2}, 0) \otimes V] \otimes [(0, \frac{1}{2}) \otimes V^*]$ to the ideal of the Poincaré algebra generated by translations is given as the product of a nonzero intertwiner from $(\frac{1}{2}, 0) \otimes (0, \frac{1}{2})$ to $(1/2, 1/2)$ by the "contraction intertwiner" from $V \otimes V^*$ to the trivial representation. On the other hand, its reduced intertwiner from $[(\frac{1}{2}, 0) \otimes V] \otimes [(\frac{1}{2}, 0) \otimes V]$ is the product of a (anti-symmetric) intertwiner from $(\frac{1}{2}, 0) \otimes (\frac{1}{2}, 0)$ to $(0,0)$ and an antisymmetric intertwiner A from N^2 to B. Conjugate it to get the corresponding case for the other half.

8.1.1 $N = 1$

B is now $u(1)$ (called R-symmetry) and V is the 1D representation of $u(1)$ with "charge" 1. A (the intertwiner defined above) would have to be zero since it is antisymmetric.

Actually, there are two versions of $N=1$ SUSY, one without the $u(1)$ (i.e. B is zero-dimensional) and the other with $u(1)$.

8.1.2 $N = 2$

B is now $su(2) \oplus u(1)$ and V is the 2D doublet representation of $su(2)$ with a zero $u(1)$ "charge". Now, A is a nonzero intertwiner to the $u(1)$ part of B.

Alternatively, V could be a 2D doublet with a nonzero $u(1)$ "charge". In this case, A would have to be zero.

Yet another possibility would be to let B be $u(1)_A \oplus u(1)_B \oplus u(1)_C$. V is invariant under $u(1)_B$ and $u(1)_C$ and decomposes into a 1D rep with $u(1)_A$ charge 1 and another 1D rep with charge -1. The intertwiner A would be complex with the real part mapping to $u(1)_B$ and the imaginary part mapping to $u(1)_C$.

Or we could have B being $su(2) \oplus u(1)_A \oplus u(1)_B$ with V being the doublet rep of $su(2)$ with zero $u(1)$ charges and A being a complex intertwiner with the real part mapping to $u(1)_A$ and the imaginary part to $u(1)_B$.

This doesn't even exhaust all the possibilities. We see that there is more than one $N = 2$ supersymmetry; likewise, the SUSYs for $N > 2$ are also not unique (in fact, it only gets worse).

8.1.3 $N = 3$

It is theoretically allowed, but the multiplet structure becomes automatically the same with that of an $N=4$ supersymmetric theory. So it is less often discussed compared to $N=1,2,4$ versions.

8.1.4 $N = 4$

This is the maximal number of supercharges in a theory without gravity.

8.1.5   SUSY in various dimensions

In $0 + 1$, $2 + 1$, $3 + 1$, $4 + 1$, $6 + 1$, $7 + 1$, $8 + 1$, $10 + 1$ dimensions, etc., a SUSY algebra is classified by a positive integer N.

In $1 + 1$, $5 + 1$, $9 + 1$ dimensions, etc., a SUSY algebra is classified by two nonnegative integers (M, N), at least one of which is nonzero. M represents the number of left-handed SUSYs and N represents the number of right-handed SUSYs.

The reason of this has to do with the reality conditions of the spinors.

Hereafter $d = 9$ means $d = 8 + 1$ in Minkowski signature, etc. The structure of supersymmetry algebra is mainly determined by the number of the fermionic generators, that is the number N times the real dimension of the spinor in d dimensions. It is because one can obtain a supersymmetry algebra of lower dimension easily from that of higher dimensionality by the use of dimensional reduction.

$d = 11$

The only example is the $N = 1$ supersymmetry with 32 supercharges.

$d = 10$

From $d = 11$, $N = 1$ susy, one obtains $N = (1, 1)$ nonchiral susy algebra, which is also called the type IIA supersymmetry. There is also $N = (2, 0)$ susy algebra, which is called the type IIB supersymmetry. Both of them have 32 supercharges.

$N = (1, 0)$ susy algebra with 16 supercharges is the minimal susy algebra in 10 dimensions. It is also called the type I supersymmetry. Type IIA / IIB / I superstring theory has the susy algebra of the corresponding name. The supersymmetry algebra for the heterotic superstrings is that of type I.

8.2   References

[1] Aitchison, Ian J R. "Supersymmetry and the MSSM: An Elementary Introduction". arXiv:hep-ph/0505105.

[2] P. van Nieuwenhuizen, Phys. Rep. 68, 189 (1981)

Chapter 9

Cartan connection

In the mathematical field of differential geometry, a **Cartan connection** is a flexible generalization of the notion of an affine connection. It may also be regarded as a specialization of the general concept of a principal connection, in which the geometry of the principal bundle is tied to the geometry of the base manifold using a solder form. Cartan connections describe the geometry of manifolds modelled on homogeneous spaces.

The theory of Cartan connections was developed by Élie Cartan, as part of (and a way of formulating) his method of moving frames (**repère mobile**).[1] The main idea is to develop a suitable notion of the connection forms and curvature using moving frames adapted to the particular geometrical problem at hand. For instance, in relativity or Riemannian geometry, orthonormal frames are used to obtain a description of the Levi-Civita connection as a Cartan connection. For Lie groups, Maurer–Cartan frames are used to view the Maurer–Cartan form of the group as a Cartan connection.

Cartan reformulated the differential geometry of (pseudo) Riemannian geometry, as well as the differential geometry of manifolds equipped with some non-metric structure, including Lie groups and homogeneous spaces. The term Cartan connection most often refers to Cartan's formulation of a (pseudo-)Riemannian, affine, projective, or conformal connection. Although these are the most commonly used Cartan connections, they are special cases of a more general concept.

Cartan's approach seems at first to be coordinate dependent because of the choice of frames it involves. However, it is not, and the notion can be described precisely using the language of principal bundles. Cartan connections induce covariant derivatives and other differential operators on certain associated bundles, hence a notion of parallel transport. They have many applications in geometry and physics: see the method of moving frames, Cartan connection applications and Einstein–Cartan theory for some examples.

9.1 Introduction

At its roots, geometry consists of a notion of *congruence* between different objects in a space. In the late 19th century, notions of congruence were typically supplied by the action of a Lie group on space. Lie groups generally act quite rigidly, and so a Cartan geometry is a generalization of this notion of congruence to allow for curvature to be present. The *flat* Cartan geometries — those with zero curvature — are locally equivalent to homogeneous spaces, hence geometries in the sense of Klein.

A Klein geometry consists of a Lie group G together with a Lie subgroup H of G. Together G and H determine a homogeneous space G/H, on which the group G acts by left-translation. Klein's aim was then to study objects living on the homogeneous space which were *congruent* by the action of G. A Cartan geometry extends the notion of a Klein geometry by attaching to each point of a manifold a copy of a Klein geometry, and to regard this copy as *tangent* to the manifold. Thus the geometry of the manifold is *infinitesimally* identical to that of the Klein geometry, but globally can be quite different. In particular, Cartan geometries no longer have a well-defined action of G on them. However, a **Cartan connection** supplies a way of connecting the infinitesimal model spaces within the manifold by means of parallel transport.

9.1.1 Motivation

Consider a smooth surface S in 3-dimensional Euclidean space $\mathbf{R}^3$. Near to any point, S can be approximated by its tangent plane at that point, which is an affine subspace of Euclidean space. The affine subspaces are *model* surfaces — they are the simplest surfaces in $\mathbf{R}^3$, and are homogeneous under the Euclidean group of the plane, hence they are *Klein geometries* in the sense of Felix Klein's Erlangen programme. Every smooth surface S has a unique affine plane tangent to it at each point. The family of all such planes in $\mathbf{R}^3$, one attached to each point of S, is called the **congruence** of tangent planes. A tangent plane can be "rolled" along S, and as it does so the point of contact traces out a curve on S. Conversely, given a curve on S, the tangent plane can be rolled along that curve. This provides a way to identify the tangent planes at different points along the curve by affine (in fact Euclidean) transformations, and is an example of a Cartan connection called an affine connection.

Another example is obtained by replacing the planes, as model surfaces, by spheres, which are homogeneous under the Möbius group of conformal transformations. There is no longer a unique sphere tangent to a smooth surface S at each point, since the radius of the sphere is undetermined. This can be fixed by supposing that the sphere has the same mean curvature as S at the point of contact. Such spheres can again be rolled along curves on S, and this equips S with another type of Cartan connection called a conformal connection.

Differential geometers in the late 19th and early 20th century were very interested in using model families such as planes or spheres to describe the geometry of surfaces. A family of model spaces attached to each point of a surface S is called a **congruence**: in the previous examples there is a canonical choice of such a congruence. A Cartan connection provides an identification between the model spaces in the congruence along any curve in S. An important feature of these identifications is that the point of contact of the model space with S *always moves* with the curve. This generic condition is characteristic of Cartan connections.

In the modern treatment of affine connections, the point of contact is viewed as the *origin* in the tangent plane (which is then a vector space), and the movement of the origin is corrected by a translation, and so Cartan connections are not needed. However, there is no canonical way to do this in general: in particular for the conformal connection of a sphere congruence, it is not possible to separate the motion of the point of contact from the rest of the motion in a natural way.

In both of these examples the model space is a homogeneous space G/H.

- In the first case, G/H is the affine plane, with $G =$ Aff$(\mathbf{R}^2)$ the affine group of the plane, and $H = $ GL(2) the corresponding general linear group.

- In the second case, G/H is the conformal (or celestial) sphere, with $G = $ O$^+(3,1)$ the (orthochronous) Lorentz group, and H the stabilizer of a null line in $\mathbf{R}^{3,1}$.

The Cartan geometry of S consists of a copy of the model space G/H at each point of S (with a marked point of contact) together with a notion of "parallel transport" along curves which identifies these copies using elements of G. This notion of parallel transport is generic in the intuitive sense that the point of contact always moves along the curve.

In general, let G be a group with a subgroup H, and M a manifold of the same dimension as G/H. Then, roughly speaking, a Cartan connection on M is a G-connection which is generic with respect to a reduction to H.

9.1.2 Affine connections

Main article: Affine connection

An **affine connection** on a manifold M is a connection (principal bundle) on the frame bundle of M (or equivalently, a connection (vector bundle) on the tangent bundle of M). A key aspect of the Cartan connection point of view is to elaborate this notion in the context of principal bundles (which could be called the "general or abstract theory of frames").

Let H be a Lie group, $\mathfrak{h}$ its Lie algebra. Then a **principal H-bundle** is a fiber bundle P over M with a smooth action of H on P which is free and transitive on the fibers. Thus P is a smooth manifold with a smooth map $\pi\colon P \to M$ which looks *locally* like the trivial bundle $M \times H \to M$. The frame bundle of M is a principal GL(n)-bundle, while if M is a Riemannian manifold, then the orthonormal frame bundle is a principal O(n)-bundle.

Let Rh denote the (right) action of $h \in $ H on P. The derivative of this action defines a **vertical vector field** on P for each element ξ of $\mathfrak{h}$: if $h(t)$ is a 1-parameter subgroup with $h(0)=e$ (the identity element) and $h'(0)=\xi$, then the corresponding vertical vector field is

$$X_\xi = \left.\frac{\mathrm{d}}{\mathrm{d}t} R_{h(t)}\right|_{t=0}.$$

A **principal H-connection** on P is a 1-form $\omega\colon TP \to \mathfrak{h}$ on P, with values in the Lie algebra $\mathfrak{h}$ of H, such that

1. $\mathrm{Ad}(h)(R_h^* \omega) = \omega$

2. for any $\xi \in \mathfrak{h}$, $\omega(X\xi) = \xi$ (identically on P).

The intuitive idea is that $\omega(X)$ provides a *vertical component* of X, using the isomorphism of the fibers of π with H to identify vertical vectors with elements of $\mathfrak{h}$.

Frame bundles have additional structure called the solder form, which can be used to extend a principal connection on P to a trivialization of the tangent bundle of P called an **absolute parallelism**.

In general, suppose that M has dimension n and H acts on $\mathbf{R}^n$ (this could be any n-dimensional real vector space). A **solder form** on a principal H-bundle P over M is an $\mathbf{R}^n$-valued 1-form $\theta\colon TP \to \mathbf{R}^n$ which is horizontal and equivariant so that it induces a bundle homomorphism from TM

to the associated bundle $P \times_H \mathbf{R}^n$. This is furthermore required to be a bundle isomorphism. Frame bundles have a (canonical or tautological) solder form which sends a tangent vector $X \in T_pP$ to the coordinates of $d\pi p(X) \in T_{\pi(p)}M$ with respect to the frame p.

The pair (ω, θ) (a principal connection and a solder form) defines a 1-form η on P, with values in the Lie algebra $\mathfrak{g}$ of the semidirect product G of H with $\mathbf{R}^n$, which provides an isomorphism of each tangent space T_pP with $\mathfrak{g}$. It induces a principal connection α on the associated principal G-bundle $P \times_H G$. This is a Cartan connection.

Cartan connections generalize affine connections in two ways.

- The action of H on $\mathbf{R}^n$ need not be effective. This allows, for example, the theory to include spin connections, in which H is the spin group $\mathrm{Spin}(n)$ rather than the orthogonal group $\mathrm{O}(n)$.

- The group G need not be a semidirect product of H with $\mathbf{R}^n$.

9.1.3 Klein geometries as model spaces

Klein's Erlangen programme suggested that geometry could be regarded as a study of homogeneous spaces: in particular, it is the study of the many geometries of interest to geometers of 19th century (and earlier). A Klein geometry consisted of a space, along with a law for motion within the space (analogous to the Euclidean transformations of classical Euclidean geometry) expressed as a Lie group of transformations. These generalized spaces turn out to be homogeneous smooth manifolds diffeomorphic to the quotient space of a Lie group by a Lie subgroup. The extra differential structure that these homogeneous spaces possess allows one to study and generalize their geometry using calculus.

The general approach of Cartan is to begin with such a *smooth Klein geometry*, given by a Lie group G and a Lie subgroup H, with associated Lie algebras $\mathfrak{g}$ and $\mathfrak{h}$, respectively. Let P be the underlying principal homogeneous space of G. A Klein geometry is the homogeneous space given by the quotient P/H of P by the right action of H. There is a right H-action on the fibres of the canonical projection

$$\pi: P \to P/H$$

given by $Rhg = gh$. Moreover, each fibre of π is a copy of H. P has the structure of a principal H-bundle over P/H.[2]

A vector field X on P is *vertical* if $d\pi(X) = 0$. Any $\xi \in \mathfrak{h}$ gives rise to a canonical vertical vector field $X\xi$ by taking the

derivative of the right action of the 1-parameter subgroup of H associated to ξ. The Maurer-Cartan form η of P is the $\mathfrak{g}$-valued one-form on P which identifies each tangent space with the Lie algebra. It has the following properties:

1. $\mathrm{Ad}(h)\, Rh^*\eta = \eta$ for all h in H

2. $\eta(X\xi) = \xi$ for all ξ in $\mathfrak{h}$

3. for all $g \in P$, η restricts a linear isomorphism of T_gP with $\mathfrak{g}$ (η is an **absolute parallelism** on P).

In addition to these properties, η satisfies the **structure (or structural) equation**

$$d\eta + \tfrac{1}{2}[\eta, \eta] = 0.$$

Conversely, one can show that given a manifold M and a principal H-bundle P over M, and a 1-form η with these properties, then P is locally isomorphic as an H-bundle to the principal homogeneous bundle $G \to G/H$. The structure equation is the integrability condition for the existence of such a local isomorphism.

A Cartan geometry is a generalization of a smooth Klein geometry, in which the structure equation is not assumed, but is instead used to define a notion of curvature. Thus the Klein geometries are said to be the **flat models** for Cartan geometries.[3]

9.2 Cartan connections and pseudogroups

Cartan connections are closely related to pseudogroup structures on a manifold. Each is thought of as *modelled on* a Klein geometry G/H, in a manner similar to the way in which Riemannian geometry is modelled on Euclidean space. On a manifold M, one imagines attaching to each point of M a copy of the model space G/H. The symmetry of the model space is then built into the Cartan geometry or pseudogroup structure by positing that the model spaces of nearby points are related by a transformation in G. The fundamental difference between a Cartan geometry and pseudogroup geometry is that the symmetry for a Cartan geometry relates *infinitesimally* close points by an *infinitesimal* transformation in G (i.e., an element of the Lie algebra of G) and the analogous notion of symmetry for a pseudogroup structure applies for points that are physically separated within the manifold.

The process of attaching spaces to points, and the attendant symmetries, can be concretely realized by using special coordinate systems.[4] To each point $p \in M$, a neighborhood

U_p of p is given along with a mapping $\varphi_p : U_p \to G/H$. In this way, the model space is attached to each point of M by realizing M locally at each point as an open subset of G/H. We think of this as a family of coordinate systems on M, parametrized by the points of M. Two such parametrized coordinate systems φ and φ' are H-related if there is an element $h_p \in H$, parametrized by p, such that

$$\varphi'_p = h_p\, \varphi_p.^{[5]}$$

This freedom corresponds roughly to the physicists' notion of a gauge.

Nearby points are related by joining them with a curve. Suppose that p and p' are two points in M joined by a curve p_t. Then p_t supplies a notion of transport of the model space along the curve.[6] Let $\tau_t : G/H \to G/H$ be the (locally defined) composite map

$$\tau_t = \varphi_{pt} \circ \varphi_{p0}^{-1}.$$

Intuitively, τ_t is the transport map. A pseudogroup structure requires that τ_t be a *symmetry of the model space* for each t: $\tau_t \in G$. A Cartan connection requires only that the derivative of τ_t be a symmetry of the model space: $\tau'_0 \in \mathfrak{g}$, the Lie algebra of G.

Typical of Cartan, one motivation for introducing the notion of a Cartan connection was to study the properties of pseudogroups from an infinitesimal point of view. A Cartan connection defines a pseudogroup precisely when the derivative of the transport map τ' can be integrated, thus recovering a true (G-valued) transport map between the coordinate systems. There is thus an integrability condition at work, and Cartan's method for realizing integrability conditions was to introduce a differential form.

In this case, τ'_0 defines a differential form at the point p as follows. For a curve $\gamma(t) = p_t$ in M starting at p, we can associate the tangent vector X, as well as a transport map τ_t^γ. Taking the derivative determines a linear map

$$X \mapsto \frac{d}{dt}\tau_t^\gamma \Big|_{t=0} = \theta(X) \in \mathfrak{g}.$$

So θ defines a $\mathfrak{g}$-valued differential 1-form on M.

This form, however, is dependent on the choice of parametrized coordinate system. If $h : U \to H$ is an H-relation between two parametrized coordinate systems φ and φ', then the corresponding values of θ are also related by

$$\theta'_p = Ad(h_p^{-1})\theta_p + h_p^*\omega_H,$$

where ωH is the Maurer-Cartan form of H.

9.3 Formal definition

A Cartan geometry modelled on a homogeneous space G/H can be viewed as a *deformation* of this geometry which allows for the presence of *curvature*. For example:

- a Riemannian manifold can be seen as a deformation of Euclidean space;
- a Lorentzian manifold can be seen as a deformation of Minkowski space;
- a conformal manifold can be seen as a deformation of the conformal sphere;
- a manifold equipped with an affine connection can be seen as a deformation of an affine space.

There are two main approaches to the definition. In both approaches, M is a smooth manifold of dimension n, H is a Lie group of dimension m, with Lie algebra $\mathfrak{h}$, and G is a Lie group G of dimension $n+m$, with Lie algebra $\mathfrak{g}$, containing H as a subgroup.

9.3.1 Definition via gauge transitions

A **Cartan connection** consists[7][8] of a coordinate atlas of open sets U in M, along with a $\mathfrak{g}$-valued 1-form θU defined on each chart such that

1. $\theta U : TU \to \mathfrak{g}$.

2. $\theta U \bmod \mathfrak{h} : T_u U \to \mathfrak{g}/\mathfrak{h}$ is a linear isomorphism for every $u \in U$.

3. For any pair of charts U and V in the atlas, there is a smooth mapping $h : U \cap V \to H$ such that

$$\theta_V = Ad(h^{-1})\theta_U + h^*\omega_H,$$

where ωH is the Maurer-Cartan form of H.

By analogy with the case when the θU came from coordinate systems, condition 3 means that φU is related to φV by h.

The curvature of a Cartan connection consists of a system of 2-forms defined on the charts, given by

$$\Omega_U = d\theta_U + \tfrac{1}{2}[\theta_U, \theta_U].$$

ΩU satisfy the compatibility condition:

If the forms θU and θV are related by a function $h : U \cap V \to H$, as above, then $\Omega V = \text{Ad}(h^{-1}) \Omega U$

The definition can be made independent of the coordinate systems by forming the quotient space

$$P = (\coprod_U U \times H)/ \sim$$

of the disjoint union over all U in the atlas. The equivalence relation $\sim$ is defined on pairs $(x,h_1) \in U_1 \times H$ and $(x, h_2) \in U_2 \times H$, by

$(x,h_1) \sim (x, h_2)$ if and only if $x \in U_1 \cap U_2$, θU_1 is related to θU_2 by h, and $h_2 = h(x)^{-1} h_1$.

Then P is a principal H-bundle on M, and the compatibility condition on the connection forms θU implies that they lift to a g-valued 1-form η defined on P (see below).

9.3.2 Definition via absolute parallelism

Let P be a principal H bundle over M. Then a **Cartan connection**[9] is a $\mathfrak{g}$-valued 1-form η on P such that

1. for all h in H, $\text{Ad}(h)Rh^* \eta = \eta$

2. for all ξ in $\mathfrak{h}$, $\eta(X\xi) = \xi$

3. for all p in P, the restriction of η defines a linear isomorphism from the tangent space T_pP to $\mathfrak{g}$.

The last condition is sometimes called the **Cartan condition**: it means that η defines an **absolute parallelism** on P. The second condition implies that η is already injective on vertical vectors and that the 1-form η mod $\mathfrak{h}$, with values in $\mathfrak{g}/\mathfrak{h}$, is horizontal. The vector space $\mathfrak{g}/\mathfrak{h}$ is a representation of H using the adjoint representation of H on $\mathfrak{g}$, and the first condition implies that η mod $\mathfrak{h}$ is equivariant. Hence it defines a bundle homomorphism from TM to the associated bundle $P \times_H \mathfrak{g}/\mathfrak{h}$. The Cartan condition is equivalent to this bundle homomorphism being an isomorphism, so that η mod $\mathfrak{h}$ is a solder form.

The **curvature** of a Cartan connection is the $\mathfrak{g}$-valued 2-form Ω defined by

$$\Omega = d\eta + \tfrac{1}{2}[\eta \wedge \eta].$$

Note that this definition of a Cartan connection looks very similar to that of a principal connection. There are several important differences, however. First, the 1-form η takes values in g, but is only equivariant under the action of H. Indeed, it cannot be equivariant under the full group G because there is no G bundle and no G action. Secondly, the 1-form is an absolute parallelism, which intuitively means that η yields information about the behavior of additional directions in the principal bundle (rather than simply being a projection operator onto the vertical space). Concretely, the existence of a solder form binds (or solders) the Cartan connection to the underlying differential topology of the manifold.

An intuitive interpretation of the Cartan connection in this form is that it determines a *fracturing* of the tautological principal bundle associated to a Klein geometry. Thus Cartan geometries are deformed analogues of Klein geometries. This deformation is roughly a prescription for attaching a copy of the model space G/H to each point of M and thinking of that model space as being *tangent* to (and *infinitesimally identical* with) the manifold at a point of contact. The fibre of the tautological bundle $G \to G/H$ of the Klein geometry at the point of contact is then identified with the fibre of the bundle P. Each such fibre (in G) carries a Maurer-Cartan form for G, and the Cartan connection is a way of assembling these Maurer-Cartan forms gathered from the points of contact into a coherent 1-form η defined on the whole bundle. The fact that only elements of H contribute to the Maurer-Cartan equation $\text{Ad}(h)Rh^* \eta = \eta$ has the intuitive interpretation that any other elements of G would move the model space away from the point of contact, and so no longer be tangent to the manifold.

From the Cartan connection, defined in these terms, one can recover a Cartan connection as a system of 1-forms on the manifold (as in the gauge definition) by taking a collection of local trivializations of P given as sections $sU : U \to P$ and letting $\theta U = s^* \eta$ be the pullbacks of the Cartan connection along the sections.

9.3.3 Cartan connections as principal connections

Another way in which to define a Cartan connection is as a principal connection on a certain principal G-bundle. From this perspective, a Cartan connection consists of

- a principal G-bundle Q over M

- a principal G-connection α on Q (the Cartan connection)

- a principal H-subbundle P of Q (i.e., a reduction of structure group)

such that the pullback η of α to P satisfies the Cartan condition.

The principal connection α on Q can be recovered from the form η by taking Q to be the associated bundle $P \times H\ G$. Conversely, the form η can be recovered from α by pulling back along the inclusion $P \subset Q$.

Since α is a principal connection, it induces a connection on any associated bundle to Q. In particular, the bundle $Q \times G\ G/H$ of homogeneous spaces over M, whose fibers are copies of the model space G/H, has a connection. The reduction of structure group to H is equivalently given by a section s of $E = Q \times G\ G/H$. The fiber of $P \times_H \mathfrak{g}/\mathfrak{h}$ over x in M may be viewed as the tangent space at $s(x)$ to the fiber of $Q \times G\ G/H$ over x. Hence the Cartan condition has the intuitive interpretation that the model spaces are tangent to M along the section s. Since this identification of tangent spaces is induced by the connection, the marked points given by s always move under parallel transport.

9.3.4 Definition by an Ehresmann connection

Yet another way to define a Cartan connection is with an Ehresmann connection on the bundle $E = Q \times G\ G/H$ of the preceding section.[10] A Cartan connection then consists of

- A fibre bundle $\pi : E \to M$ with fibre G/H and vertical space $VE \subset TE$.

- A section $s : M \to E$.

- A G-connection $\theta : TE \to VE$ such that

  $s^* \theta_x : T_x M \to V_{s(x)}E$ is a linear isomorphism of vector spaces for all $x \in M$.

This definition makes rigorous the intuitive ideas presented in the introduction. First, the preferred section s can be thought of as identifying a point of contact between the manifold and the tangent space. The last condition, in particular, means that the tangent space of M at x is isomorphic to the tangent space of the model space at the point of contact. So the model spaces are, in this way, tangent to the manifold.

This definition also brings prominently into focus the idea of development. If x_t is a curve in M, then the Ehresmann connection on E supplies an associated parallel transport map $\tau_t : E_{x_t} \to E_{x_0}$ from the fibre over the endpoint of the curve to the fibre over the initial point. In particular, since E is equipped with a preferred section s, the points $s(x_t)$ transport back to the fibre over x_0 and trace out a curve in E_{x_0}. This curve is then called the *development* of the curve x_t.

To show that this definition is equivalent to the others above, one must introduce a suitable notion of a moving

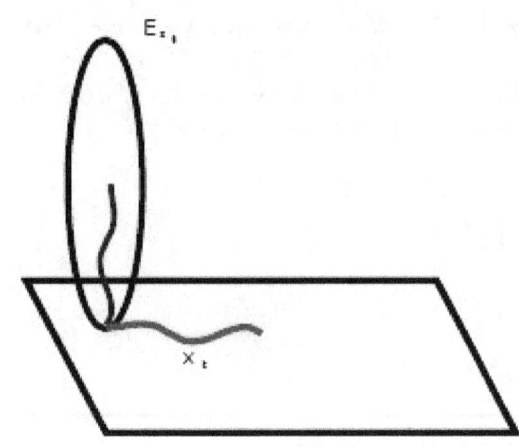

Development of a curve into the model space at x_0

frame for the bundle E. In general, this is possible for any G-connection on a fibre bundle with structure group G. See Ehresmann connection#Associated bundles for more details.

9.4 Special Cartan connections

9.4.1 Reductive Cartan connections

Let P be a principal H-bundle on M, equipped with a Cartan connection $\eta : TP \to \mathfrak{g}$. If $\mathfrak{g}$ is a reductive module for H, meaning that $\mathfrak{g}$ admits an $\mathrm{Ad}(H)$-invariant splitting of vector spaces $\mathfrak{g} = \mathfrak{h} \oplus \mathfrak{m}$, then the $\mathfrak{m}$-component of η generalizes the solder form for an affine connection.[11] In detail, η splits into $\mathfrak{h}$ and $\mathfrak{m}$ components:

$$\eta = \eta_{\mathfrak{h}} + \eta_{\mathfrak{m}}.$$

Note that the 1-form $\eta_{\mathfrak{h}}$ is a principal H-connection on the original Cartan bundle P. Moreover, the 1-form $\eta_{\mathfrak{m}}$ satisfies:

$\eta_{\mathfrak{m}}(X) = 0$ for every vertical vector $X \in TP$. ($\eta_{\mathfrak{m}}$ is *horizontal*.)

$R_h^* \eta_{\mathfrak{m}} = \mathrm{Ad}(h^{-1}) \eta_{\mathfrak{m}}$ for every $h \in H$. ($\eta_{\mathfrak{m}}$ is *equivariant* under the right H-action.)

In other words, η is a solder form for the bundle P.

Hence, P equipped with the form $\eta_{\mathfrak{m}}$ defines a (first order) H-structure on M. The form $\eta_{\mathfrak{h}}$ defines a connection on the H-structure.

9.4.2 Parabolic Cartan connections

If **g** is a semisimple Lie algebra with parabolic subalgebra **p** (i.e., **p** contains a maximal solvable subalgebra of **g**) and G and P are associated Lie groups, then a Cartan connection modelled on $(G,P,\mathbf{g},\mathbf{p})$ is called a **parabolic Cartan geometry**, or simply a **parabolic geometry**. A distinguishing feature of parabolic geometries is a Lie algebra structure on its cotangent spaces: this arises because the perpendicular subspace $\mathbf{p}^\perp$ of **p** in **g** with respect to the Killing form of **g** is a subalgebra of **p**, and the Killing form induces a natural duality between $\mathbf{p}^\perp$ and **g**/**p**. Thus the bundle associated to $\mathbf{p}^\perp$ is isomorphic to the cotangent bundle.

Parabolic geometries include many of those of interest in research and applications of Cartan connections, such as the following examples:

- Conformal connections: Here $G = SO(p+1,q+1)$, and P is the stabilizer of a null ray in $\mathbf{R}^{n+2}$.

- Projective connections: Here $G = PGL(n+1)$ and P is the stabilizer of a point in $\mathbf{RP}^n$.

- CR structures and Cartan-Chern-Tanaka connections: $G = PSU(p+1,q+1)$, P = stabilizer of a point on the projective null hyperquadric.

- Contact projective connections:[12] Here $G = SP(2n+2)$ and P is the stabilizer of the ray generated by the first standard basis vector in $\mathbf{R}^{n+2}$.

- Generic rank 2 distributions on 5-manifolds: Here $G = Aut(\mathbf{O}_s)$ is the automorphism group of the algebra $\mathbf{O}_s$ of split octonions, a closed subgroup of $SO(3,4)$, and P is the intersection of G with the stabilizer of the isotropic line spanned by the first standard basis vector in $\mathbf{R}^7$ viewed as the purely imaginary split octonions (orthogonal complement of the unit element in $\mathbf{O}_s$).[13]

9.5 Associated differential operators

9.5.1 Covariant differentiation

Suppose that M is a Cartan geometry modelled on G/H, and let (Q,α) be the principal G-bundle with connection, and (P,η) the corresponding reduction to H with η equal to the pullback of α. Let V a representation of G, and form the vector bundle $\mathbf{V} = Q \times_G V$ over M. Then the principal G-connection α on Q induces a covariant derivative on $\mathbf{V}$, which is a first order linear differential operator

$$\nabla : \Omega^0_M(\mathbf{V}) \to \Omega^1_M(\mathbf{V}),$$

where $\Omega^k_M(\mathbf{V})$ denotes the space of k-forms on M with values in $\mathbf{V}$ so that $\Omega^0_M(\mathbf{V})$ is the space of sections of $\mathbf{V}$ and $\Omega^1_M(\mathbf{V})$ is the space of sections of Hom(TM,$\mathbf{V}$). For any section v of $\mathbf{V}$, the contraction of the covariant derivative ∇v with a vector field X on M is denoted ∇Xv and satisfies the following Leibniz rule:

$$\nabla_X(fv) = df(X)v + f\nabla_X v$$

for any smooth function f on M.

The covariant derivative can also be constructed from the Cartan connection η on P. In fact, constructing it in this way is slightly more general in that V need not be a fully fledged representation of G.[14] Suppose instead that that V is a ($\mathfrak{g}$, H)-module: a representation of the group H with a compatible representation of the Lie algebra $\mathfrak{g}$. Recall that a section v of the induced vector bundle $\mathbf{V}$ over M can be thought of as an H-equivariant map $P \to V$. This is the point of view we shall adopt. Let X be a vector field on M. Choose any right-invariant lift $\bar{X}$ to the tangent bundle of P. Define

$$\nabla_X v = dv(\bar{X}) + \eta(\bar{X}) \cdot v$$

In order to show that ∇v is well defined, it must:

1. be independent of the chosen lift $\bar{X}$

2. be equivariant, so that it descends to a section of the bundle $\mathbf{V}$.

For (1), the ambiguity in selecting a right-invariant lift of X is a transformation of the form $X \mapsto X + X_\xi$ where X_ξ is the right-invariant vertical vector field induced from $\xi \in \mathfrak{h}$. So, calculating the covariant derivative in terms of the new lift $\bar{X} + X_\xi$, one has

$$\nabla_X v = dv(\bar{X} + X_\xi) + \eta(\bar{X} + X_\xi)) \cdot v$$

$$= dv(\bar{X}) + dv(X_\xi) + \eta(\bar{X}) \cdot v + \xi \cdot v$$

$$= dv(\bar{X}) + \eta(\bar{X}) \cdot v$$

since $\xi \cdot v + dv(X_\xi) = 0$ by taking the differential of the equivariance property $h \cdot R_h^* v = v$ at h equal to the identity element.

For (2), observe that since v is equivariant and $\bar{X}$ is right-invariant, $dv(\bar{X})$ is equivariant. On the other hand, since η is also equivariant, it follows that $\eta(\bar{X}) \cdot v$ is equivariant as well.

9.5.2 The fundamental or universal derivative

Suppose that V is only a representation of the subgroup H and not necessarily the larger group G. Let $\Omega^k(P, V)$ be the space of V-valued differential k-forms on P. In the presence of a Cartan connection, there is a canonical isomorphism

$$\varphi\colon \Omega^k(P, V) \cong \Omega^0(P, V \otimes \textstyle\bigwedge^k \mathfrak{g}^*)$$

given by $\varphi(\beta)(\xi_1, \xi_2, \ldots, \xi_k) = \beta(\eta^{-1}(\xi_1), \ldots, \eta^{-1}(\xi_k))$ where $\beta \in \Omega^k(P, V)$ and $\xi_j \in \mathfrak{g}$.

For each k, the exterior derivative is a first order operator differential operator

$$d\colon \Omega^k(P, V) \to \Omega^{k+1}(P, V)$$

and so, for $k=0$, it defines a differential operator

$$\varphi \circ d\colon \Omega^0(P, V) \to \Omega^0(P, V \otimes \mathfrak{g}^*).$$

Because η is equivariant, if v is equivariant, so is $Dv := \varphi(dv)$. It follows that this composite descends to a first order differential operator D from sections of $\mathbf{V} = P \times_H V$ to sections of the bundle $P \times_H (\mathbf{V} \otimes \mathfrak{g}^*)$. This is called the fundamental or universal derivative, or fundamental D-operator.

9.6 Notes

[1] Although Cartan only began formalizing this theory in particular cases in the 1920s (Cartan 1926), he made much use of the general idea much earlier. In particular, the high point of his remarkable 1910 paper on Pfaffian systems in five variables is the construction of a Cartan connection modelled on a 5-dimensional homogeneous space for the exceptional Lie group G_2, which he and Engels had discovered independently in 1894.

[2] Chevalley 1946, p. 110.

[3] See R. Hermann (1983), Appendix 1–3 to Cartan (1951).

[4] This appears to be Cartan's way of viewing the connection. Cf. Cartan 1923, p. 362; Cartan 1924, p. 208 especially *..un repère définissant un système de coordonnées projectives...*; Cartan 1951, p. 34. Modern readers can arrive at various interpretations of these statements, cf. Hermann's 1983 notes in Cartan 1951, pp. 384–385, 477.

[5] More precisely, h_p is required to be in the isotropy group of $\varphi_p(p)$, which is a group in G isomorphic to H.

[6] In general, this is not the rolling map described in the motivation, although it is related.

[7] Sharpe 1997.

[8] Lumiste 2001a.

[9] This is the standard definition. Cf. Hermann (1983), Appendix 2 to Cartan 1951; Kobayashi 1970, p. 127; Sharpe 1997; Slovák 1997.

[10] Ehresmann 1950, Kobayashi 1957, Lumiste 2001b.

[11] For a treatment of affine connections from this point of view, see Kobayashi & Nomizu (1996, Volume 1).

[12] See, for example, Fox (2005).

[13] Sagerschnig 2006; Cap & Sagerschnig 2007.

[14] See, for instance, Čap & Gover (2002, Definition 2.4).

9.7 References

• Čap, Andreas; Gover, A. Rod (2002), "Tractor calculi for parabolic geometries]" (PDF), *Trans. Amer. Math. Soc.* **354** (04): 1511–1548, doi:10.1090/S0002-9947-01-02909-9.

• Čap, A.; Sagerschnig, K. (2007), *On Nurowski's Conformal Structure Associated to a Generic Rank Two Distribution in Dimension Five*, ESI Preprint 1963. arXiv:0710.2208.

• Cartan, Élie (1910), "Les systèmes de Pfaff à cinq variables et les équations aux dérivées partielles du second ordre", *Annales scientifiques de l'École Normale Supérieure* **27**: 109–192.

• Cartan, Élie (1923), "Sur les variétés à connexion affine et la théorie de la relativité généralisée (première partie)", *Annales Scientifiques de l'École Normale Supérieure* **40**: 325–412.

• Cartan, Élie (1924), "Sur les variétés à connexion projective", *Bulletin de la Société Mathématique* **52**: 205–241.

• Cartan, Élie (1926), "Espaces à connexion affine, projective et conforme", *Acta Math.* **48**: 1–42, doi:10.1007/BF02629755.

• Cartan, Élie (1951), with appendices by Robert Hermann, ed., *Geometry of Riemannian Spaces* (translation by James Glazebrook of *Leçons sur la géométrie des espaces de Riemann*, 2nd ed.), Math Sci Press, Massachusetts (published 1983), ISBN 978-0-915692-34-7.

- Chevalley, C. (1946), *The Theory of Lie Groups*, Princeton University Press, ISBN 0-691-08052-6.

- Ehresmann, C. (1950), "Les connexions infinitésimales dans un espace fibré différentiel", *Colloque de Topologie, Bruxelles*: 29–55, MR 0042768.

- Fox, D.J.F. (2005), "Contact projective structures", *Indiana Univ. Math. J.* **54** (6): 1547–1598, doi:10.1512/iumj.2005.54.2603.

- Griffiths, Phillip (1974), "On Cartan's method of Lie groups and moving frames as applied to uniqueness and existence questions in differential geometry", *Duke Mathematical Journal* **41** (4): 775–814, doi:10.1215/S0012-7094-74-04180-5.

- Kobayashi, Shoshichi; Nomizu, Katsumi (1996), *Foundations of Differential Geometry, Vol. 1 & 2* (New ed.), Wiley-Interscience, ISBN 0-471-15733-3.

- Kobayashi, Shoshichi (1970), *Transformation Groups in Differential Geometry* (1st ed.), Springer, ISBN 3-540-05848-6.

- Kobayashi, Shoshichi (1957), "Theory of Connections", *Ann. Mat. Pura Appl.* **43**: 119–194, doi:10.1007/BF02411907.

- Lumiste, Ü. (2001a), "Conformal connection", in Hazewinkel, Michiel, *Encyclopaedia of Mathematics*, Kluwer Academic Publishers, ISBN 978-1-55608-010-4.

- Lumiste, Ü. (2001b), "Connections on a manifold", in Hazewinkel, Michiel, *Encyclopaedia of Mathematics*, Kluwer Academic Publishers, ISBN 978-1-55608-010-4.

- Sagerschnig, K. (2006), "Split octonions and generic rank two distributions in dimension five", *Arch. Math. (Brno)*, 42 Suppl.: 329–339.

- Sharpe, R.W. (1997), *Differential Geometry: Cartan's Generalization of Klein's Erlangen Program*, Springer-Verlag, New York, ISBN 0-387-94732-9.

- Slovák, Jan (1997), *Parabolic Geometries*, Research Lecture Notes, Part of DrSc-dissertation, Masaryk University.

9.8 Books

- Kobayashi, Shoshichi (1972), *Transformations Groups in Differential Geometry* (Classics in Mathematics 1995 ed.), Springer-Verlag, Berlin, ISBN 978-3-540-58659-3.

The section 3. **Cartan Connections** [pages 127-130] treats conformal and projective connections in a unified manner.

9.9 External links

- Ü. Lumiste (2001), "Affine connection", in Hazewinkel, Michiel, *Encyclopedia of Mathematics*, Springer, ISBN 978-1-55608-010-4

Chapter 10

Graviton

This article is about the hypothetical particle. For other uses, see Graviton (disambiguation).

In physics, the **graviton** is a hypothetical elementary particle that mediates the force of gravitation in the framework of quantum field theory. If it exists, the graviton is expected to be massless (because the gravitational force appears to have unlimited range) and must be a spin-2 boson. The spin follows from the fact that the source of gravitation is the stress–energy tensor, a second-rank tensor (compared to electromagnetism's spin-1 photon, the source of which is the four-current, a first-rank tensor). Additionally, it can be shown that any massless spin-2 field would give rise to a force indistinguishable from gravitation, because a massless spin-2 field would couple to the stress–energy tensor in the same way that gravitational interactions do. Seeing as the graviton is hypothetical, its discovery would unite quantum theory with gravity.[4] This result suggests that, if a massless spin-2 particle is discovered, it must be the graviton, so that the only experimental verification needed for the graviton may simply be the discovery of a massless spin-2 particle.[5]

10.1 Theory

The three other known forces of nature are mediated by elementary particles: electromagnetism by the photon, the strong interaction by the gluons, and the weak interaction by the W and Z bosons. The hypothesis is that the gravitational interaction is likewise mediated by an – as yet undiscovered – elementary particle, dubbed as *the graviton*. In the classical limit, the theory would reduce to general relativity and conform to Newton's law of gravitation in the weak-field limit.[6][7][8]

10.1.1 Gravitons and renormalisation

When describing graviton interactions, the classical theory of Feynman diagrams, and semiclassical corrections such as one-loop diagrams behave normally. But, Feynman diagrams with at least two loops lead to ultraviolet divergences. These infinite results cannot be removed because quantized general relativity is not perturbatively renormalizable, unlike quantum electrodynamics models such as Yang-Mills. Therefore, incalculable answers are found from the perturbation method by which physicists calculate the probability of a particle to emit or absorb gravitons; and the theory loses predictive veracity. Those problems and the complementary approximation framework are grounds to show that a theory more unified than quantized general relativity suffices to describe the behavior near the Planck scale.

10.1.2 Comparison with other forces

Unlike the force carriers of the other forces, gravitation plays a special role in general relativity in defining the spacetime in which events take place. In some descriptions, matter modifies the 'shape' of spacetime itself, and gravity is a result of this shape, an idea which at first glance may appear hard to match with the idea of a force acting between particles.[9] Because the diffeomorphism invariance of the theory does not allow any particular space-time background to be singled out as the "true" space-time background, general relativity is said to be background independent. In contrast, the Standard Model is *not* background independent, with Minkowski space enjoying a special status as the fixed background space-time.[10] A theory of quantum gravity is needed in order to reconcile these differences.[11] Whether this theory should be background independent is an open question. The answer to this question will determine our understanding of what specific role gravitation plays in the fate of the universe.[12]

10.1.3 Gravitons in speculative theories

String theory predicts the existence of gravitons and their well-defined interactions. A graviton in perturbative string theory is a closed string in a very particular low-energy vibrational state. The scattering of gravitons in string theory can also be computed from the correlation functions in conformal field theory, as dictated by the AdS/CFT correspondence, or from matrix theory.

A feature of gravitons in string theory is that, as closed strings without endpoints, they would not be bound to branes and could move freely between them. If we live on a brane (as hypothesized by brane theories) this "leakage" of gravitons from the brane into higher-dimensional space could explain why gravitation is such a weak force, and gravitons from other branes adjacent to our own could provide a potential explanation for dark matter. However, if gravitons were to move completely freely between branes this would dilute gravity too much, causing a violation of Newton's inverse square law. To combat this, Lisa Randall found that a three-brane (such as ours) would have a gravitational pull of its own, preventing gravitons from drifting freely, possibly resulting in the diluted gravity we observe while roughly maintaining Newton's inverse square law.[13] See brane cosmology.

A theory by Ahmed Farag Ali and Saurya Das adds quantum mechanical corrections (using Bohm trajectories) to general relativistic geodesics. If gravitons are given a small but non-zero mass, it could explain the cosmological constant without need for dark energy and solve the smallness problem.[14] The theory received an Honorable Mention in the 2014 Essay Competition of the Gravity Research Foundation for explaining the smallness of cosmological constant.[15] Also the theory received an Honorable Mention in the 2015 Essay Competition of the Gravity Research Foundation for naturally explaining the observed large scale homogeneity and isotropy of the universe due to the proposed quantum corrections.[16]

10.2   Experimental observation

Unambiguous detection of individual gravitons, though not prohibited by any fundamental law, is impossible with any physically reasonable detector.[17] The reason is the extremely low cross section for the interaction of gravitons with matter. For example, a detector with the mass of Jupiter and 100% efficiency, placed in close orbit around a neutron star, would only be expected to observe one graviton every 10 years, even under the most favorable conditions. It would be impossible to discriminate these events from the background of neutrinos, since the dimensions of the required neutrino shield would ensure collapse into a black hole.[17]

LIGO and Virgo collaborations' observations have directly detected of gravitational waves.[18][19][20] Others have postulated that graviton scattering yields gravitational waves as particle interactions yield coherent states.[21] Although

these experiments cannot detect individual gravitons, they might provide information about certain properties of the graviton.[22] For example, if gravitational waves were observed to propagate slower than c (the speed of light in a vacuum), that would imply that the graviton has mass (however, gravitational waves must propagate slower than c in a region with non-zero mass density if they are to be detectable).[23] Recent observations of gravitational waves have put an upper bound of 1.2×10^{-22} eV/c^2 on the graviton's mass.[19] Astronomical observations of the kinematics of galaxies, especially the galaxy rotation problem and modified Newtonian dynamics, might point toward gravitons having non-zero mass.[24]

10.3   Difficulties and outstanding issues

Most theories containing gravitons suffer from severe problems. Attempts to extend the Standard Model or other quantum field theories by adding gravitons run into serious theoretical difficulties at energies close to or above the Planck scale. This is because of infinities arising due to quantum effects; technically, gravitation is not renormalizable. Since classical general relativity and quantum mechanics seem to be incompatible at such energies, from a theoretical point of view, this situation is not tenable. One possible solution is to replace particles with strings. String theories are quantum theories of gravity in the sense that they reduce to classical general relativity plus field theory at low energies, but are fully quantum mechanical, contain a graviton, and are thought to be mathematically consistent.[25]

10.4   See also

- Gravitation
- Gravitational wave
- Gravitino
- Gravitomagnetism
- Multiverse
- Planck mass
- Static forces and virtual-particle exchange

10.5   References

[1] G is used to avoid confusion with gluons (symbol g)

[2] Rovelli, C. (2001). "Notes for a brief history of quantum gravity". arXiv:gr-qc/0006061 [gr-qc].

[3] Blokhintsev, D. I.; Gal'perin, F. M. (1934). "Gipoteza neitrino i zakon sokhraneniya energii" [Neutrino hypothesis and conservation of energy]. *Pod Znamenem Marxisma* (in Russian) **6**: 147–157.

[4] Lightman, A. P.; Press, W. H.; Price, R. H.; Teukolsky, S. A. (1975). "Problem 12.16". *Problem book in Relativity and Gravitation*. Princeton University Press. ISBN 0-691-08162-X.

[5] For a comparison of the geometric derivation and the (non-geometric) spin-2 field derivation of general relativity, refer to box 18.1 (and also 17.2.5) of Misner, C. W.; Thorne, K. S.; Wheeler, J. A. (1973). *Gravitation*. W. H. Freeman. ISBN 0-7167-0344-0.

[6] Feynman, R. P.; Morinigo, F. B.; Wagner, W. G.; Hatfield, B. (1995). *Feynman Lectures on Gravitation*. Addison-Wesley. ISBN 0-201-62734-5.

[7] Zee, A. (2003). *Quantum Field Theory in a Nutshell*. Princeton University Press. ISBN 0-691-01019-6.

[8] Randall, L. (2005). *Warped Passages: Unraveling the Universe's Hidden Dimensions*. Ecco Press. ISBN 0-06-053108-8.

[9] See the other articles on General relativity, Gravitational field, Gravitational wave, etc

[10] Colosi, D.; et al. (2005). "Background independence in a nutshell: The dynamics of a tetrahedron". *Classical and Quantum Gravity* **22** (14): 2971. arXiv:gr-qc/0408079. Bibcode:2005CQGra..22.2971C. doi:10.1088/0264-9381/22/14/008.

[11] Witten, E. (1993). "Quantum Background Independence In String Theory". arXiv:hep-th/9306122 [hep-th].

[12] Smolin, L. (2005). "The case for background independence". arXiv:hep-th/0507235 [hep-th].

[13] Kaku, Michio (2006). *Parallel Worlds - The science of alternative universes and our future in the Cosmos*. pp. 218–221.

[14] Ali, Ahmed Farag (2014). "Cosmology from quantum potential". *Physical Letters B* **741**: 276–279. arXiv:1404.3093v3. doi:10.1016/j.physletb.2014.12.057.

[15] Das, Saurya (2014). "Cosmic coincidence or graviton mass?". *International Journal of Modern Physics D* **23**: 1442017. arXiv:1405.4011. doi:10.1142/S0218271814420176.

[16] Das, Saurya (2015). "Bose–Einstein condensation as an alternative to inflation". *International Journal of Modern Physics D* **24**: 1544001. arXiv:1509.02658. doi:10.1142/S0218271815440010.

[17] Rothman, T.; Boughn, S. (2006). "Can Gravitons be Detected?". *Foundations of Physics* **36** (12): 1801–1825. arXiv:gr-qc/0601043. Bibcode:2006FoPh...36.1801R. doi:10.1007/s10701-006-9081-9.

[18] Castelvecchi, Davide; Witze, Witze (February 11, 2016). "Einstein's gravitational waves found at last". *Nature News*. doi:10.1038/nature.2016.19361. Retrieved 2016-02-11.

[19] B. P. Abbott et al. (LIGO Scientific Collaboration and Virgo Collaboration) (2016). "Observation of Gravitational Waves from a Binary Black Hole Merger". *Physical Review Letters* **116** (6). doi:10.1103/PhysRevLett.116.061102.

[20] "Gravitational waves detected 100 years after Einstein's prediction | NSF - National Science Foundation". *www.nsf.gov*. Retrieved 2016-02-11.

[21] Senatore, L., Silverstein, E., & Zaldarriaga, M. (2014). New sources of gravitational waves during inflation. Journal of Cosmology and Astroparticle Physics, 2014(08), 016.

[22] Dyson, Freeman (8 October 2013). "Is a graviton detectable?". *International Journal of Modern Physics A* **28** (25): 1330041-1–1330035-14. Bibcode:2013IJMPA..2830041D. doi:10.1142/S0217751X1330041X.

[23] Will, C. M. (1998). "Bounding the mass of the graviton using gravitational-wave observations of inspiralling compact binaries". *Physical Review D* **57** (4): 2061–2068. arXiv:gr-qc/9709011. Bibcode:1998PhRvD..57.2061W. doi:10.1103/PhysRevD.57.2061.

[24] Trippe, S. (2013), "A Simplified Treatment of Gravitational Interaction on Galactic Scales", J. Kor. Astron. Soc. **46**, 41. arXiv:1211.4692

[25] Sokal, A. (July 22, 1996). "Don't Pull the String Yet on Superstring Theory". *The New York Times*. Retrieved March 26, 2010.

10.6   External links

-

- Graviton on *In Our Time* at the BBC. (listen now)

Chapter 11

Superpartner

In particle physics, a **Superpartner** (also **sparticle**) is a hypothetical elementary particle. Supersymmetry is one of the synergistic theories in current high-energy physics that predicts the existence of these "shadow" particles.[1][2]

The word *sparticle* features the *s-* prefix which is used to form names of superpartners of the individual fermions, e.g. the stop squark.

11.1 Theoretical predictions

According to the supersymmetry theory, each fermion should have a partner boson, the fermion's superpartner, and each boson should have a partner fermion. Exact *unbroken* supersymmetry would predict that a particle and its superpartners would have the same mass. No superpartners of the Standard Model particles have yet been found. This may indicate that supersymmetry is incorrect, or it may also be the result of the fact that supersymmetry is not an exact, *unbroken* symmetry of nature. If superpartners are found, their masses would indicate the scale at which supersymmetry is broken.[1][3]

For particles that are real scalars (such as an axion), there is a fermion superpartner as well as a second, real scalar field. For axions, these particles are often referred to as axinos and saxions.

In extended supersymmetry there may be more than one superparticle for a given particle. For instance, with two copies of supersymmetry in four dimensions, a photon would have two fermion superpartners and a scalar superpartner.

In zero dimensions it is possible to have supersymmetry, but no superpartners. However, this is the only situation where supersymmetry does not imply the existence of superpartners.

11.2 Recreating superpartners

If the supersymmetry theory is correct, it should be possible to recreate these particles in high-energy particle accelerators. Doing so will not be an easy task; these particles may have masses up to a thousand times greater than their corresponding "real" particles.[1]

Some researchers have hoped the Large Hadron Collider at CERN might produce evidence for the existence of superpartner particles.[1] However, as of 2013, no such evidence has been found.[4]

11.3 See also

- Chargino
- Gluino
- Gravitino as a superpartner of the hypothetical graviton
- Neutralino
- Sfermion
- Higgsino

11.4 References

[1] Langacker, Paul (November 22, 2010). Sprouse, Gene D., ed. "Meet a superpartner at the LHC". *Physics* (New York: American Physical Society) 3 (98). Bibcode:2010PhyOJ...3...98L. doi:10.1103/Physics.3.98. ISSN 1943-2879. OCLC 233971234. Archived from the original on 2011-02-22. Retrieved 21 February 2011.

[2] Overbye, Dennis (May 15, 2007). "A Giant Takes On Physics' Biggest Questions". *The New York Times* (Manhattan, New York: Arthur Ochs Sulzberger, Jr.). p. F1. ISSN 0362-4331. OCLC 1645522. Retrieved 21 February 2011.

[3] Quigg, Chris (January 17, 2008). "Sidebar: Solving the Higgs Puzzle". *Scientific American* (Nature Publishing Group). ISSN 0036-8733. OCLC 1775222. Archived from the original on 2011-02-22. Retrieved 21 February 2011.

[4] Jamieson, Valerie (13 December 2013). "Higgs Nobel bash: I was at the party of the universe". *New Scientist*. Retrieved 20 December 2013. So far the Higgs hasn't given many supersymmetric clues.

11.5   External links

- Argonne National Laboratory

- Large Hadron Collider

- CERN homepage

Chapter 12

Gravitino

In supergravity theories combining general relativity and supersymmetry, the **gravitino** (G) is the gauge fermion supersymmetric partner of the hypothesized graviton. It has been suggested as a candidate for dark matter.

If it exists, it is a fermion of spin $\frac{3}{2}$ and therefore obeys the Rarita-Schwinger equation. The gravitino field is conventionally written as $\psi\mu\alpha$ with $\mu = 0,1,2,3$ a four-vector index and $\alpha = 1,2$ a spinor index. For $\mu = 0$ one would get negative norm modes, as with every massless particle of spin 1 or higher. These modes are unphysical, and for consistency there must be a gauge symmetry which cancels these modes: $\delta\psi\mu\alpha = \partial\mu\epsilon\alpha$ where $\epsilon\alpha(x)$ is a spinor function of spacetime. This gauge symmetry is a local supersymmetry transformation, and the resulting theory is supergravity.

Thus the gravitino is the fermion mediating supergravity interactions, just as the photon is mediating electromagnetism, and the graviton is presumably mediating gravitation. Whenever supersymmetry is broken in supergravity theories, it acquires a mass which is determined by the scale at which supersymmetry is broken. This varies greatly between different models of supersymmetry breaking, but if supersymmetry is to solve the hierarchy problem of the Standard Model, the gravitino cannot be more massive than about 1 TeV/c².

12.1 Gravitino cosmological problem

If the gravitino indeed has a mass of the order of TeV, then it creates a problem in the standard model of cosmology, at least naïvely.[1][2][3][4]

One option is that the gravitino is stable. This would be the case if the gravitino is the lightest supersymmetric particle and R-parity is conserved (or nearly so). In this case the gravitino is a candidate for dark matter; as such gravitinos will have been created in the very early universe. However, one may calculate the density of gravitinos and it turns out to be much higher than the observed dark matter density.

The other option is that the gravitino is unstable. Thus the gravitinos mentioned above would decay and will not contribute to the observed dark matter density. However, since they decay only through gravitational interactions, their lifetime would be very long, of the order of Mpl^2 / m^3 in natural units, where Mpl is the Planck mass and m is the mass of a gravitino. For a gravitino mass of the order of TeV this would be 10^5 s, much later than the era of nucleosynthesis. At least one possible channel of decay must include either a photon, a charged lepton or a meson, each of which would be energetic enough to destroy a nucleus if it strikes one. One can show that enough such energetic particles will be created in the decay as to destroy almost all the nuclei created in the era of nucleosynthesis, in contrast with observations. In fact, in such a case the universe would have been made of hydrogen alone, and star formation would probably be impossible.

One possible solution to the cosmological gravitino problem is the split supersymmetry model, where the gravitino mass is much higher than the TeV scale, but other fermionic supersymmetric partners of standard model particles already appear at this scale.

Another solution is that R-parity is slightly violated and the gravitino is the lightest supersymmetric particle. This causes almost all supersymmetric particles in the early Universe to decay into Standard Model particles via R-parity violating interactions well before the synthesis of primordial nuclei; a small fraction however decay into gravitinos, whose half-life is orders of magnitude greater than the age of the Universe due to the suppression of the decay rate by the Planck scale and the small R-parity violating couplings.[5]

12.2 See also

- Supersymmetry

12.3   References

[1] T. Moroi, H. Murayama Cosmological constraints on the light stable gravitino Phys.Lett.B303:289–294,1993

[2] N. Okada, O. Seto A brane world cosmological solution to the gravitino problem Phys.Rev.D71:023517,2005

[3] A. de Gouvea, T. Moroi, H. Murayama Cosmology of Supersymmetric Models with Low-energy Gauge Mediation Phys.Rev.D56:1281–1299,1997

[4] M. Endo Moduli Stabilization and Moduli-Induced Gravitino Problem talk given at SUSY'06, 12 June 2006

[5] F. Takayama and M. Yamaguchi, Phys. Lett. B 485 (2000)

Chapter 13

Cosmological constant

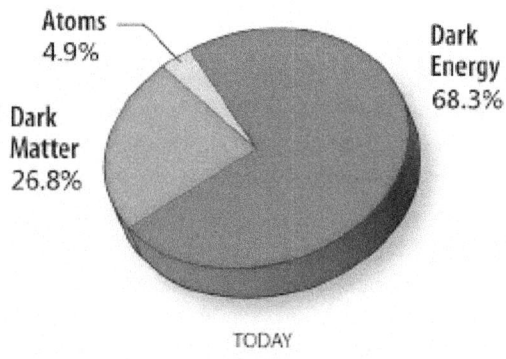

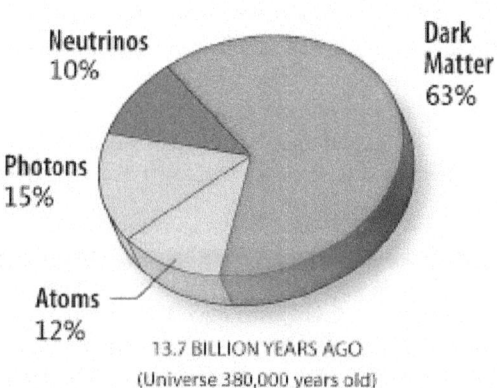

Estimated ratios of dark matter and dark energy (which may be the cosmological constant) in the universe. According to current theories of physics, dark energy now dominates as the largest source of energy of the universe, in contrast to earlier epochs when it was insignificant.

In cosmology, the **cosmological constant** (usually denoted by the Greek capital letter lambda: Λ) is the value of the energy density of the vacuum of space. It was originally introduced by Albert Einstein in 1917[1] as an addition to his theory of general relativity to "hold back gravity" and achieve a static universe, which was the accepted view at the time. Einstein abandoned the concept after Hubble's 1929 discovery that all galaxies outside the Local Group

(the group that contains the Milky Way Galaxy) are moving away from each other, implying an overall expanding universe. From 1929 until the early 1990s, most cosmology researchers assumed the cosmological constant to be zero.

Since the 1990s, several developments in observational cosmology, especially the discovery of the accelerating universe from distant supernovae in 1998 (in addition to independent evidence from the cosmic microwave background and large galaxy redshift surveys), have shown that around 68% of the mass–energy density of the universe can be attributed to dark energy.[2] While dark energy is poorly understood at a fundamental level, the main required properties of dark energy are that it functions as a type of antigravity, it dilutes much more slowly than matter as the universe expands, and it clusters much more weakly than matter, or perhaps not at all. The cosmological constant is the simplest possible form of dark energy since it is constant in both space and time, and this leads to the current standard model of cosmology known as the Lambda-CDM model, which provides a good fit to many cosmological observations as of 2016.

13.1 Equation

The cosmological constant Λ appears in Einstein's field equation in the form of

$$R_{\mu\nu} - \frac{1}{2}R\,g_{\mu\nu} + \Lambda\,g_{\mu\nu} = \frac{8\pi G}{c^4}T_{\mu\nu},$$

where R and g describe the structure of spacetime, T pertains to matter and energy affecting that structure, and G and c are conversion factors that arise from using traditional units of measurement. When Λ is zero, this reduces to the original field equation of general relativity. When T is zero, the field equation describes empty space (the vacuum).

The cosmological constant has the same effect as an intrinsic energy density of the vacuum, ϱ_{vac} (and an associated pressure). In this context, it is commonly moved onto the right-hand side of the equation, and defined with a proportionality factor of 8π: $\Lambda = 8\pi\varrho_{vac}$, where unit conventions of general relativity are used (otherwise factors of G and c would also appear, i.e. $\Lambda = 8\pi\,(G/c^2)\varrho_{vac} = \kappa\,\varrho_{vac}$, where κ is Einstein's constant). It is common to quote values of energy density directly, though still using the name "cosmological constant", with convention $8\pi\,G = 1$. (In fact, the true dimension of Λ is a length^{-2} and it has the value of $\sim 1\ 10^{-52}$ m^{-2} or in reduced Planck units : $\sim 3\ 10^{-122}$, calculated with the best present (2015) values of $\Omega\Lambda = 0.6911 \pm 0.0062$ and H$_o = 67.74 \pm 0.46$ km/s / Mpc $= 2.195 \pm 0.015\ 10^{-18}$ s^{-1}).

A positive vacuum energy density resulting from a cosmological constant implies a negative pressure, and vice versa. If the energy density is positive, the associated negative pressure will drive an accelerated expansion of the universe, as observed. (See dark energy and cosmic inflation for details.)

13.1.1 $\Omega\Lambda$ (Omega Lambda)

Instead of the cosmological constant itself, cosmologists often refer to the ratio between the energy density due to the cosmological constant and the critical density of the universe, the tipping point for a sufficient density to stop the universe from expanding forever. This ratio is usually denoted $\Omega\Lambda$, and is estimated to be 0.6911 ± 0.0062, according to the recent Planck results released in 2015.[3] In a flat universe $\Omega\Lambda$ is the fraction of the energy of the universe due to the cosmological constant, i.e., what we would intuitively call the fraction of the universe that is made up of dark energy. Note that this value changes over time: the critical density changes with cosmological time, but the energy density due to the cosmological constant remains unchanged throughout the history of the universe: the amount of dark energy increases as the universe grows, while the amount of matter does not.

13.1.2 Equation of state

Another ratio that is used by scientists is the equation of state, usually denoted w, which is the ratio of pressure that dark energy puts on the universe to the energy per unit volume.[4] This ratio is $w = -1$ for a true cosmological constant, and is generally different for alternative time-varying forms of vacuum energy such as quintessence.

13.2 History

Einstein included the cosmological constant as a term in his field equations for general relativity because he was dissatisfied that otherwise his equations did not allow, apparently, for a static universe: gravity would cause a universe that was initially at dynamic equilibrium to contract. To counteract this possibility, Einstein added the cosmological constant.[5] However, soon after Einstein developed his static theory, observations by Edwin Hubble indicated that the universe appears to be expanding; this was consistent with a cosmological solution to the *original* general relativity equations that had been found by the mathematician Friedmann, working on the Einstein equations of general relativity. Einstein later reputedly referred to his failure to accept the validation of his equations—when they had predicted the expansion of the universe in theory, before it was demonstrated in observation of the cosmological red shift—as the "biggest blunder" of his life.[6][7]

In fact, adding the cosmological constant to Einstein's equations does not lead to a static universe at equilibrium because the equilibrium is unstable: if the universe expands slightly, then the expansion releases vacuum energy, which causes yet more expansion. Likewise, a universe that contracts slightly will continue contracting.[8]:59

However, the cosmological constant remained a subject of theoretical and empirical interest. Empirically, the onslaught of cosmological data in the past decades strongly suggests that our universe has a positive cosmological constant.[5] The explanation of this small but positive value is an outstanding theoretical challenge (*see the section below*).

Finally, it should be noted that some early generalizations of Einstein's gravitational theory, known as classical unified field theories, either introduced a cosmological constant on theoretical grounds or found that it arose naturally from the mathematics. For example, Sir Arthur Stanley Eddington claimed that the cosmological constant version of the vacuum field equation expressed the "epistemological" property that the universe is "self-gauging", and Erwin Schrödinger's pure-affine theory using a simple variational principle produced the field equation with a cosmological term.

13.3 Positive value

Observations announced in 1998 of distance–redshift relation for Type Ia supernovae[9][10] indicated that the expansion of the universe is accelerating. When combined with measurements of the cosmic microwave background radiation these implied a value of $\Omega\Lambda \approx 0.7$,[11] a result which has been supported and refined by more recent measurements. There are other possible causes of an accelerating universe, such as quintessence, but the cosmological constant is in most respects the simplest solution. Thus, the current standard model of cosmology, the Lambda-CDM model, includes the cosmological constant, which is measured to be on the order of 10^{-52} m^{-2}, in metric units. Multiplied by other constants that appear in the equations, it is often expressed as 10^{-52} m^{-2}, 10^{-35} s^{-2}, 10^{-47} GeV^4, 10^{-29} g/cm^3.[12] In terms of Planck units, and as a natural dimensionless value, the cosmological constant, Λ, is on the order of 10^{-122}.[13]

As was only recently seen, by works of 't Hooft, Susskind[14] and others, a positive cosmological constant has surprising consequences, such as a finite maximum entropy of the observable universe (see the holographic principle).

13.4 Predictions

13.4.1 Quantum field theory

See also: Vacuum catastrophe

A major outstanding problem is that most quantum field theories predict a huge value for the quantum vacuum. A common assumption is that the quantum vacuum is equivalent to the cosmological constant. Although no theory exists that supports this assumption, arguments can be made in its favor.[15]

Such arguments are usually based on dimensional analysis and effective field theory. If the universe is described by an effective local quantum field theory down to the Planck scale, then we would expect a cosmological constant of the order of M_{pl}^4. As noted above, the measured cosmological constant is smaller than this by a factor of 10^{-120}. This discrepancy has been called "the worst theoretical prediction in the history of physics!".[16]

Some supersymmetric theories require a cosmological constant that is exactly zero, which further complicates things. This is the *cosmological constant problem*, the worst problem of fine-tuning in physics: there is no known natural way to derive the tiny cosmological constant used in cosmology from particle physics.

13.4.2 Anthropic principle

One possible explanation for the small but non-zero value was noted by Steven Weinberg in 1987 following the anthropic principle.[17] Weinberg explains that if the vacuum energy took different values in different domains of the universe, then observers would necessarily measure values similar to that which is observed: the formation of life-supporting structures would be suppressed in domains where the vacuum energy is much larger. Specifically, if the vacuum energy is negative and its absolute value is substantially larger than it appears to be in the observed universe (say, a factor of 10 larger), holding all other variables (e.g. matter density) constant, that would mean that the universe is closed; furthermore, its lifetime would be shorter than the age of our universe, possibly too short for intelligent life to form. On the other hand, a universe with a large positive cosmological constant would expand too fast, preventing galaxy formation. According to Weinberg, domains where the vacuum energy is compatible with life would be comparatively rare. Using this argument, Weinberg predicted that the cosmological constant would have a value of less than a hundred times the currently accepted value.[18] In 1992, Weinberg refined this prediction of the cosmological constant to 5 to 10 times the matter density.[19]

This argument depends on a lack of a variation of the distribution (spatial or otherwise) in the vacuum energy density, as would be expected if dark energy were the cosmological constant. There is no evidence that the vacuum energy does vary, but it may be the case if, for example, the vacuum energy is (even in part) the potential of a scalar field such as the residual inflaton (also see quintessence). Another theoretical approach that deals with the issue is that of multiverse theories, which predict a large number of "parallel" universes with different laws of physics and/or values of fundamental constants. Again, the anthropic principle states that we can only live in one of the universes that is compatible with some form of intelligent life. Critics claim that these theories, when used as an explanation for fine-tuning, commit the inverse gambler's fallacy.

In 1995, Weinberg's argument was refined by Alexander Vilenkin to predict a value for the cosmological constant that was only ten times the matter density,[20] i.e. about three times the current value since determined.

13.4.3 Cyclic model

More recent work has suggested the problem may be indirect evidence of a cyclic universe possibly as allowed by string theory. With every cycle of the universe (Big Bang then eventually a Big Crunch) taking about a trillion (10^{12}) years, "the amount of matter and radiation in the universe is reset, but the cosmological constant is not. Instead, the cosmological constant gradually diminishes over many cycles to the small value observed today."[21] Critics respond that, as the authors acknowledge in their paper, the model "entails ... the same degree of tuning required in any cosmological model".[22]

13.5   See also

- Higgs mechanism

- Lambdavacuum solution

- Naturalness (physics)

- Quantum electrodynamics

- de Sitter relativity

- Unruh effect

13.6   References

[1] Einstein, A (1917). "Kosmologische Betrachtungen zur allgemeinen Relativitaetstheorie". *Sitzungsberichte der Königlich Preussischen Akademie der Wissenschaften Berlin.* part 1: 142–152.

[2] What is Dark Energy?, Space.com, 1 May 2013

[3] Collaboration, Planck, PAR Ade, N Aghanim, C Armitage-Caplan, M Arnaud, et al., Planck 2015 results. XIII. Cosmological parameters. arXiv preprint 1502.1589v2 , 6 Feb 2015.

[4] Hogan, Jenny (2007). "Welcome to the Dark Side". *Nature* **448** (7151): 240–245. Bibcode:2007Natur.448..240H. doi:10.1038/448240a. PMID 17637630.

[5] Urry, Meg (2008). *The Mysteries of Dark Energy*. Yale Science. Yale University.

[6] Gamov, George (1970). *My World Line*. Viking Press. p. 44. ISBN 978-0670503766

[7] Rosen, Rebecca J. "Einstein Likely Never Said One of His Most Oft-Quoted Phrases". *The Atlantic*. The Atlantic Media Company. Retrieved 10 August 2013.

[8] Barbara Sue Ryden (2003). *Introduction to cosmology*. Addison-Wesley. ISBN 978-0-8053-8912-8.

[9] Riess, A.; et al. (September 1998). "Observational Evidence from Supernovae for an Accelerating Universe and a Cosmological Constant". *The Astronomical Journal* **116** (3): 1009–1038. arXiv:astro-ph/9805201. Bibcode:1998AJ....116.1009R. doi:10.1086/300499.

[10] Perlmutter, S.; et al. (June 1999). "Measurements of Omega and Lambda from 42 High-Redshift Supernovae". *The Astrophysical Journal* **517** (2): 565–586. arXiv:astro-ph/9812133. Bibcode:1999ApJ...517..565P. doi:10.1086/307221.

[11] See e.g. Baker, Joanne C.; et al. (1999). "Detection of cosmic microwave background structure in a second field with the Cosmic Anisotropy Telescope". *Monthly Notices of the Royal Astronomical Society* **308** (4): 1173–1178. arXiv:astro-ph/9904415. Bibcode:1999MNRAS.308.1173B. doi:10.1046/j.1365-8711.1999.02829.x.

[12] Tegmark, Max; et al. (2004). "Cosmological parameters from SDSS and WMAP". *Physical Review D* **69** (103501): 103501. arXiv:astro-ph/0310723. Bibcode:2004PhRvD..69j3501T. doi:10.1103/PhysRevD.69.103501.

[13] John D. Barrow The Value of the Cosmological Constant

[14] Lisa Dyson, Matthew Kleban, Leonard Susskind: "Disturbing Implications of a Cosmological Constant"

[15] Rugh, S; Zinkernagel, H. (2001). "The Quantum Vacuum and the Cosmological Constant Problem". *Studies in History and Philosophy of Modern Physics* **33** (4): 663–705. doi:10.1016/S1355-2198(02)00033-3.

[16] MP Hobson, GP Efstathiou & AN Lasenby (2006). *General Relativity: An introduction for physicists* (Reprinted with corrections 2007 ed.). Cambridge University Press. p. 187. ISBN 978-0-521-82951-9.

[17] Weinberg, S (1987). "Anthropic Bound on the Cosmological Constant". *Phys. Rev. Lett.* **59** (22): 2607–2610. Bibcode:1987PhRvL..59.2607W. doi:10.1103/PhysRevLett.59.2607. PMID 10035596.

[18] Alexander Vilenkin, *Many Worlds in One: The Search for Other Universes*, ISBN 978-0-8090-9523-0, pp. 138–9

[19] Weinberg, Steven (1993). *Dreams of a Final Theory: the search for the fundamental laws of nature*. Vintage Press. p. 182. ISBN 0-09-922391-0.

[20] Alexander Vilenkin, *Many Worlds in One: The Search for Other universes*, ISBN 978-0-8090-9523-0, p. 146, which references Vilenkin' *Predictions from quantum cosmology*, Physical Review Letters, vol 74, p. 846 (1995)

[21] 'Cyclic universe' can explain cosmological constant, NewScientistSpace, 4 May 2006

[22] Steinhardt, P. J.; Turok, N. (2002-04-25). "A Cyclic Model of the Universe". *Science* **296** (5572): 1436–1439. arXiv:hep-th/0111030v2. Bibcode:2002Sci...296.1436S. doi:10.1126/science.1070462. PMID 11976408. Retrieved 2012-04-29.

13.7   Further reading

- Michael, E., University of Colorado, Department of Astrophysical and Planetary Sciences, "The Cosmological Constant"

- Ferguson, Kitty (1991). *Stephen Hawking: Quest For A Theory of Everything*, Franklin Watts. ISBN 0-553-29895-X.

- John D. Barrow and John K. Webb (June 2005). "Inconstant Constants". *Scientific American.*

- *Beyond the Cosmological Standard Model*[1] (2014)

13.8 External links

- Cosmological constant (astronomy) at *Encyclopædia Britannica*

- Carroll, Sean M., *"The Cosmological Constant"* (short), *"The Cosmological Constant"* (extended).

- News story: More evidence for dark energy being the cosmological constant

- Cosmological constant article from Scholarpedia

- Copeland, Ed; Merrifield, Mike. "Λ – Cosmological Constant". *Sixty Symbols.* Brady Haran for the University of Nottingham.

[1] Austin Joyce, Bhuvnesh Jain, Justin Khoury, Mark Trodden (2014). "Beyond the Cosmological Standard Model".

Chapter 14

Quantum field theory

"Relativistic quantum field theory" redirects here. For other uses, see Relativity.

In theoretical physics, **quantum field theory (QFT)** is a theoretical framework for constructing quantum mechanical models of subatomic particles in particle physics and quasiparticles in condensed matter physics. A QFT treats particles as excited states of an underlying physical field, so these are called field quanta.

In quantum field theory, quantum mechanical interactions between particles are described by interaction terms between the corresponding underlying quantum fields. These interactions are conveniently visualized by Feynman diagrams, that also serve as a formal tool to evaluate various processes.

Historically the development began in the 1920s with the quantization of the electromagnetic field, the quantization being based on an analogy with the eigenmode expansion of a vibrating string with fixed endpoints. In Weinberg (2005), QFT is brought forward as an unavoidable consequence of the reconciliation of quantum mechanics with special relativity.

14.1 Definition

Quantum electrodynamics (QED) has one electron field and one photon field; quantum chromodynamics (QCD) has one field for each type of quark; and, in condensed matter, there is an atomic displacement field that gives rise to phonon particles.

14.1.1 Dynamics

See also: Relativistic dynamics

Ordinary quantum mechanical systems have a fixed number of particles, with each particle having a finite number of degrees of freedom. In contrast, the excited states of a quantum field can represent any number of particles. This makes quantum field theories especially useful for describing systems where the particle count/number may change over time, a crucial feature of relativistic dynamics.

14.1.2 States

QFT interaction terms are similar in spirit to those between charges with electric and magnetic fields in Maxwell's equations. However, unlike the classical fields of Maxwell's theory, fields in QFT generally exist in quantum superpositions of states and are subject to the laws of quantum mechanics.

Because the fields are continuous quantities over space, there exist excited states with arbitrarily large numbers of particles in them, providing QFT systems with an effectively infinite number of degrees of freedom. Infinite degrees of freedom can easily lead to divergences of calculated quantities (e.g., the quantities become infinite). Techniques such as renormalization of QFT parameters or discretization of spacetime, as in lattice QCD, are often used to avoid such infinities so as to yield physically plausible results.

14.1.3 Fields and radiation

The gravitational field and the electromagnetic field are the only two fundamental fields in nature that have infinite range and a corresponding classical low-energy limit, which greatly diminishes and hides their "particle-like" excitations. Albert Einstein in 1905, attributed "particle-like" and discrete exchanges of momenta and energy, characteristic of "field quanta", to the electromagnetic field. Originally, his principal motivation was to explain the thermodynamics of radiation. Although the photoelectric effect and Compton scattering strongly suggest the existence of the photon, it might alternately be explained by a mere quantization of emission; more definitive evidence of the quantum nature of radiation is now taken up into modern quantum optics as in the antibunching effect.[1]

14.2 Theories

There is currently no complete quantum theory of the remaining fundamental force, gravity. Many of the proposed theories to describe gravity as a QFT postulate the existence of a graviton particle that mediates the gravitational force. Presumably, the as yet unknown correct quantum field-theoretic treatment of the gravitational field will behave like Einstein's general theory of relativity in the low-energy limit. Quantum field theory of the fundamental forces itself has been postulated to be the low-energy effective field theory limit of a more fundamental theory such as superstring theory.

Most theories in standard particle physics are formulated as **relativistic quantum field theories**, such as QED, QCD, and the Standard Model. QED, the quantum field-theoretic description of the electromagnetic field, approximately reproduces Maxwell's theory of electrodynamics in the low-energy limit, with small non-linear corrections to the Maxwell equations required due to virtual electron–positron pairs.

In the perturbative approach to quantum field theory, the full field interaction terms are approximated as a perturbative expansion in the number of particles involved. Each term in the expansion can be thought of as forces between particles being mediated by other particles. In QED, the electromagnetic force between two electrons is caused by an exchange of photons. Similarly, intermediate vector bosons mediate the weak force and gluons mediate the strong force in QCD. The notion of a force-mediating particle comes from perturbation theory, and does not make sense in the context of non-perturbative approaches to QFT, such as with bound states.

14.3 The History of QFT

14.3.1 The Early Development

The first achievement of QFT, namely quantum electrodynamics, is "still the paradigmatic example of a successful quantum field theory" according to Weinberg (2005). Ordinary QM cannot give an account of photons, which constitute the prime case of relativistic 'particles'. Since photons have rest mass zero, and correspondingly travel in the vacuum at the speed c, a non-relativistic theory such as ordinary QM cannot give even an approximate description. Photons are implicit in the emission and absorption processes which

Max Born (1882–1970), one of the founders of quantum field theory.
He is also known for the Born rule that introduced the probabilistic interpretation in quantum mechanics. He received the 1954 Nobel Prize in Physics together with Walther Bothe.

have to be postulated, for instance, when one of an atom's electrons makes a transition between energy levels. The formalism of QFT is needed for an explicit description of photons. In fact most topics in the early development of quantum theory (1900–1927) were related to the interaction of radiation and matter and thus should be treated by quantum field theoretical methods. However, quantum mechanics as formulated by Dirac, Heisenberg, and Schrödinger (1926-1927) started from atomic spectra and did not focus much on problems of radiation.

As soon as the conceptual framework of quantum mechanics was developed, a small group of theoreticians tried to extend quantum methods to electromagnetic fields. A good example is the famous paper by Born, Jordan & Heisenberg (1926). P. Jordan was especially acquainted with the literature on light quanta and made important contributions to QFT. The basic idea was that in QFT the electromagnetic field should be represented by matrices in the same way that position and momentum were represented in QM by matrices in matrix mechanics. The ideas of QM were extended to systems having an infinite number of degrees of freedom.

The inception of QFT is usually considered to be Dirac's

famous 1927 paper on "The quantum theory of the emission and absorption of radiation".[2] Here Dirac coined the name "quantum electrodynamics" (QED) for the part of QFT that was developed first. Dirac supplied a systematic procedure for transferring the characteristic quantum phenomenon of discreteness of physical quantities from the quantum-mechanical treatment of particles to a corresponding treatment of fields. Employing the theory of the quantum harmonic oscillator, Dirac gave a theoretical description of how photons appear in the quantization of the electromagnetic radiation field. Later, Dirac's procedure became a model for the quantization of other fields as well. These first approaches to QFT were further developed during the following three years. P. Jordan introduced creation and annihilation operators for fields obeying Fermi–Dirac statistics. These differ from the corresponding operators for Bose–Einstein statistics in that the former satisfy *anticommutation relations* while the latter satisfy commutation relations.

The methods of QFT could be applied to derive equations resulting from the quantum-mechanical (field-like) treatment of particles, e.g. the Dirac equation, the Klein–Gordon equation and the Maxwell equations. Schweber points out[3] that the idea and procedure of second quantization goes back to Jordan, in a number of papers from 1927,[4] while the expression itself was coined by Dirac. Some difficult problems concerning commutation relations, statistics, and Lorentz invariance were eventually solved. The first comprehensive account of a general theory of quantum fields, in particular the method of canonical quantization, was presented by Heisenberg & Pauli in 1929. Whereas Jordan's second quantization procedure applied to the coefficients of the normal modes of the field, Heisenberg & Pauli started with the fields themselves and subjected them to the canonical procedure. Heisenberg and Pauli thus established the basic structure of QFT as presented in modern introductions to QFT. Fermi and Dirac, as well as Fock and Podolski, presented different formulations which played a heuristic role in the following years.

Quantum electrodynamics rests on two pillars, see e.g., the short and lucid "Historical Introduction" of Scharf (2014). The first pillar is the quantization of the electromagnetic field, i.e., it is about photons as the quantized excitations or 'quanta' of the electromagnetic field. This procedure will be described in some more detail in the section on the particle interpretation. As Weinberg points out the "photon is the only particle that was known as a field before it was detected as a particle" so that it is natural that QED began with the analysis of the radiation field.[5] The second pillar of QED consists of the relativistic theory of the electron, centered on the Dirac equation.

14.3.2   The Emergence of Infinities

Pascual Jordan (1902–1980), doctoral student of Max Born, was a pioneer in quantum field theory, coauthoring a number of seminal papers with Born and Heisenberg.
Jordan algebras were introduced by him to formalize the notion of an algebra of observables in quantum mechanics. He was awarded the Max Planck medal 1954.

Quantum field theory started with a theoretical framework that was built in analogy to quantum mechanics. Although there was no unique and fully developed theory, quantum field theoretical tools could be applied to concrete processes. Examples are the scattering of radiation by free electrons, Compton scattering, the collision between relativistic electrons or the production of electron-positron pairs by photons. Calculations to the first order of approximation were quite successful, but most people working in the field thought that QFT still had to undergo a major change. On the one side some calculations of effects for cosmic rays clearly differed from measurements. On the other side and, from a theoretical point of view more threatening, calculations of higher orders of the perturbation series led to infinite results. The self-energy of the electron as well as vacuum fluctuations of the electromagnetic field seemed to be infinite. The perturbation expansions did not converge to a finite sum and even most individual terms were divergent.

The various forms of infinities suggested that the divergences were more than failures of specific calculations. Many physicists tried to avoid the divergences by formal tricks (truncating the integrals at some value of momentum, or even ignoring infinite terms) but such rules were not reliable, violated the requirements of relativity and were not considered as satisfactory. Others came up with first ideas of coping with infinities by a redefinition of the parameters of the theory and using a measured finite value, for example of the charge of the electron, instead of the infinite 'bare' value. This process is called renormalization.

From the point of view of philosophy of science it is remarkable that these divergences did not give enough reason to discard the theory. The years from 1930 to the beginning of World War II were characterized by a variety of attitudes towards QFT. Some physicists tried to circumvent the infinities by more-or-less arbitrary prescriptions, others worked on transformations and improvements of the theoretical framework. Most of the theoreticians believed that QED would break down at high energies. There was also a considerable number of proposals in favour of alternative approaches. These proposals included changes in the basic concepts e.g. negative probabilities and interactions at a distance instead of a field theoretical approach, and a methodological change to phenomenological methods that focusses on relations between observable quantities without an analysis of the microphysical details of the interaction, the so-called S-matrix theory where the basic elements are amplitudes for various scattering processes.

Despite the feeling that QFT was imperfect and lacking rigour, its methods were extended to new areas of applications. In 1933 Fermi's theory of the beta decay started with conceptions describing the emission and absorption of photons, transferred them to beta radiation and analyzed the creation and annihilation of electrons and neutrinos described by the weak interaction. Further applications of QFT outside of quantum electrodynamics succeeded in nuclear physics with the strong interaction. In 1934 a new type of fields (scalar fields), described by the Klein–Gordon equation, could be quantized. This is another example of second quantization. This new theory for matter fields could be applied a decade later when new particles, pions, were detected.

14.3.3 The Taming of Infinities

After the end of World War II more reliable and effective methods for dealing with infinities in QFT were developed, namely coherent and systematic rules for performing relativistic field theoretical calculations, and a general renormalization theory. On three famous conferences, the Shelter Island Conference 1947, the Pocono Conference

Werner Heisenberg (1901–1976), doctoral student of Arnold Sommerfeld, was one of the founding fathers of quantum mechanics. In particular, he introduced the version of quantum mechanics known as matrix mechanics, but is now more known for the Heisenberg uncertainty relations. He was awarded the Nobel prize in physics 1932 together with Erwin Schrödinger and Paul Dirac.

1948, and the 1949 Oldstone Conference, developments in theoretical physics were confronted with relevant new experimental results. In the late forties there were two different ways to address the problem of divergences. One of these was discovered by Richard Feynman, the other one (based on an operator formalism) by Julian Schwinger and independently by Sin-Itiro Tomonaga. In 1949 Freeman Dyson showed that the two approaches are in fact equivalent. Thus, Freeman Dyson, Feynman, Schwinger and Tomonaga became the inventors of renormalization theory. The most spectacular experimental successes of renormalization theory were the calculations of the anomalous magnetic moment of electron and the Lamb shift in the spectrum of hydrogen. These successes were so outstanding because the theoretical results were in better agreement with high precision experiments than anything in physics before. Nevertheless, mathematical problems lingered on and prompted a search for rigorous formulations (to be discussed in the main article).

The basic idea of renormalization is to avoid divergences that appear in physical predictions by shifting them into a part of the theory where they do not influence empirical propositions. Dyson could show that a rescaling of charge and mass ('renormalization') is sufficient to remove all divergences in QED to all orders of perturbation theory. In general, a QFT is called renormalizable, if all infinities can be absorbed into a redefinition of a finite number of coupling constants and masses. A consequence is that the physical charge and mass of the electron must be measured and cannot be computed from first principles. Perturbation theory gives well defined predictions only in renormalizable quantum field theories, and luckily QED, the first fully developed QFT, belonged to this class of renormalizable theories. There are various technical procedures to renormalize a theory. One way is to cut off the integrals in the calculations at a certain value Λ of the momentum which is large but finite. This cut-off procedure is successful if, after taking the limit $\Lambda \to \infty$, the resulting quantities are independent of Λ. Part II of Peskin & Schroeder (1995) gives an extensive description of renormalization. https://upload.wikimedia.org/wikipedia/en/4/42/

Feynman's formulation of QED is of special interest from a philosophical point of view. His so-called space-time approach is visualized by the famous Feynman diagrams that look like depicting paths of particles. Feynman's method of calculating scattering amplitudes is based on the functional integral formulation of field theory.[6] A set of graphical rules can be derived so that the probability of a specific scattering process can be calculated by drawing a diagram of that process and then using the diagram to write down the mathematical expressions for calculating its amplitude. The diagrams provide an effective way to organize and visualize the various terms in the perturbation series, and they seem to display the flow of electrons and photons during the scattering process. External lines in the diagrams represent incoming and outgoing particles, internal lines are connected with virtual particles and vertices with interactions. Each of these graphical elements is associated with mathematical expressions that contribute to the amplitude of the respective process. The diagrams are part of Feynman's very efficient and elegant algorithm for computing the probability of scattering processes. The idea of particles travelling from one point to another was heuristically useful in constructing the theory. This heuristics, based on Huygen's principle, is useful for concrete calculations and actually give the correct particle propagators as derived more rigorously.[7] Nevertheless, an analysis of the theoretical justification of the space-time approach shows that its success does not imply that particle paths have to be taken seriously. General arguments against a particle interpretation of QFT clearly

Richard Feynman (1918–1988)
His 1945 PhD thesis developed the path integral formulation of ordinary quantum mechanics. This was later generalized to field theory.

exclude that the diagrams represent paths of particles in the interaction area. Feynman himself was not particularly interested in ontological questions.

14.3.4 The Standard Model of Particle Physics

In the beginning of the 1950s QED had become a reliable theory which no longer counted as preliminary. It took two decades from writing down the first equations until QFT could be applied to interesting physical problems in a systematic way. The new developments made it possible to apply QFT to new particles and new interactions. In the following decades QFT was extended to describe not only the electromagnetic force, but also weak and strong interaction so that new Lagrangians had to be found which contain new classes of 'particles' or quantum fields. The research aimed at a more comprehensive theory of matter and in the end at a unified theory of all interactions.

New theoretical concepts had to be introduced, mainly connected with non-Abelian gauge theories and spontaneous symmetry breaking. See also the entry on symmetry and symmetry breaking. Today there are trustworthy theories of the strong, weak, and electromagnetic interactions of elementary particles which have a similar structure as QED. A combined theory associated with the gauge group SU(3) ⊗ SU(2) ⊗ U(1) is considered as 'the standard model' of elementary particle physics which was achieved by Sheldon Glashow, Steven Weinberg and Abdul Salam in 1962.

According to the standard model there are, on the one side, six types of leptons (e.g. the electron and its neutrino) and six types of quarks, where the members of both group are all fermions with spin 1/2. On the other side, there are spin 1 particles (thus bosons) that mediate the interaction between elementary particles and the fundamental forces, namely the photon for electromagnetic interaction, two W and one Z-boson for weak interaction, and the gluon for strong interaction. Altogether there is good agreement with experimental data, for example the masses of W+ and W− bosons (detected in 1983) confirmed the theoretical prediction within one per cent deviation.

Further Reading. The first chapter in Weinberg (1995) is a very good short description of the earlier history of QFT. Detailed accounts of the historical development of QFT can be found, e.g., in Darrigol 1986, Schweber (1994) and Cao 1997a. Various historical and conceptual studies of the standard model are gathered in Hoddeson et al. 1997 and of renormalization theory in Brown 1993.

14.4 History

Main article: History of quantum field theory

14.4.1 Foundations

The early development of the field involved Dirac, Fock, Pauli, Heisenberg and Bogolyubov. This phase of development culminated with the construction of the theory of quantum electrodynamics in the 1950s.

14.4.2 Gauge theory

Gauge theory was formulated and quantized, leading to the **unification of forces** embodied in the standard model of particle physics. This effort started in the 1950s with the work of Yang and Mills, was carried on by Martinus Veltman and a host of others during the 1960s and completed by the 1970s through the work of Gerard 't Hooft, Frank Wilczek, David Gross and David Politzer.

14.4.3 Grand synthesis

Parallel developments in the understanding of phase transitions in condensed matter physics led to the study of the renormalization group. This in turn led to the grand synthesis of theoretical physics, which unified theories of particle and condensed matter physics through quantum field theory. This involved the work of Michael Fisher and Leo Kadanoff in the 1970s, which led to the seminal reformulation of quantum field theory by Kenneth G. Wilson in 1975.

14.5 Principles

14.5.1 Classical and quantum fields

Main article: Classical field theory

A classical field is a function defined over some region of space and time.[8] Two physical phenomena which are described by classical fields are Newtonian gravitation, described by Newtonian gravitational field $\mathbf{g}(\mathbf{x}, t)$, and classical electromagnetism, described by the electric and magnetic fields $\mathbf{E}(\mathbf{x}, t)$ and $\mathbf{B}(\mathbf{x}, t)$. Because such fields can in principle take on distinct values at each point in space, they are said to have infinite degrees of freedom.[8]

Classical field theory does not, however, account for the quantum-mechanical aspects of such physical phenomena. For instance, it is known from quantum mechanics that certain aspects of electromagnetism involve discrete particles—photons—rather than continuous fields. The business of *quantum* field theory is to write down a field that is, like a classical field, a function defined over space and time, but which also accommodates the observations of quantum mechanics. This is a *quantum field*.

It is not immediately clear *how* to write down such a quantum field, since quantum mechanics has a structure very unlike a field theory. In its most general formulation, quantum mechanics is a theory of abstract operators (observables) acting on an abstract state space (Hilbert space), where the observables represent physically observable quantities and the state space represents the possible states of the system under study.[9] For instance, the fundamental observables associated with the motion of a single quantum mechanical particle are the position and momentum operators $\hat{x}$ and $\hat{p}$. Field theory, in contrast, treats x as a way to index the field rather than as an operator.[10]

There are two common ways of developing a quantum field: the path integral formalism and canonical quantization.[11] The latter of these is pursued in this article.

Lagrangian formalism

Quantum field theory frequently makes use of the Lagrangian formalism from classical field theory. This formalism is analogous to the Lagrangian formalism used in classical mechanics to solve for the motion of a particle under the influence of a field. In classical field theory, one writes down a Lagrangian density, $\mathcal{L}$, involving a field, $\varphi(\mathbf{x},t)$, and possibly its first derivatives ($\partial\varphi/\partial t$ and $\nabla\varphi$), and then applies a field-theoretic form of the Euler–Lagrange equation. Writing coordinates $(t, \mathbf{x}) = (x^0, x^1, x^2, x^3) = x^\mu$, this form of the Euler–Lagrange equation is[8]

$$\frac{\partial}{\partial x^\mu}\left[\frac{\partial\mathcal{L}}{\partial(\partial\varphi/\partial x^\mu)}\right] - \frac{\partial\mathcal{L}}{\partial\varphi} = 0,$$

where a sum over μ is performed according to the rules of Einstein notation.

By solving this equation, one arrives at the "equations of motion" of the field.[8] For example, if one begins with the Lagrangian density

$$\mathcal{L}(\varphi, \nabla\varphi) = -\rho(t, \mathbf{x})\,\varphi(t, \mathbf{x}) - \frac{1}{8\pi G}|\nabla\varphi|^2,$$

and then applies the Euler–Lagrange equation, one obtains the equation of motion

$$4\pi G\rho(t, \mathbf{x}) = \nabla^2\varphi.$$

This equation is Newton's law of universal gravitation, expressed in differential form in terms of the gravitational potential $\varphi(t, \mathbf{x})$ and the mass density $\rho(t, \mathbf{x})$. Despite the nomenclature, the "field" under study is the gravitational potential, φ, rather than the gravitational field, $\mathbf{g}$. Similarly, when classical field theory is used to study electromagnetism, the "field" of interest is the electromagnetic four-potential $(V/c, \mathbf{A})$, rather than the electric and magnetic fields $\mathbf{E}$ and $\mathbf{B}$.

Quantum field theory uses this same Lagrangian procedure to determine the equations of motion for quantum fields. These equations of motion are then supplemented by commutation relations derived from the canonical quantization procedure described below, thereby incorporating quantum mechanical effects into the behavior of the field.

14.5.2   Single- and many-particle quantum mechanics

Main articles: Quantum mechanics and First quantization

In non-relativistic quantum mechanics, a particle (such as an electron or proton) is described by a complex wavefunction, $\psi(x, t)$, whose time-evolution is governed by the Schrödinger equation:

$$-\frac{\hbar^2}{2m}\frac{\partial^2}{\partial x^2}\psi(x, t) + V(x)\psi(x, t) = i\hbar\frac{\partial}{\partial t}\psi(x, t).$$

Here m is the particle's mass and $V(x)$ is the applied potential. Physical information about the behavior of the particle is extracted from the wavefunction by constructing expected values for various quantities; for example, the expected value of the particle's position is given by integrating $\psi^*(x)\,x\,\psi(x)$ over all space, and the expected value of the particle's momentum is found by integrating $-i\hbar\psi^*(x)d\psi/dx$. The quantity $\psi^*(x)\psi(x)$ is itself in the Copenhagen interpretation of quantum mechanics interpreted as a probability density function. This treatment of quantum mechanics, where a particle's wavefunction evolves against a classical background potential $V(x)$, is sometimes called *first quantization*.

This description of quantum mechanics can be extended to describe the behavior of multiple particles, so long as the number and the type of particles remain fixed. The particles are described by a wavefunction $\psi(x_1, x_2, ..., xN, t)$, which is governed by an extended version of the Schrödinger equation.

Often one is interested in the case where N particles are all of the same type (for example, the 18 electrons orbiting a neutral argon nucleus). As described in the article on identical particles, this implies that the state of the entire system must be either symmetric (bosons) or antisymmetric (fermions) when the coordinates of its constituent particles are exchanged. This is achieved by using a Slater determinant as the wavefunction of a fermionic system (and a Slater permanent for a bosonic system), which is equivalent to an element of the symmetric or antisymmetric subspace of a tensor product.

For example, the general quantum state of a system of N bosons is written as

$$|\phi_1\cdots\phi_N\rangle = \sqrt{\frac{\prod_j N_j!}{N!}}\sum_{p\in S_N}|\phi_{p(1)}\rangle\otimes\cdots\otimes|\phi_{p(N)}\rangle,$$

where $|\phi_i\rangle$ are the single-particle states, Nj is the number of particles occupying state j, and the sum is taken over all possible permutations p acting on N elements. In general, this is a sum of $N!$ (N factorial) distinct terms. $\sqrt{\frac{\prod_j N_j!}{N!}}$ is a normalizing factor.

There are several shortcomings to the above description of quantum mechanics, which are addressed by quantum field theory. First, it is unclear how to extend quantum mechanics to include the effects of special relativity.[12] Attempted replacements for the Schrödinger equation, such as the Klein–Gordon equation or the Dirac equation, have many unsatisfactory qualities; for instance, they possess energy eigenvalues that extend to $-\infty$, so that there seems to be no easy definition of a ground state. It turns out that such inconsistencies arise from relativistic wavefunctions not having a well-defined probabilistic interpretation in position space, as probability conservation is not a relativistically covariant concept. The second shortcoming, related to the first, is that in quantum mechanics there is no mechanism to describe particle creation and annihilation;[13] this is crucial for describing phenomena such as pair production, which result from the conversion between mass and energy according to the relativistic relation $E = mc^2$.

14.5.3 Second quantization

Main article: Second quantization

In this section, we will describe a method for constructing a quantum field theory called second quantization. This basically involves choosing a way to index the quantum mechanical degrees of freedom in the space of multiple identical-particle states. It is based on the Hamiltonian formulation of quantum mechanics.

Several other approaches exist, such as the Feynman path integral,[14] which uses a Lagrangian formulation. For an overview of some of these approaches, see the article on quantization.

Bosons

For simplicity, we will first discuss second quantization for bosons, which form perfectly symmetric quantum states. Let us denote the mutually orthogonal single-particle states which are possible in the system by $|\phi_1\rangle, |\phi_2\rangle, |\phi_3\rangle$, and so on. For example, the 3-particle state with one particle in state $|\phi_1\rangle$ and two in state $|\phi_2\rangle$ is

$$\frac{1}{\sqrt{3}} \left[|\phi_1\rangle|\phi_2\rangle|\phi_2\rangle + |\phi_2\rangle|\phi_1\rangle|\phi_2\rangle + |\phi_2\rangle|\phi_2\rangle|\phi_1\rangle \right].$$

The first step in second quantization is to express such quantum states in terms of **occupation numbers**, by listing the number of particles occupying each of the single-particle states $|\phi_1\rangle, |\phi_2\rangle$, etc. This is simply another way of labelling the states. For instance, the above 3-particle state is denoted as

$$|1, 2, 0, 0, 0, \ldots\rangle.$$

An N-particle state belongs to a space of states describing systems of N particles. The next step is to combine the individual N-particle state spaces into an extended state space, known as Fock space, which can describe systems of any number of particles. This is composed of the state space of a system with no particles (the so-called vacuum state, written as $|0\rangle$), plus the state space of a 1-particle system, plus the state space of a 2-particle system, and so forth. States describing a definite number of particles are known as Fock states: a general element of Fock space will be a linear combination of Fock states. There is a one-to-one correspondence between the occupation number representation and valid boson states in the Fock space.

At this point, the quantum mechanical system has become a quantum field in the sense we described above. The field's elementary degrees of freedom are the occupation numbers, and each occupation number is indexed by a number j indicating which of the single-particle states $|\phi_1\rangle, |\phi_2\rangle, \ldots, |\phi_j\rangle, \ldots$ it refers to:

$$|N_1, N_2, N_3, \ldots, N_j, \ldots\rangle.$$

The properties of this quantum field can be explored by defining creation and annihilation operators, which add and subtract particles. They are analogous to ladder operators in the quantum harmonic oscillator problem, which added and subtracted energy quanta. However, these operators literally create and annihilate particles of a given quantum state. The bosonic annihilation operator a_2 and creation operator $a_2^\dagger$ are easily defined in the occupation number representation as having the following effects:

$$a_2|N_1, N_2, N_3, \ldots\rangle = \sqrt{N_2} \mid N_1, (N_2 - 1), N_3, \ldots\rangle,$$

$$a_2^\dagger|N_1, N_2, N_3, \ldots\rangle = \sqrt{N_2 + 1} \mid N_1, (N_2+1), N_3, \ldots\rangle.$$

It can be shown that these are operators in the usual quantum mechanical sense, i.e. linear operators acting on the Fock space. Furthermore, they are indeed Hermitian conjugates, which justifies the way we have written them. They can be shown to obey the commutation relation

$$[a_i, a_j] = 0 \quad , \quad \left[a_i^\dagger, a_j^\dagger\right] = 0 \quad , \quad \left[a_i, a_j^\dagger\right] = \delta_{ij},$$

where δ stands for the Kronecker delta. These are precisely the relations obeyed by the ladder operators for an infinite set of independent quantum harmonic oscillators, one for each single-particle state. Adding or removing bosons from each state is therefore analogous to exciting or de-exciting a quantum of energy in a harmonic oscillator.

Applying an annihilation operator a_k followed by its corresponding creation operator $a_k^\dagger$ returns the number N_k of particles in the kth single-particle eigenstate:

$$a_k^\dagger a_k | \ldots, N_k, \ldots \rangle = N_k | \ldots, N_k, \ldots \rangle.$$

The combination of operators $a_k^\dagger a_k$ is known as the number operator for the kth eigenstate.

The Hamiltonian operator of the quantum field (which, through the Schrödinger equation, determines its dynamics) can be written in terms of creation and annihilation operators. For instance, for a field of free (non-interacting) bosons, the total energy of the field is found by summing the energies of the bosons in each energy eigenstate. If the kth single-particle energy eigenstate has energy E_k and there are N_k bosons in this state, then the total energy of these bosons is $E_k N_k$. The energy in the *entire* field is then a sum over k :

$$E_{\text{tot}} = \sum_k E_k N_k$$

This can be turned into the Hamiltonian operator of the field by replacing N_k with the corresponding number operator, $a_k^\dagger a_k$. This yields

$$H = \sum_k E_k a_k^\dagger a_k.$$

Fermions

It turns out that a different definition of creation and annihilation must be used for describing fermions. According to the Pauli exclusion principle, fermions cannot share quantum states, so their occupation numbers Ni can only take on the value 0 or 1. The fermionic annihilation operators c and creation operators $c^\dagger$ are defined by their actions on a Fock state thus

$$c_j | N_1, N_2, \ldots, N_j = 0, \ldots \rangle = 0$$

$$c_j | N_1, N_2, \ldots, N_j = 1, \ldots \rangle = (-1)^{(N_1 + \cdots + N_{j-1})} | N_1, N_2, \ldots, N_j = 0, \ldots \rangle$$

$$c_j^\dagger | N_1, N_2, \ldots, N_j = 0, \ldots \rangle = (-1)^{(N_1 + \cdots + N_{j-1})} | N_1, N_2, \ldots, N_j = 1, \ldots \rangle$$

$$c_j^\dagger | N_1, N_2, \ldots, N_j = 1, \ldots \rangle = 0.$$

These obey an anticommutation relation:

$$\{c_i, c_j\} = 0 \quad , \quad \{c_i^\dagger, c_j^\dagger\} = 0 \quad , \quad \{c_i, c_j^\dagger\} = \delta_{ij}.$$

One may notice from this that applying a fermionic creation operator twice gives zero, so it is impossible for the particles to share single-particle states, in accordance with the exclusion principle.

Field operators

We have previously mentioned that there can be more than one way of indexing the degrees of freedom in a quantum field. Second quantization indexes the field by enumerating the single-particle quantum states. However, as we have discussed, it is more natural to think about a "field", such as the electromagnetic field, as a set of degrees of freedom indexed by position.

To this end, we can define *field operators* that create or destroy a particle at a particular point in space. In particle physics, these operators turn out to be more convenient to work with, because they make it easier to formulate theories that satisfy the demands of relativity.

Single-particle states are usually enumerated in terms of their momenta (as in the particle in a box problem.) We can construct field operators by applying the Fourier transform to the creation and annihilation operators for these states. For example, the bosonic field annihilation operator $\phi(\mathbf{r})$ is

$$\phi(\mathbf{r}) \overset{\text{def}}{=} \sum_j e^{i \mathbf{k}_j \cdot \mathbf{r}} a_j.$$

The bosonic field operators obey the commutation relation

$$[\phi(\mathbf{r}), \phi(\mathbf{r}')] = 0 \quad , \quad [\phi^\dagger(\mathbf{r}), \phi^\dagger(\mathbf{r}')] = 0 \quad , \quad [\phi(\mathbf{r}), \phi^\dagger(\mathbf{r}')] = \delta$$

where $\delta(x)$ stands for the Dirac delta function. As before, the fermionic relations are the same, with the commutators replaced by anticommutators.

The field operator is not the same thing as a single-particle wavefunction. The former is an operator acting on the Fock space, and the latter is a quantum-mechanical amplitude for finding a particle in some position. However, they are closely related, and are indeed commonly denoted with the same symbol. If we have a Hamiltonian with a space representation, say

$$H = -\frac{\hbar^2}{2m} \sum_i \nabla_i^2 + \sum_{i<j} U(|\mathbf{r}_i - \mathbf{r}_j|)$$

where the indices i and j run over all particles, then the field theory Hamiltonian (in the non-relativistic limit and for negligible self-interactions) is

$$H = -\frac{\hbar^2}{2m} \int d^3r \, \phi^\dagger(\mathbf{r}) \nabla^2 \phi(\mathbf{r}) + \frac{1}{2} \int d^3r \int d^3r' \, \phi^\dagger($$

$$\mathbf{r})\phi^\dagger(\mathbf{r}')U(|\mathbf{r}-\mathbf{r}'|)\phi(\mathbf{r}')\phi(\mathbf{r}).$$

This looks remarkably like an expression for the expectation value of the energy, with ϕ playing the role of the wavefunction. This relationship between the field operators and wavefunctions makes it very easy to formulate field theories starting from space-projected Hamiltonians.

14.5.4 Dynamics

Once the Hamiltonian operator is obtained as part of the canonical quantization process, the time dependence of the state is described with the Schrödinger equation, just as with other quantum theories. Alternatively, the Heisenberg picture can be used where the time dependence is in the operators rather than in the states.

14.5.5 Implications

Unification of fields and particles

The "second quantization" procedure that we have outlined in the previous section takes a set of single-particle quantum states as a starting point. Sometimes, it is impossible to define such single-particle states, and one must proceed directly to quantum field theory. For example, a quantum theory of the electromagnetic field *must* be a quantum field theory, because it is impossible (for various reasons) to define a wavefunction for a single photon.[15] In such situations, the quantum field theory can be constructed by examining the mechanical properties of the classical field and guessing the corresponding quantum theory. For free (non-interacting) quantum fields, the quantum field theories obtained in this way have the same properties as those obtained using second quantization, such as well-defined creation and annihilation operators obeying commutation or anticommutation relations.

Quantum field theory thus provides a unified framework for describing "field-like" objects (such as the electromagnetic field, whose excitations are photons) and "particle-like" objects (such as electrons, which are treated as excitations of an underlying electron field), so long as one can treat interactions as "perturbations" of free fields. There are still unsolved problems relating to the more general case of interacting fields that may or may not be adequately described by perturbation theory. For more on this topic, see Haag's theorem.

Physical meaning of particle indistinguishability

The second quantization procedure relies crucially on the particles being identical. We would not have been able to construct a quantum field theory from a distinguishable many-particle system, because there would have been no way of separating and indexing the degrees of freedom.

Many physicists prefer to take the converse interpretation, which is that *quantum field theory explains what identical particles are*. In ordinary quantum mechanics, there is not much theoretical motivation for using symmetric (bosonic) or antisymmetric (fermionic) states, and the need for such states is simply regarded as an empirical fact. From the point of view of quantum field theory, particles are identical if and only if they are excitations of the same underlying quantum field. Thus, the question "why are all electrons identical?" arises from mistakenly regarding individual electrons as fundamental objects, when in fact it is only the electron field that is fundamental.

Particle conservation and non-conservation

During second quantization, we started with a Hamiltonian and state space describing a fixed number of particles (N), and ended with a Hamiltonian and state space for an arbitrary number of particles. Of course, in many common situations N is an important and perfectly well-defined quantity, e.g. if we are describing a gas of atoms sealed in a box. From the point of view of quantum field theory, such situations are described by quantum states that are eigenstates of the number operator $\hat{N}$, which measures the total number of particles present. As with any quantum mechanical observable, $\hat{N}$ is conserved if it commutes with the Hamiltonian. In that case, the quantum state is trapped in the N-particle subspace of the total Fock space, and the situation could equally well be described by ordinary N-particle quantum mechanics. (Strictly speaking, this is only true in the noninteracting case or in the low energy density limit of renormalized quantum field theories)

For example, we can see that the free-boson Hamiltonian described above conserves particle number. Whenever the Hamiltonian operates on a state, each particle destroyed by an annihilation operator a_k is immediately put back by the creation operator $a_k^\dagger$.

On the other hand, it is possible, and indeed common, to encounter quantum states that are *not* eigenstates of $\hat{N}$, which do not have well-defined particle numbers. Such states are difficult or impossible to handle using ordinary quantum mechanics, but they can be easily described in quantum field theory as quantum superpositions of states having different values of N. For example, suppose we have a bosonic field whose particles can be created or destroyed by interactions with a fermionic field. The Hamiltonian of the combined system would be given by the Hamiltonians of the free boson and free fermion fields, plus a "potential energy" term such as

$$H_I = \sum_{k,q} V_q(a_q + a_{-q}^\dagger)c_{k+q}^\dagger c_k,$$

where $a_k^\dagger$ and a_k denotes the bosonic creation and annihilation operators, $c_k^\dagger$ and c_k denotes the fermionic creation and annihilation operators, and V_q is a parameter that describes the strength of the interaction. This "interaction term" describes processes in which a fermion in state k either absorbs or emits a boson, thereby being kicked into a different eigenstate $k + q$. (In fact, this type of Hamiltonian is used to describe interaction between conduction electrons and phonons in metals. The interaction between electrons and photons is treated in a similar way, but is a little more complicated because the role of spin must be taken into account.) One thing to notice here is that even if we start out with a fixed number of bosons, we will typically end up with a superposition of states with different numbers of bosons at later times. The number of fermions, however, is conserved in this case.

In condensed matter physics, states with ill-defined particle numbers are particularly important for describing the various superfluids. Many of the defining characteristics of a superfluid arise from the notion that its quantum state is a superposition of states with different particle numbers. In addition, the concept of a coherent state (used to model the laser and the BCS ground state) refers to a state with an ill-defined particle number but a well-defined phase.

14.5.6 Axiomatic approaches

The preceding description of quantum field theory follows the spirit in which most physicists approach the subject. However, it is not mathematically rigorous. Over the past several decades, there have been many attempts to put quantum field theory on a firm mathematical footing by formulating a set of axioms for it. These attempts fall into two broad classes.

The first class of axioms, first proposed during the 1950s, include the Wightman, Osterwalder–Schrader, and Haag–Kastler systems. They attempted to formalize the physicists' notion of an "operator-valued field" within the context of functional analysis, and enjoyed limited success. It was possible to prove that any quantum field theory satisfying these axioms satisfied certain general theorems, such as the spin-statistics theorem and the CPT theorem. Unfortunately, it proved extraordinarily difficult to show that any

realistic field theory, including the Standard Model, satisfied these axioms. Most of the theories that could be treated with these analytic axioms were physically trivial, being restricted to low-dimensions and lacking interesting dynamics. The construction of theories satisfying one of these sets of axioms falls in the field of constructive quantum field theory. Important work was done in this area in the 1970s by Segal, Glimm, Jaffe and others.

During the 1980s, a second set of axioms based on geometric ideas was proposed. This line of investigation, which restricts its attention to a particular class of quantum field theories known as topological quantum field theories, is associated most closely with Michael Atiyah and Graeme Segal, and was notably expanded upon by Edward Witten, Richard Borcherds, and Maxim Kontsevich. However, most of the physically relevant quantum field theories, such as the Standard Model, are not topological quantum field theories; the quantum field theory of the fractional quantum Hall effect is a notable exception. The main impact of axiomatic topological quantum field theory has been on mathematics, with important applications in representation theory, algebraic topology, and differential geometry.

Finding the proper axioms for quantum field theory is still an open and difficult problem in mathematics. One of the Millennium Prize Problems—proving the existence of a mass gap in Yang–Mills theory—is linked to this issue.

14.6   Associated phenomena

In the previous part of the article, we described the most general features of quantum field theories. Some of the quantum field theories studied in various fields of theoretical physics involve additional special ideas, such as renormalizability, gauge symmetry, and supersymmetry. These are described in the following sections.

14.6.1   Renormalization

Main article: Renormalization

Early in the history of quantum field theory, it was found that many seemingly innocuous calculations, such as the perturbative shift in the energy of an electron due to the presence of the electromagnetic field, give infinite results. The reason is that the perturbation theory for the shift in an energy involves a sum over all other energy levels, and there are infinitely many levels at short distances that each give a finite contribution which results in a divergent series.

Many of these problems are related to failures in classical electrodynamics that were identified but unsolved in the

19th century, and they basically stem from the fact that many of the supposedly "intrinsic" properties of an electron are tied to the electromagnetic field that it carries around with it. The energy carried by a single electron—its self energy—is not simply the bare value, but also includes the energy contained in its electromagnetic field, its attendant cloud of photons. The energy in a field of a spherical source diverges in both classical and quantum mechanics, but as discovered by Weisskopf with help from Furry, in quantum mechanics the divergence is much milder, going only as the logarithm of the radius of the sphere.

The solution to the problem, presciently suggested by Stueckelberg, independently by Bethe after the crucial experiment by Lamb, implemented at one loop by Schwinger, and systematically extended to all loops by Feynman and Dyson, with converging work by Tomonaga in isolated postwar Japan, comes from recognizing that all the infinities in the interactions of photons and electrons can be isolated into redefining a finite number of quantities in the equations by replacing them with the observed values: specifically the electron's mass and charge: this is called renormalization. The technique of renormalization recognizes that the problem is essentially purely mathematical, that extremely short distances are at fault. In order to define a theory on a continuum, first place a cutoff on the fields, by postulating that quanta cannot have energies above some extremely high value. This has the effect of replacing continuous space by a structure where very short wavelengths do not exist, as on a lattice. Lattices break rotational symmetry, and one of the crucial contributions made by Feynman, Pauli and Villars, and modernized by 't Hooft and Veltman, is a symmetry-preserving cutoff for perturbation theory (this process is called regularization). There is no known symmetrical cutoff outside of perturbation theory, so for rigorous or numerical work people often use an actual lattice.

On a lattice, every quantity is finite but depends on the spacing. When taking the limit of zero spacing, we make sure that the physically observable quantities like the observed electron mass stay fixed, which means that the constants in the Lagrangian defining the theory depend on the spacing. Hopefully, by allowing the constants to vary with the lattice spacing, all the results at long distances become insensitive to the lattice, defining a continuum limit.

The renormalization procedure only works for a certain class of quantum field theories, called **renormalizable quantum field theories**. A theory is **perturbatively renormalizable** when the constants in the Lagrangian only diverge at worst as logarithms of the lattice spacing for very short spacings. The continuum limit is then well defined in perturbation theory, and even if it is not fully well defined non-perturbatively, the problems only show up at distance scales that are exponentially small in the inverse coupling for weak couplings. The Standard Model of particle physics

is perturbatively renormalizable, and so are its component theories (quantum electrodynamics/electroweak theory and quantum chromodynamics). Of the three components, quantum electrodynamics is believed to not have a continuum limit, while the asymptotically free SU(2) and SU(3) weak hypercharge and strong color interactions are nonperturbatively well defined.

The renormalization group describes how renormalizable theories emerge as the long distance low-energy effective field theory for any given high-energy theory. Because of this, renormalizable theories are insensitive to the precise nature of the underlying high-energy short-distance phenomena. This is a blessing because it allows physicists to formulate low energy theories without knowing the details of high energy phenomenon. It is also a curse, because once a renormalizable theory like the standard model is found to work, it gives very few clues to higher energy processes. The only way high energy processes can be seen in the standard model is when they allow otherwise forbidden events, or if they predict quantitative relations between the coupling constants.

14.6.2 Haag's theorem

See also: Haag's theorem

From a mathematically rigorous perspective, there exists no interaction picture in a Lorentz-covariant quantum field theory. This implies that the perturbative approach of Feynman diagrams in QFT is not strictly justified, despite producing vastly precise predictions validated by experiment. This is called Haag's theorem, but most particle physicists relying on QFT largely shrug it off.

14.6.3 Gauge freedom

A gauge theory is a theory that admits a symmetry with a local parameter. For example, in every quantum theory the global phase of the wave function is arbitrary and does not represent something physical. Consequently, the theory is invariant under a global change of phases (adding a constant to the phase of all wave functions, everywhere); this is a global symmetry. In quantum electrodynamics, the theory is also invariant under a *local* change of phase, that is – one may shift the phase of all wave functions so that the shift may be different at every point in space-time. This is a *local* symmetry. However, in order for a well-defined derivative operator to exist, one must introduce a new field, the gauge field, which also transforms in order for the local change of variables (the phase in our example) not to affect the derivative. In quantum electrodynamics this gauge field is the electromagnetic field. The change of local gauge of variables is termed gauge transformation. It is worth noting that by Noether's theorem, for every such symmetry there exists an associated conserved current. The aforementioned symmetry of the wavefunction under global phase changes implies the conservation of electric charge.

In quantum field theory the excitations of fields represent particles. The particle associated with excitations of the gauge field is the gauge boson, which is the photon in the case of quantum electrodynamics.

The degrees of freedom in quantum field theory are local fluctuations of the fields. The existence of a gauge symmetry reduces the number of degrees of freedom, simply because some fluctuations of the fields can be transformed to zero by gauge transformations, so they are equivalent to having no fluctuations at all, and they therefore have no physical meaning. Such fluctuations are usually called "non-physical degrees of freedom" or *gauge artifacts*; usually some of them have a negative norm, making them inadequate for a consistent theory. Therefore, if a classical field theory has a gauge symmetry, then its quantized version (i.e. the corresponding quantum field theory) will have this symmetry as well. In other words, a gauge symmetry cannot have a quantum anomaly. If a gauge symmetry is anomalous (i.e. not kept in the quantum theory) then the theory is non-consistent: for example, in quantum electrodynamics, had there been a gauge anomaly, this would require the appearance of photons with longitudinal polarization and polarization in the time direction, the latter having a negative norm, rendering the theory inconsistent; another possibility would be for these photons to appear only in intermediate processes but not in the final products of any interaction, making the theory non-unitary and again inconsistent (see optical theorem).

In general, the gauge transformations of a theory consist of several different transformations, which may not be commutative. These transformations are together described by a mathematical object known as a gauge group. Infinitesimal gauge transformations are the gauge group generators. Therefore, the number of gauge bosons is the group dimension (i.e. number of generators forming a basis).

All the fundamental interactions in nature are described by gauge theories. These are:

- Quantum chromodynamics, whose gauge group is SU(3). The gauge bosons are eight gluons.

- The electroweak theory, whose gauge group is U(1) × SU(2), (a direct product of U(1) and SU(2)).

- Gravity, whose classical theory is general relativity, admits the equivalence principle, which is a form

of gauge symmetry. However, it is explicitly non-renormalizable.

14.6.4   Multivalued gauge transformations

The gauge transformations which leave the theory invariant involve, by definition, only single-valued gauge functions $\Lambda(x_i)$ which satisfy the Schwarz integrability criterion

$$\partial_{x_i x_j}\Lambda = \partial_{x_j x_i}\Lambda.$$

An interesting extension of gauge transformations arises if the gauge functions $\Lambda(x_i)$ are allowed to be multivalued functions which violate the integrability criterion. These are capable of changing the physical field strengths and are therefore not proper symmetry transformations. Nevertheless, the transformed field equations describe correctly the physical laws in the presence of the newly generated field strengths. See the textbook by H. Kleinert cited below for the applications to phenomena in physics.

14.6.5   Supersymmetry

Main article: Supersymmetry

Supersymmetry assumes that every fundamental fermion has a superpartner that is a boson and vice versa. It was introduced in order to solve the so-called Hierarchy Problem, that is, to explain why particles not protected by any symmetry (like the Higgs boson) do not receive radiative corrections to its mass driving it to the larger scales (GUT, Planck...). It was soon realized that supersymmetry has other interesting properties: its gauged version is an extension of general relativity (Supergravity), and it is a key ingredient for the consistency of string theory.

The way supersymmetry protects the hierarchies is the following: since for every particle there is a superpartner with the same mass, any loop in a radiative correction is cancelled by the loop corresponding to its superpartner, rendering the theory UV finite.

Since no superpartners have yet been observed, if supersymmetry exists it must be broken (through a so-called soft term, which breaks supersymmetry without ruining its helpful features). The simplest models of this breaking require that the energy of the superpartners not be too high; in these cases, supersymmetry is expected to be observed by experiments at the Large Hadron Collider. The Higgs particle has been detected at the LHC, and no such superparticles have been discovered.

14.7 See also

- Abraham–Lorentz force
- Basic concepts of quantum mechanics
- Common integrals in quantum field theory
- Einstein–Maxwell–Dirac equations
- Form factor (quantum field theory)
- Green–Kubo relations
- Green's function (many-body theory)
- Invariance mechanics
- List of quantum field theories
- Quantization of a field
- Quantum electrodynamics
- Quantum field theory in curved spacetime
- Quantum flavordynamics
- Quantum hydrodynamics
- Quantum triviality
- Relation between Schrödinger's equation and the path integral formulation of quantum mechanics
- Relationship between string theory and quantum field theory
- Schwinger–Dyson equation
- Static forces and virtual-particle exchange
- Symmetry in quantum mechanics
- Theoretical and experimental justification for the Schrödinger equation
- Ward–Takahashi identity
- Wheeler–Feynman absorber theory
- Wigner's classification
- Wigner's theorem

14.8 Remarks

14.9 Notes

[1] Thorn et al. 2004

[2] Dirac 1927

[3] Schweber 1994, p. 28

[4] See references in Schweber (1994, pp. 695f)

[5] Weinberg 2005, p. 15

[6] Peskin & Schroeder (1995, Chapter 4)

[7] Greiner & Reinhardt 1996

[8] Tong 2015, Chapter 1

[9] Srednicki 2007, p. 19

[10] Srednicki 2007, pp. 25–26

[11] Zee 2010, p. 61

[12] Tong 2015, Introduction

[13] Zee 2010, p. 3

[14] Pais 1994 Pais recounts how his astonishment at the rapidity with which Feynman could calculate using his method. Feynman's method is now part of the standard methods for physicists.

[15] Newton & Wigner 1949, pp. 400–406

14.10 References

Historical references

- Born, M.; Jordan, P.; Heisenberg, W. (1926). "Zur quantenmechanic II" [On Quantum mechanics II]. *Zeitschrift für Physik* (in German) (Springer Verlag) **35** (8). doi:10.1007/BF01379806. ISSN 0044-3328. (subscription required (help)).

- Dirac, P. A. M. (1927). "The quantum theory of the emission and absorption of radiation". *Proc. R. Soc. Lond. A* (Royal Society Publishing) **114** (767): 243–265. doi:10.1098/rspa.1927.0039. (subscription required (help)).

General readers

- Pais, A. (1994) [1986]. *Inward Bound: Of Matter and Forces in the Physical World* (reprint ed.). Oxford, New York, Toronto: Oxford University Press. ISBN 978-0198519973.

- Schweber, S. S. (1994). *QED and the Men Who Made It: Dyson, Feynman, Schwinger, and Tomonaga.* Princeton University Press. ISBN 9780691033273.

Articles

- Newton, T. D.; Wigner, E.P. (1949). "Localized states for elementary systems". *Rev. Mod. Phys.* (APS) **21** (3). Bibcode:1949RvMP...21..400N. doi:10.1103/RevModPhys.21.400. ISSN 0034-6861.

- Thorn, J. J.; Neel, M. S.; Donato, W. V.; Bergreen, G. S.; Davies, R. E.; Beck, M.. (2004). "Observing the quantum behavior of light in an undergraduate laboratory" (PDF). *Am. J. Phys.* (American Association of Physics Teachers) **72** (1210): 243–265. doi:10.1119/1.1737397.

Introductory texts

- Greiner, W.; Reinhardt, J. (1996). *Field Quantization.* Springer Publishing. ISBN 3-540-59179-6.

- Peskin, M.; Schroeder, D. (1995). *An Introduction to Quantum Field Theory.* Westview Press. ISBN 0-201-50397-2.

- Scharf, Günter (2014) [1989]. *Finite Quantum Electrodynamics: The Causal Approach* (third ed.). Dover Publications. ISBN 978-0486492735.

- Srednicki, M. (2007). *Quantum Field Theory.* Cambridge University Press. ISBN 978-0521-8644-97.

- Tong, David (2015). "Lectures on Quantum Field Theory". Retrieved 2016-02-09.

- Zee, Anthony (2010). *Quantum Field Theory in a Nutshell* (2nd ed.). Princeton University Press. ISBN 978-0691140346.

Advanced texts

- Weinberg, S. (2005). *The Quantum Theory of Fields* **1**. Cambridge University Press. ISBN 978-0521670531.

14.11   Further reading

General readers

- Feynman, R.P. (2001) [1964]. *The Character of Physical Law.* MIT Press. ISBN 0-262-56003-8.

- Feynman, R.P. (2006) [1985]. *QED: The Strange Theory of Light and Matter.* Princeton University Press. ISBN 0-691-12575-9.

- Gribbin, J. (1998). *Q is for Quantum: Particle Physics from A to Z.* Weidenfeld & Nicolson. ISBN 0-297-81752-3.

- Schumm, Bruce A. (2004) *Deep Down Things.* Johns Hopkins Univ. Press. Chpt. 4.

Introductory texts

- McMahon, D. (2008). *Quantum Field Theory.* McGraw-Hill. ISBN 978-0-07-154382-8.

- Bogoliubov, N.; Shirkov, D. (1982). *Quantum Fields.* Benjamin-Cummings. ISBN 0-8053-0983-7.

- Frampton, P.H. (2000). *Gauge Field Theories. Frontiers in Physics (2nd ed.).* Wiley.

- Greiner, W; Müller, B. (2000). *Gauge Theory of Weak Interactions.* Springer. ISBN 3-540-67672-4.

- Itzykson, C.; Zuber, J.-B. (1980). *Quantum Field Theory.* McGraw-Hill. ISBN 0-07-032071-3.

- Kane, G.L. (1987). *Modern Elementary Particle Physics.* Perseus Books. ISBN 0-201-11749-5.

- Kleinert, H.; Schulte-Frohlinde, Verena (2001). *Critical Properties of φ^4-Theories.* World Scientific. ISBN 981-02-4658-7.

- Kleinert, H. (2008). *Multivalued Fields in Condensed Matter, Electrodynamics, and Gravitation* (PDF). World Scientific. ISBN 978-981-279-170-2.

- Loudon, R (1983). *The Quantum Theory of Light.* Oxford University Press. ISBN 0-19-851155-8.

- Mandl, F.; Shaw, G. (1993). *Quantum Field Theory.* John Wiley & Sons. ISBN 978-0-471-94186-6.

- Ryder, L.H. (1985). *Quantum Field Theory.* Cambridge University Press. ISBN 0-521-33859-X.

- Schwartz, M.D. (2014). *Quantum Field Theory and the Standard Model.* Cambridge University Press. ISBN 978-1107034730.

- Ynduráin, F.J. (1996). *Relativistic Quantum Mechanics and Introduction to Field Theory* (1st ed.). Springer. ISBN 978-3-540-60453-2.

Advanced texts

- Brown, Lowell S. (1994). *Quantum Field Theory*. Cambridge University Press. ISBN 978-0-521-46946-3.

- Bogoliubov, N.; Logunov, A.A.; Oksak, A.I.; Todorov, I.T. (1990). *General Principles of Quantum Field Theory*. Kluwer Academic Publishers. ISBN 978-0-7923-0540-8.

Articles

- 't Hooft, Gerard (2007). Butterfield, J.; Earman, John, eds. *Philosophy of Physics*. Part A. The Conceptual Basis of Quantum Field Theory: Elsevier. pp. 661–730 – via ScienceDirect. (subscription required (help)). On web at 't Hooft's university website

- Wilczek, frank (1999). "Quantum field theory". *Rev. Mod. Phys* **71** (S85–S95). arXiv:hep-th/9803075v2. doi:10.1103/RevModPhys.71.S85.

14.12 External links

- Hazewinkel, Michiel, ed. (2001), "Quantum field theory", *Encyclopedia of Mathematics*, Springer, ISBN 978-1-55608-010-4

- *Stanford Encyclopedia of Philosophy*: "Quantum Field Theory", by Meinard Kuhlmann.

- Siegel, Warren, 2005. *Fields*. A free text, also available from arXiv:hep-th/9912205.

- Quantum Field Theory by P. J. Mulders

Chapter 15

Conformal field theory

A **conformal field theory** (**CFT**) is a quantum field theory that is invariant under conformal transformations. In two dimensions, there is an infinite-dimensional algebra of local conformal transformations, and conformal field theories can sometimes be exactly solved or classified.

Conformal field theory has important applications[1] to string theory, statistical mechanics, and condensed matter physics. Statistical and condensed matter systems are indeed often conformally invariant at their thermodynamic or quantum critical points.

15.1 Scale invariance vs. conformal invariance

While it is possible for a quantum field theory to be scale invariant but not conformally-invariant, examples are rare.[2] For this reason, the terms are often used interchangeably in the context of quantum field theory, even though the scale symmetry group is smaller.

In some particular cases it is possible to prove that scale invariance implies conformal invariance in a quantum field theory, for example in unitary compact conformal field theories in two dimensions.

15.2 Dimensional considerations

15.2.1 Two dimensions

There are two versions of 2D CFT: 1) Euclidean, and 2) Lorentzian. The former applies to statistical mechanics, and the latter to quantum field theory. The two versions are related by a Wick rotation.

Two-dimensional CFTs are (in some way) invariant under an *infinite-dimensional symmetry group*. For example, consider a CFT on the Riemann sphere. It has the Möbius transformations as the conformal group, which is isomorphic to

(the finite-dimensional) PSL(2,C).

However, the infinitesimal conformal transformations[3] form an **infinite-dimensional algebra**, called the Witt algebra, but this infinity of conformal transformations do not have global inverses on C. Only the primary fields (or chiral fields) are invariant with respect to this full infinitesimal conformal group. Its generators are indexed by integers n,

$$L_n = \oint_{z=0} \frac{dz}{2\pi i} z^{n+1} T_{zz} ,$$

where T_{zz} is the holomorphic part of the non-trace piece of the energy momentum tensor of the theory. E.g., for a free scalar field,

$$T_{zz} = \tfrac{1}{2}(\partial_z \phi)^2 .$$

In most conformal field theories, a conformal anomaly, also known as a Weyl anomaly, arises in the quantum theory. This results in the appearance of a nontrivial central charge, and the Witt algebra is extended to the Virasoro algebra.

In Euclidean CFT, one has both a holomorphic and an antiholomorphic copy of the Virasoro algebra. In Lorentzian CFT, one has a left-moving and a right moving copy of the Virasoro algebra (spacetime is a cylinder, with space being a circle, and time a line).

This symmetry makes it possible to classify two-dimensional CFTs much more precisely than in higher dimensions. In particular, it is possible to relate the spectrum of primary operators in a theory to the value of the central charge, c.

The Hilbert space of physical states is a unitary module of the Virasoro algebra corresponding to a fixed value of c. Stability requires that the energy spectrum of the Hamiltonian be nonnegative. The modules of interest are the highest weight modules of the Virasoro algebra.

A chiral field is a holomorphic field $W(z)$ which transforms as

$$L_n W(z) = -z^{n+1} \frac{\partial}{\partial z} W(z) - (n+1)\Delta z^n W(z)$$

and

$$\bar{L}_n W(z) = 0 \,.$$

Analogously, mutatis mutandis, for an antichiral field. Δ is called the *conformal weight* of the chiral field W.

Furthermore, it was shown by Alexander Zamolodchikov that there exists a function, C, which decreases monotonically under the renormalization group flow of a two-dimensional quantum field theory, and is equal to the central charge for a two-dimensional conformal field theory. This is known as the Zamolodchikov C-theorem, and tells us that renormalization group flow in two dimensions is irreversible.

Frequently, we are not just interested in the operators, but we are also interested in the vacuum state, or in statistical mechanics, the thermal state. Unless $c=0$, there can't possibly be any state which leaves the entire infinite dimensional conformal symmetry unbroken. The best we can come up with is a state which is invariant under L_{-1}, L_0, L_1, L_i, $i > 1$. This contains the Möbius subgroup. The rest of the conformal group is spontaneously broken.

Two-dimensional conformal field theories play an important role in statistical mechanics, where they describe critical points of many lattice models.

15.2.2 More than two dimensions

In $d > 2$ dimensions, the conformal group is isomorphic to $SO(d+1, 1)$ in Euclidean signature, or $SO(d, 2)$ in Minkowski space.

Higher-dimensional conformal field theories are prominent in the AdS/CFT correspondence, in which a gravitational theory in anti-de Sitter space (AdS) is equivalent to a conformal field theory on the AdS boundary. Notable examples are d=4 N = 4 supersymmetric Yang–Mills theory, which is dual to Type IIB string theory on AdS_5 x S^5, and d=3 N=6 super-Chern–Simons theory, which is dual to M-theory on AdS_4 x S^7. (The prefix "super" denotes supersymmetry, N denotes the degree of extended supersymmetry possessed by the theory, and d the number of space-time dimensions on the boundary.)

15.3 Conformal symmetry

Conformal symmetry is a symmetry under scale invariance

and under the special conformal transformations having the following relations.

$$[P_\mu, P_\nu] = 0,$$

$$[D, K_\mu] = -K_\mu,$$

$$[D, P_\mu] = P_\mu,$$

$$[K_\mu, K_\nu] = 0,$$

$$[K_\mu, P_\nu] = \eta_{\mu\nu} D - i M_{\mu\nu},$$

where P generates translations, D generates scaling transformations as a scalar and K_μ generates the special conformal transformations as a covariant vector under Lorentz transformation.

Main articles: Conformal symmetry and Conformal Killing equation

15.4 See also

- Logarithmic conformal field theory

- AdS/CFT correspondence

- Operator product expansion

- Vertex operator algebra

- WZW model

- Critical point

- Boundary conformal field theory

- Primary field

- Superconformal algebra

- Conformal algebra

- Conformal bootstrap

15.5 References

[1] Paul Ginsparg (1989), *Applied Conformal Field Theory*. arXiv:hep-th/9108028. Published in *Ecole d'Eté de Physique Théorique: Champs, cordes et phénomènes critiques/Fields, strings and critical phenomena* (Les Houches), ed. by E. Brézin and J. Zinn-Justin, Elsevier Science Publishers B.V.

[2] One physical example is the theory of elasticity in two and three dimensions (also known as the theory of a vector field without gauge invariance). See Riva V, Cardy J (2005). "Scale and conformal invariance in field theory: a physical counterexample". *Phys. Lett.* **B 622**: 339–342. arXiv:hep-th/0504197.   Bibcode:2005PhLB..622..339R. doi:10.1016/j.physletb.2005.07.010.

[3] Since the conformal Killing equations in two dimensions, $\partial_\mu \xi_\nu + \partial_\nu \xi_\mu = \partial \cdot \xi \eta_{\mu\nu}$, reduce to just the Cauchy-Riemann equations, $\partial_{\bar{z}} \xi(z) = 0 = \partial_z \xi(\bar{z})$, the infinity of modes of arbitrary analytic coordinate transformations $\xi(z)$ yield the infinity of Killing vector fields $z^n \, \partial z$.

15.6   Further reading

- Martin Schottenloher, *A Mathematical Introduction to Conformal Field Theory*, Springer-Verlag, Berlin, Heidelberg, 1997. ISBN 3-540-61753-1, 2nd edition 2008, ISBN 978-3-540-68625-5.

- P. Di Francesco, P. Mathieu, and D. Sénéchal, *Conformal Field Theory*, Springer-Verlag, New York, 1997. ISBN 0-387-94785-X.

- Conformal Field Theory page in String Theory Wiki lists books and reviews.

- Slava Rychkov, Lectures on Conformal Field Theory in D≥ 3 Dimensions, 2012

Chapter 16

Orbifold

This terminology should not be blamed on me. It was obtained by a democratic process in my course of 1976–77. An orbifold is something with many folds; unfortunately, the word "manifold" already has a different definition. I tried "foldamani", which was quickly displaced by the suggestion of "manifolded". After two months of patiently saying "no, not a manifold, a manifol*dead*," we held a vote, and "orbifold" won.

Thurston (1980, section 13.2) explaining the origin of the word "orbifold"

In the mathematical disciplines of topology, geometry, and geometric group theory, an **orbifold** (for "orbit-manifold") is a generalization of a manifold. It is a topological space (called the *underlying space*) with an orbifold structure (see below).

The underlying space locally looks like the quotient space of a Euclidean space under the linear action of a finite group. Definitions of orbifold have been given several times: by Satake in the context of automorphic forms in the 1950s under the name *V-manifold*;[1] by Thurston in the context of the geometry of 3-manifolds in the 1970s[2] when he coined the name *orbifold*, after a vote by his students; and by Haefliger in the 1980s in the context of Gromov's programme on CAT(k) spaces under the name *orbihedron*.[3] The definition of Thurston will be described here: it is the most widely used and is applicable in all cases.

Mathematically, orbifolds arose first as surfaces with singular points long before they were formally defined.[4] One of the first classical examples arose in the theory of modular forms[5] with the action of the modular group $SL(2,\mathbf{Z})$ on the upper half-plane: a version of the Riemann–Roch theorem holds after the quotient is compactified by the addition of two orbifold cusp points. In 3-manifold theory, the theory of Seifert fiber spaces, initiated by Seifert, can be phrased in terms of 2-dimensional orbifolds.[6] In geometric group theory, post-Gromov, discrete groups have been studied in terms of the local curvature properties of orbihedra and their covering spaces.[7]

In string theory, the word "orbifold" has a slightly different meaning,[8] discussed in detail below. In conformal field theory, a mathematical part of string theory, it is often used to refer to the theory attached to the fixed point subalgebra of a vertex algebra under the action of a finite group of automorphisms.

The main example of underlying space is a quotient space of a manifold under the properly discontinuous action of a possibly infinite group of diffeomorphisms with finite isotropy subgroups.[9] In particular this applies to any action of a finite group; thus a manifold with boundary carries a natural orbifold structure, since it is the quotient of its double by an action of $\mathbf{Z}_2$. Similarly the quotient space of a manifold by a smooth proper action of S^1 carries the structure of an orbifold.

Orbifold structure gives a natural stratification by open manifolds on its underlying space, where one stratum corresponds to a set of singular points of the same type.

It should be noted that one topological space can carry many different orbifold structures. For example, consider the orbifold O associated with a factor space of the 2-sphere along a rotation by π; it is homeomorphic to the 2-sphere, but the natural orbifold structure is different. It is possible to adopt most of the characteristics of manifolds to orbifolds and these characteristics are usually different from correspondent characteristics of underlying space. In the above example, the *orbifold fundamental group* of O is $\mathbf{Z}_2$ and its *orbifold Euler characteristic* is 1.

16.1   Formal definitions

Like a manifold, an orbifold is specified by local conditions; however, instead of being locally modelled on open subsets of $\mathbf{R}^n$, an orbifold is locally modelled on quotients of open subsets of $\mathbf{R}^n$ by finite group actions. The structure of an orbifold encodes not only that of the underlying quotient space, which need not be a manifold, but also that of the isotropy subgroups.

An *n*-dimensional **orbifold** is a Hausdorff topological space

X, called the **underlying space**, with a covering by a collection of open sets Ui, closed under finite intersection. For each Ui, there is

- an open subset Vi of $\mathbf{R}^n$, invariant under a faithful linear action of a finite group Γi

- a continuous map φi of Vi onto Ui invariant under Γi, called an **orbifold chart**, which defines a homeomorphism between $Vi / \Gamma i$ and Ui.

The collection of orbifold charts is called an **orbifold atlas** if the following properties are satisfied:

- for each inclusion $Ui \subset Uj$ there is an injective group homomorphism $fij : \Gamma i \to \Gamma j$

- for each inclusion $Ui \subset Uj$ there is a Γi-equivariant homeomorphism ψij, called a **gluing map**, of Vi onto an open subset of Vj

- the gluing maps are compatible with the charts, i.e. $\varphi j{\cdot}\psi ij = \varphi i$

- the gluing maps are unique up to composition with group elements, i.e. any other possible gluing map from Vi to Vj has the form $g{\cdot}\psi ij$ for a unique g in Γj

The orbifold atlas defines the **orbifold structure** completely: two orbifold atlases of X give the same orbifold structure if they can be consistently combined to give a larger orbifold atlas. Note that the orbifold structure determines the isotropy subgroup of any point of the orbifold up to isomorphism: it can be computed as the stabilizer of the point in any orbifold chart. If $Ui \subset Uj \subset Uk$, then there is a unique *transition element* g_{ijk} in Γk such that

$$gijk{\cdot}\psi ik = \psi jk{\cdot}\psi ij$$

These transition elements satisfy

$$(\text{Ad } gijk){\cdot}fik = fjk{\cdot}fij$$

as well as the *cocycle relation* (guaranteeing associativity)

$$fkm(gijk){\cdot}gikm = gijm{\cdot}gjkm.$$

More generally, attached to an open covering of an orbifold by orbifold charts, there is the combinatorial data of a so-called *complex of groups* (see below).

Exactly as in the case of manifolds, differentiability conditions can be imposed on the gluing maps to give a definition

of a **differentiable orbifold**. It will be a *Riemannian orbifold* if in addition there are invariant Riemannian metrics on the orbifold charts and the gluing maps are isometries.

For applications in geometric group theory, it is often convenient to have a slightly more general notion of orbifold, due to Haefliger. An **orbispace** is to topological spaces what an orbifold is to manifolds. An orbispace is a topological generalization of the orbifold concept. It is defined by replacing the model for the orbifold charts by a locally compact space with a *rigid* action of a finite group, i.e. one for which points with trivial isotropy are dense. (This condition is automatically satisfied by faithful linear actions, because the points fixed by any non-trivial group element form a proper linear subspace.) It is also useful to consider metric space structures on an orbispace, given by invariant metrics on the orbispace charts for which the gluing maps preserve distance. In this case each orbispace chart is usually required to be a length space with unique geodesics connecting any two points.

16.1.1 Examples

- Any manifold without boundary is trivially an orbifold. Each of the groups Γi is the trivial group.

- If N is a compact manifold with boundary, its **double** M can formed by gluing together a copy of N and its mirror image along their common boundary. There is natural *reflection* action of $\mathbf{Z}_2$ on the manifold M fixing the common boundary; the quotient space can be identified with N, so that N has a natural orbifold structure.

- If M is a Riemannian n-manifold with a cocompact proper isometric action of a discrete group Γ, then the orbit space $X = M/\Gamma$ has natural orbifold structure: for each x in X take a representative m in M and an open neighbourhood Vm of m invariant under the stabiliser Γm, identified equivariantly with a Γm-subset of TmM under the exponential map at m; finitely many neighbourhoods cover X and each of their finite intersections, if non-empty, is covered by an intersection of Γ-translates $gm{\cdot}Vm$ with corresponding group $gm \Gamma gm^{-1}$. Orbifolds that arise in this way are called *developable* or *good*.

- A classical theorem of Henri Poincaré constructs Fuchsian groups as hyperbolic reflection groups generated by reflections in the edges of a geodesic triangle in the hyperbolic plane for the Poincaré metric. If the triangle has angles π / ni for positive integers ni, the triangle is a fundamental domain and naturally a 2-dimensional orbifold. The corresponding group is an example of a hyperbolic triangle group. Poincaré also gave a 3-dimensional version of this result for Kleinian groups: in this case the Kleinian group Γ is generated by hyperbolic reflections and the orbifold is $\mathbf{H}^3 / \Gamma$.

- If M is a closed 2-manifold, new orbifold structures can be defined on Mi by removing finitely many disjoint closed discs from M and gluing back copies of discs $D/ \Gamma i$ where D is the closed unit disc and Γi is a finite cyclic group of rotations. This generalises Poincaré's construction.

16.2   Orbifold fundamental group

There are several ways to define the **orbifold fundamental group**. More sophisticated approaches use orbifold covering spaces or classifying spaces of groupoids. The simplest approach (adopted by Haefliger and known also to Thurston) extends the usual notion of loop used in the standard definition of the fundamental group.

An **orbifold path** is a path in the underlying space provided with an explicit piecewise lift of path segments to orbifold charts and explicit group elements identifying paths in overlapping charts; if the underlying path is a loop, it is called an **orbifold loop**. Two orbifold paths are identified if they are related through multiplication by group elements in orbifold charts. The orbifold fundamental group is the group formed by homotopy classes of orbifold loops.

If the orbifold arises as the quotient of a simply connected manifold M by a proper rigid action of a discrete group Γ, the orbifold fundamental group can be identified with Γ. In general it is an extension of Γ by $\pi_1 M$.

The orbifold is said to be *developable* or *good* if it arises as the quotient by a finite group action; otherwise it is called *bad*. A *universal covering orbifold* can be constructed for an orbifold by direct analogy with the construction of the universal covering space of a topological space, namely as the space of pairs consisting of points of the orbifold and homotopy classes of orbifold paths joining them to the basepoint. This space is naturally an orbifold.

Note that if an orbifold chart on a contractible open subset corresponds to a group Γ, then there is a natural *local homomorphism* of Γ into the orbifold fundamental group.

In fact the following conditions are equivalent:

- The orbifold is developable.

- The orbifold structure on the universal covering orbifold is trivial.

- The local homomorphisms are all injective for a covering by contractible open sets.

16.3   Non-positively curved orbispaces

As explained above, an **orbispace** is basically a generalization of the orbifold concept applied to topological spaces. Let then X be an orbispace endowed with a metric space structure for which the charts are geodesic length spaces. The preceding definitions and results for orbifolds can be generalized to give definitions of *orbispace fundamental group* and *universal covering orbispace*, with analogous criteria for developability. The distance functions on the orbispace charts can be used to define the length of an orbispace path in the universal covering orbispace. If the distance function in each chart is non-positively curved, then the Birkhoff curve shortening argument can be used to prove that any orbispace path with fixed endpoints is homotopic to a unique geodesic. Applying this to constant paths in an orbispace chart, it follows that each local homomorphism is injective and hence:

- every non-positively curved orbispace is developable (i.e. *good*).

16.4   Complexes of groups

Every orbifold has associated with it an additional combinatorial structure given by a *complex of groups*.

16.4.1   Definition

A **complex of groups** (Y,f,g) on an abstract simplicial complex Y is given by

- a finite group $\Gamma\sigma$ for each simplex σ of Y

- an injective homomorphism $f\sigma\tau : \Gamma\tau \to \Gamma\sigma$ whenever $\sigma \subset \tau$

- for every inclusion $\rho \subset \sigma \subset \tau$, a group element $g_\rho\sigma\tau$ in Γ_ρ such that $(\mathrm{Ad}\, g_\rho\sigma\tau)\cdot f_\rho\tau = f_\rho\sigma\cdot f\sigma\tau$ (here Ad denotes the adjoint action by conjugation)

The group elements must in addition satisfy the cocycle condition

$$f\pi_\rho(g_\rho\sigma\tau)\, g\pi_\rho\tau = g\pi\sigma\tau\, g\pi_\rho\sigma$$

for every chain of simplices $\pi \subset \rho \subset \sigma \subset \tau$. (This condition is vacuous if Y has dimension 2 or less.)

Any choice of elements $h\sigma\tau$ in $\Gamma\sigma$ yields an *equivalent* complex of groups by defining

- $f'\sigma\tau = (\text{Ad } h\sigma\tau)\cdot f\sigma\tau$

- $g'_\rho\sigma\tau = h_\rho\sigma\cdot f_\rho\sigma(h\sigma\tau)\cdot g_\rho\sigma\tau\cdot h_\rho\tau^{-1}$

A complex of groups is called **simple** whenever $g_\rho\sigma\tau = 1$ everywhere.

- An easy inductive argument shows that every complex of groups on a *simplex* is equivalent to a complex of groups with $g_\rho\sigma\tau = 1$ everywhere.

It is often more convenient and conceptually appealing to pass to the barycentric subdivision of Y. The vertices of this subdivision correspond to the simplices of Y, so that each vertex has a group attached to it. The edges of the barycentric subdivision are naturally oriented (corresponding to inclusions of simplices) and each directed edge gives an inclusion of groups. Each triangle has a transition element attached to it belonging to the group of exactly one vertex; and the tetrahedra, if there are any, give cocycle relations for the transition elements. Thus a complex of groups involves only the 3-skeleton of the barycentric subdivision; and only the 2-skeleton if it is simple.

16.4.2   Example

If X is an orbifold (or orbispace), choose a covering by open subsets from amongst the orbifold charts $fi : Vi \to Ui$. Let Y be the abstract simplicial complex given by the nerve of the covering: its vertices are the sets of the cover and its n-simplices correspond to *non-empty* intersections $U\alpha = Ui_1 \cap \cdots \cap Uin$. For each such simplex there is an associated group $\Gamma\alpha$ and the homomorphisms fij become the homomorphisms $f\sigma\tau$. For every triple $\rho \subset \sigma \subset \tau$ corresponding to intersections

$$Ui \supset Ui \cap Uj \supset Ui \cap Uj \cap Uk$$

there are charts $\varphi i : Vi \to Ui$, $\varphi ij : Vij \to Ui \cap Uj$ and $\varphi ijk : Vijk \to Ui \cap Uj \cap Uk$ and gluing maps $\psi : Vij \to Vi$, $\psi' : Vijk \to Vij$ and $\psi'' : Vijk \to Vi$.

There is a unique transition element $g_\rho\sigma\tau$ in Γi such that $g_\rho\sigma\tau\cdot\psi'' = \psi\cdot\psi'$. The relations satisfied by the transition elements of an orbifold imply those required for a complex of groups. In this way a complex of groups can be canonically associated to the nerve of an open covering by orbifold (or orbispace) charts. In the language of non-commutative sheaf theory and gerbes, the complex of groups in this case arises as a sheaf of groups associated to the covering Ui; the data $g_\rho\sigma\tau$ is a 2-cocycle in non-commutative sheaf cohomology and the data $h\sigma\tau$ gives a 2-coboundary perturbation.

16.4.3   Edge-path group

The **edge-path group** of a complex of groups can be defined as a natural generalisation of the edge path group of a simplicial complex. In the barycentric subdivision of Y, take generators eij corresponding to edges from i to j where $i \to j$, so that there is an injection $\psi ij : \Gamma i \to \Gamma j$. Let Γ be the group generated by the eij and Γk with relations

$$eij^{-1} \cdot g \cdot eij = \psi ij(g)$$

for g in Γi and

$$eik = ejk\cdot eij\cdot gijk$$

if $i \to j \to k$.

For a fixed vertex i_0, the edge-path group $\Gamma(i_0)$ is defined to be the subgroup of Γ generated by all products

$$g_0 \cdot ei_0 i_1 \cdot g_1 \cdot ei_1 i_2 \cdots \cdot gn \cdot eini_0$$

where $i_0, i_1, ..., in, i_0$ is an edge-path, gk lies in Γik and $eji = eij^{-1}$ if $i \to j$.

16.4.4   Developable complexes

A simplicial proper action of a discrete group Γ on a simplicial complex X with finite quotient is said to be **regular** if it satisfies one of the following equivalent conditions (see Bredon 1972):

- X admits a finite subcomplex as fundamental domain;

- the quotient $Y = X/\Gamma$ has a natural simplicial structure;

- the quotient simplicial structure on orbit-representatives of vertices is consistent;

- if $(v_0, ..., vk)$ and $(g_0\cdot v_0, ..., gk\cdot vk)$ are simplices, then $g\cdot vi = gi\cdot vi$ for some g in Γ.

The fundamental domain and quotient $Y = X / \Gamma$ can naturally be identified as simplicial complexes in this case, given by the stabilisers of the simplices in the fundamental domain. A complex of groups Y is said to be **developable** if it arises in this way.

- A complex of groups is developable if and only if the homomorphisms of $\Gamma\sigma$ into the edge-path group are injective.

- A complex of groups is developable if and only if for each simplex σ there is an injective homomorphism θσ from Γσ into a fixed discrete group Γ such that θτ·fστ = θσ. In this case the simplicial complex X is canonically defined: it has k-simplices (σ, xΓσ) where σ is a k-simplex of Y and x runs over Γ / Γσ. Consistency can be checked using the fact that the restriction of the complex of groups to a *simplex* is equivalent to one with trivial cocycle $g_\rho \sigma \tau$.

The action of Γ on the barycentric subdivision X' of X always satisfies the following condition, weaker than regularity:

- whenever σ and g·σ are subsimplices of some simplex τ, they are equal, i.e. σ = g·σ

Indeed, simplices in X' correspond to chains of simplices in X, so that a subsimplices, given by subchains of simplices, is uniquely determined by the *sizes* of the simplices in the subchain. When an action satisfies this condition, then g necessarily fixes all the vertices of σ. A straightforward inductive argument shows that such an action becomes regular on the barycentric subdivision; in particular

- the action on the second barycentric subdivision X" is regular;

- Γ is naturally isomorphic to the edge-path group defined using edge-paths and vertex stabilisers for the barycentric subdivision of the fundamental domain in X".

There is in fact no need to pass to a *third* barycentric subdivision: as Haefliger observes using the language of category theory, in this case the 3-skeleton of the fundamental domain of X" already carries all the necessary data – including transition elements for triangles – to define an edge-path group isomorphic to Γ.

In two dimensions this is particularly simple to describe. The fundamental domain of X" has the same structure as the barycentric subdivision Y' of a complex of groups Y, namely:

- a finite 2-dimensional simplicial complex Z;

- an orientation for all edges $i \to j$;

- if $i \to j$ and $j \to k$ are edges, then $i \to k$ is an edge and (i, j, k) is a triangle;

- finite groups attached to vertices, inclusions to edges and transition elements, describing compatibility, to triangles.

An edge-path group can then be defined. A similar structure is inherited by the barycentric subdivision Z' and its edge-path group is isomorphic to that of Z.

16.5 Orbihedra

If a countable discrete group acts by a *regular simplicial* proper action on a simplicial complex, the quotient can be given not only the structure of a complex of groups, but also that of an orbispace. This leads more generally to the definition of "orbihedron", the simplicial analogue of an orbifold.

16.5.1 Definition

Let X be a finite simplicial complex with barycentric subdivision X'. An **orbihedron** structure consists of:

- for each vertex i of X', a simplicial complex Li' endowed with a rigid simplicial action of a finite group Γi.

- a simplicial map φi of Li' onto the link Li of i in X', identifying the quotient Li' / Γi with Li.

This action of Γi on Li' extends to a simplicial action on the simplicial cone Ci over Li' (the simplicial join of i and Li'), fixing the centre i of the cone. The map φi extends to a simplicial map of Ci onto the star St(i) of i, carrying the centre onto i; thus φi identifies Ci / Γi, the quotient of the star of i in Ci, with St(i) and gives an *orbihedron chart* at i.

- for each directed edge $i \to j$ of X', an injective homomorphism fij of Γi into Γj.

- for each directed edge $i \to j$, a Γi equivariant simplicial *gluing map* ψij of Ci into Cj.

- the gluing maps are compatible with the charts, i.e. φj·ψij = φi.

- the gluing maps are unique up to composition with group elements, i.e. any other possible gluing map from Vi to Vj has the form g·ψij for a unique g in Γj.

If $i \to j \to k$, then there is a unique *transition element* g_{ijk} in Γk such that

$$g_{ijk} \cdot \psi ik = \psi jk \cdot \psi ij$$

These transition elements satisfy

(Ad *gijk*)·*fik* = *fjk*·*fij*

as well as the cocycle relation

ψ*km*(*gijk*)·*gikm* = *gijm*·*gjkm*.

16.5.2   Main properties

- The group theoretic data of an orbihedron gives a complex of groups on X, because the vertices i of the barycentric subdivision X' correspond to the simplices in X.

- Every complex of groups on X is associated with an essentially unique orbihedron structure on X. This key fact follows by noting that the star and link of a vertex i of X', corresponding to a simplex σ of X, have natural decompositions: the star is isomorphic to the abstract simplicial complex given by the join of σ and the barycentric subdivision σ' of σ; and the link is isomorphic to join of the link of σ in X and the link of the barycentre of σ in σ'. Restricting the complex of groups to the link of σ in X, all the groups Γτ come with injective homomorphisms into Γσ. Since the link of i in X' is canonically covered by a simplicial complex on which Γσ acts, this defines an orbihedron structure on X.

- The orbihedron fundamental group is (tautologically) just the edge-path group of the associated complex of groups.

- Every orbihedron is also naturally an orbispace: indeed in the geometric realization of the simplicial complex, orbispace charts can be defined using the interiors of stars.

- The orbihedron fundamental group can be naturally identified with the orbispace fundamental group of the associated orbispace. This follows by applying the simplicial approximation theorem to segments of an orbispace path lying in an orbispace chart: it is a straightforward variant of the classical proof that the fundamental group of a polyhedron can be identified with its edge-path group.

- The orbispace associated to an orbihedron has a *canonical metric structure*, coming locally from the length metric in the standard geometric realization in Euclidean space, with vertices mapped to an orthonormal basis. Other metric structures are also used, involving length metrics obtained by realizing the simplices in hyperbolic space, with simplices identified isometrically along common boundaries.

- The orbispace associated to an orbihedron is non-positively curved if and only if the link in each orbihedron chart has girth greater than or equal to 6, i.e. any closed circuit in the link has length at least 6. This condition, well known from the theory of Hadamard spaces, depends only on the underlying complex of groups.

- When the universal covering orbihedron is non-positively curved the fundamental group is infinite and is generated by isomorphic copies of the isotropy groups. This follows from the corresponding result for orbispaces.

16.6   Triangles of groups

Historically one of the most important applications of orbifolds in geometric group theory has been to *triangles of groups*. This is the simplest 2-dimensional example generalising the 1-dimensional "interval of groups" discussed in Serre's lectures on trees, where amalgamated free products are studied in terms of actions on trees. Such triangles of groups arise any time a discrete group acts simply transitively on the triangles in the affine Bruhat-Tits building for $SL_3(\mathbf{Q}_p)$; in 1979 Mumford discovered the first example for $p = 2$ (see below) as a step in producing an algebraic surface not isomorphic to projective space, but having the same Betti numbers. Triangles of groups were worked out in detail by Gersten and Stallings, while the more general case of complexes of groups, described above, was developed independently by Haefliger. The underlying geometric method of analysing finitely presented groups in terms of metric spaces of non-positive curvature is due to Gromov. In this context triangles of groups correspond to non-positively curved 2-dimensional simplicial complexes with the regular action of a group, *transitive on triangles*.

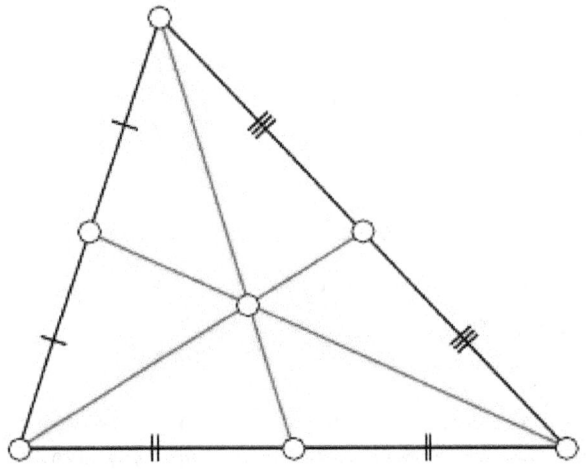

A **triangle of groups** is a *simple* complex of groups consisting of a triangle with vertices **A**, **B**, **C**. There are groups

- ΓA, ΓB, ΓC at each vertex

- ΓBC, ΓCA, ΓAB for each edge

- ΓABC for the triangle itself.

There is an injective homomorphisms of ΓABC into all the other groups and of an edge group ΓXY into ΓX and ΓY. The three ways of mapping ΓABC into a vertex group all agree. (Often ΓABC is the trivial group.) The Euclidean metric structure on the corresponding orbispace is non-positively curved if and only if the link of each of the vertices in the orbihedron chart has girth at least 6.

This girth at each vertex is always even and, as observed by Stallings, can be described at a vertex **A**, say, as the length of the smallest word in the kernel of the natural homomorphism into ΓA of the amalgamated free product over ΓABC of the edge groups ΓAB and ΓAC:

$$\Gamma_{AB} \star_{\Gamma_{ABC}} \Gamma_{AC} \to \Gamma_A.$$

The result using the Euclidean metric structure is not optimal. Angles α, β, γ at the vertices **A**, **B** and **C** were defined by Stallings as 2π divided by the girth. In the Euclidean case α, β, γ ≤ π/3. However, if it is only required that α + β + γ ≤ π, it is possible to identify the triangle with the corresponding geodesic triangle in the hyperbolic plane with the Poincaré metric (or the Euclidean plane if equality holds). It is a classical result from hyperbolic geometry that the hyperbolic medians intersect in the hyperbolic barycentre,[10] just as in the familiar Euclidean case. The barycentric subdivision and metric from this model yield a non-positively curved metric structure on the corresponding orbispace. Thus, if α+β+γ≤π,

- the orbispace of the triangle of groups is developable;

- the corresponding edge-path group, which can also be described as the colimit of the triangle of groups, is infinite;

- the homomorphisms of the vertex groups into the edge-path group are injections.

16.6.1 Mumford's example

Let $\alpha = \sqrt{-7}$ be given by the binomial expansion of $(1 - 8)^{1/2}$ in $\mathbf{Q}_2$ and set $K = \mathbf{Q}(\alpha) \subset \mathbf{Q}_2$. Let

$$\zeta = \exp 2\pi i/7$$

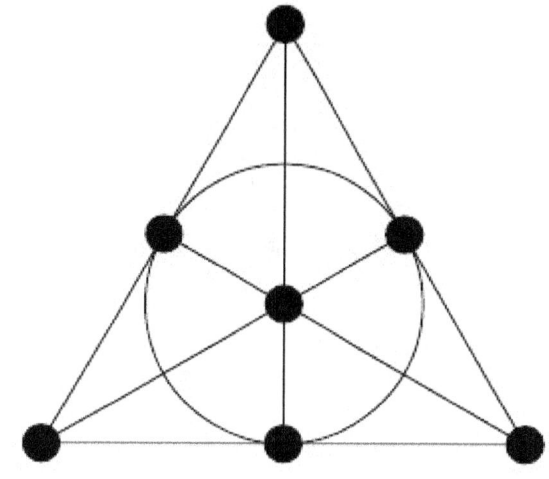

The Fano plane

$$\lambda = (\alpha - 1)/2 = \zeta + \zeta^2 + \zeta^4$$

$$\mu = \lambda/\lambda^*.$$

Let $E = \mathbf{Q}(\zeta)$, a 3-dimensional vector space over K with basis 1, ζ and ζ^2. Define K-linear operators on E as follows:

- σ is the generator of the Galois group of E over K, an element of order 3 given by $\sigma(\zeta) = \zeta^2$

- τ is the operator of multiplication by ζ on E, an element of order 7

- ρ is the operator given by $\rho(\zeta) = 1$, $\rho(\zeta^2) = \zeta$ and $\rho(1) = \mu \cdot \zeta^2$, so that ρ^3 is scalar multiplication by μ.

The elements ρ, σ and τ generate a discrete subgroup of $GL_3(K)$ which acts properly on the affine Bruhat–Tits building corresponding to $SL_3(\mathbf{Q}_2)$. This group acts *transitively* on all vertices, edges and triangles in the building. Let

$$\sigma_1 = \sigma, \ \sigma_2 = \rho\sigma\rho^{-1}, \ \sigma_3 = \rho^2\sigma\rho^{-2}.$$

Then

- σ_1, σ_2 and σ_3 generate a subgroup Γ of $SL_3(K)$.

- Γ is the smallest subgroup generated by σ and τ, invariant under conjugation by ρ.

- Γ acts simply transitively on the triangles in the building.

- There is a triangle Δ such that the stabiliser of its edges are the subgroups of order 3 generated by the σi's.

- The stabiliser of a vertices of Δ is the Frobenius group of order 21 generated by the two order 3 elements stabilising the edges meeting at the vertex.

- The stabiliser of Δ is trivial.

The elements σ and τ generate the stabiliser of a vertex. The link of this vertex can be identified with the spherical building of $SL_3(\mathbf{F}_2)$ and the stabiliser can be identified with the collineation group of the Fano plane generated by a 3-fold symmetry σ fixing a point and a cyclic permutation τ of all 7 points, satisfying $\sigma\tau = \tau^2\sigma$. Identifying $\mathbf{F}_8{}^*$ with the Fano plane, σ can be taken to be the restriction of the Frobenius automorphism $\sigma(x) = x^{22}$ of $\mathbf{F}_8$ and τ to be multiplication by any element not in the prime field $\mathbf{F}_2$, i.e. an order 7 generator of the cyclic multiplicative group of $\mathbf{F}_8$. This Frobenius group acts simply transitively on the 21 flags in the Fano plane, i.e. lines with marked points. The formulas for σ and τ on E thus "lift" the formulas on $\mathbf{F}_8$.

Mumford also obtains an action simply transitive on the vertices of the building by passing to a subgroup of $\Gamma_1 = <\rho, \sigma, \tau, -I>$. The group Γ_1 preserves the $\mathbf{Q}(\alpha)$-valued hermitian form

$$f(x,y) = xy^* + \sigma(xy^*) + \sigma^2(xy^*)$$

on $\mathbf{Q}(\zeta)$ and can be identified with $U_3(f) \cap GL_3(S)$ where $S = \mathbf{Z}[\alpha,\frac{1}{2}]$. Since $S/(\alpha) = \mathbf{F}_7$, there is a homomorphism of the group Γ_1 into $GL_3(\mathbf{F}_7)$. This action leaves invariant a 2-dimensional subspace in $\mathbf{F}_7{}^3$ and hence gives rise to a homomorphism Ψ of Γ_1 into $SL_2(\mathbf{F}_7)$, a group of order $16 \cdot 3 \cdot 7$. On the other hand, the stabiliser of a vertex is a subgroup of order 21 and Ψ is injective on this subgroup. Thus if the congruence subgroup Γ_0 is defined as the inverse image under Ψ of the 2-Sylow subgroup of $SL_2(\mathbf{F}_7)$, the action of Γ_0 on vertices must be simply transitive.

16.6.2   Generalizations

Other examples of triangles or 2-dimensional complexes of groups can be constructed by variations of the above example.

Cartwright et al. consider actions on buildings that are *simply transitive on vertices*. Each such action produces a bijection (or modified duality) between the points x and lines x^* in the flag complex of a finite projective plane and a collection of oriented triangles of points (x,y,z), invariant under cyclic permutation, such that x lies on z^*, y lies on x^* and z lies on y^* and any two points uniquely determine the third. The groups produced have generators x, labelled by

points, and relations $xyz = 1$ for each triangle. Generically this construction will not correspond to an action on a classical affine building.

More generally, as shown by Ballmann and Brin, similar algebraic data encodes all actions that are simply transitively on the vertices of a non-positively curved 2-dimensional simplicial complex, provided the link of each vertex has girth at least 6. This data consists of:

- a generating set S containing inverses, but not the identity;

- a set of relations $g\,h\,k = 1$, invariant under cyclic permutation.

The elements g in S label the vertices $g{\cdot}v$ in the link of a fixed vertex v; and the relations correspond to edges $(g^{-1}{\cdot}v, h{\cdot}v)$ in that link. The graph with vertices S and edges (g, h), for $g^{-1}h$ in S, must have girth at least 6. The original simplicial complex can be reconstructed using complexes of groups and the second barycentric subdivision.

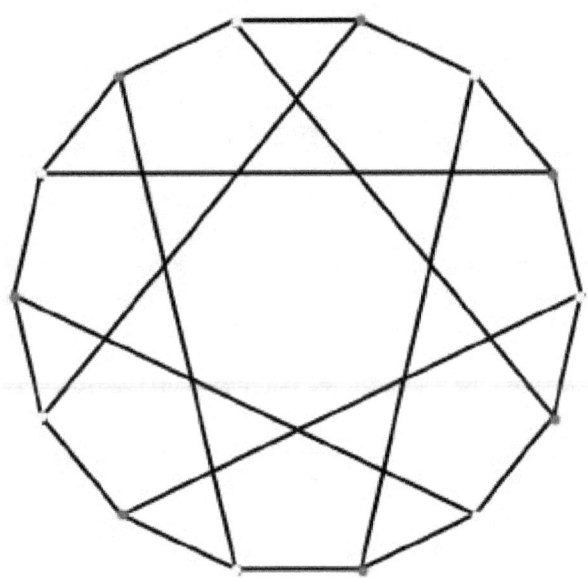

The bipartite Heawood graph

Further examples of non-positively curved 2-dimensional complexes of groups have been constructed by Swiatkowski based on actions *simply transitive on oriented edges* and inducing a 3-fold symmetry on each triangle; in this case too the complex of groups is obtained from the regular action on the second barycentric subdivision. The simplest example, discovered earlier with Ballmann, starts from a finite group H with a symmetric set of generators S, not containing the identity, such that the corresponding Cayley graph has girth at least 6. The associated group is generated by H and an involution τ subject to $(\tau g)^3 = 1$ for each g in S.

In fact, if Γ acts in this way, fixing an edge (v, w), there is an involution τ interchanging v and w. The link of v is made up of vertices $g \cdot w$ for g in a symmetric subset S of $H = \Gamma_v$, generating H if the link is connected. The assumption on triangles implies that

$$\tau \cdot (g \cdot w) = g^{-1} \cdot w$$

for g in S. Thus, if $\sigma = \tau g$ and $u = g^{-1} \cdot w$, then

$$\sigma(v) = w, \quad \sigma(w) = u, \quad \sigma(u) = w.$$

By simple transitivity on the triangle (v, w, u), it follows that $\sigma^3 = 1$.

The second barycentric subdivision gives a complex of groups consisting of singletons or pairs of barycentrically subdivided triangles joined along their large sides: these pairs are indexed by the quotient space $S/\sim$ obtained by identifying inverses in S. The single or "coupled" triangles are in turn joined along one common "spine". All stabilisers of simplices are trivial except for the two vertices at the ends of the spine, with stabilisers H and $\langle\tau\rangle$, and the remaining vertices of the large triangles, with stabiliser generated by an appropriate σ. Three of the smaller triangles in each large triangle contain transition elements.

When all the elements of S are involutions, none of the triangles need to be doubled. If H is taken to be the dihedral group D_7 of order 14, generated by an involution a and an element b of order 7 such that

$$ab = b^{-1}a,$$

then H is generated by the 3 involutions a, ab and ab^5. The link of each vertex is given by the corresponding Cayley graph, so is just the bipartite Heawood graph, i.e. exactly the same as in the affine building for $SL_3(\mathbf{Q}_2)$. This link structure implies that the corresponding simplicial complex is necessarily a Euclidean building. At present, however, it seems to be unknown whether any of these types of action can in fact be realised on a classical affine building: Mumford's group Γ_1 (modulo scalars) is only simply transitive on edges, not on oriented edges.

16.7   2-dimensional orbifolds

In two dimensions, there are three singular point types of an orbifold:

- A boundary point

- An elliptic point or gyration point of order n, such as the origin of $\mathbf{R}^2$ quotiented out by a cyclic group of order n of rotations.

- A corner reflector of order n: the origin of $\mathbf{R}^2$ quotiented out by a dihedral group of order $2n$.

A compact 2-dimensional orbifold has an **Euler characteristic** X given by

$$X = X(X_0) - \Sigma(1 - 1/ni)/2 - \Sigma(1 - 1/mi)$$

where $X(X_0)$ is the Euler characteristic of the underlying topological manifold X_0, and ni are the orders of the corner reflectors, and mi are the orders of the elliptic points.

A 2-dimensional compact connected orbifold has a hyperbolic structure if its Euler characteristic is less than 0, a Euclidean structure if it is 0, and if its Euler characteristic is positive it is either **bad** or has an elliptic structure (an orbifold is called bad if it does not have a manifold as a covering space). In other words, its universal covering space has a hyperbolic, Euclidean, or spherical structure.

The compact 2-dimensional connected orbifolds that are not hyperbolic are listed in the table below. The 17 parabolic orbifolds are the quotients of the plane by the 17 wallpaper groups.

16.8   3-dimensional orbifolds

A 3-manifold is said to be *small* if it is closed, irreducible and does not contain any incompressible surfaces.

Orbifold Theorem. Let M be a small 3-manifold. Let φ be a non-trivial periodic orientation-preserving diffeomorphism of M. Then M admits a φ-invariant hyperbolic or Seifert fibered structure.

This theorem is a special case of Thurston's orbifold theorem, announced without proof in 1981; it forms part of his geometrization conjecture for 3-manifolds. In particular it implies that if X is a compact, connected, orientable, irreducible, atoroidal 3-orbifold with non-empty singular locus, then M has a geometric structure (in the sense of orbifolds). A complete proof of the theorem was published by Boileau, Leeb & Porti in 2005.[11]

16.9   Orbifolds in string theory

In string theory, the word "orbifold" has a slightly new meaning. For mathematicians, an orbifold is a generalization of the notion of manifold that allows the presence of the points whose neighborhood is diffeomorphic to a quotient of $\mathbf{R}^n$ by a finite group, i.e. $\mathbf{R}^n/\Gamma$. In physics, the notion of an orbifold usually describes an object that can be globally written as an orbit space M/G where M is a manifold (or a theory), and G is a group of its isometries (or symmetries) — not necessarily all of them. In string theory, these symmetries do not have to have a geometric interpretation.

A quantum field theory defined on an orbifold becomes singular near the fixed points of G. However string theory requires us to add new parts of the closed string Hilbert space — namely the twisted sectors where the fields defined on the closed strings are periodic up to an action from G. Orbifolding is therefore a general procedure of string theory to derive a new string theory from an old string theory in which the elements of G have been identified with the identity. Such a procedure reduces the number of states because the states must be invariant under G, but it also increases the number of states because of the extra twisted sectors. The result is usually a perfectly smooth, new string theory.

D-branes propagating on the orbifolds are described, at low energies, by gauge theories defined by the quiver diagrams. Open strings attached to these D-branes have no twisted sector, and so the number of open string states is reduced by the orbifolding procedure.

More specifically, when the orbifold group G is a discrete subgroup of spacetime isometries, then if it has no fixed point, the result is usually a compact smooth space; the twisted sector consists of closed strings wound around the compact dimension, which are called *winding states*.

When the orbifold group G is a discrete subgroup of spacetime isometries, and it has fixed points, then these usually have conical singularities, because $\mathbf{R}^n/\mathbf{Z}k$ has such a singularity at the fixed point of Zk. In string theory, gravitational singularities are usually a sign of extra degrees of freedom which are located at a locus point in spacetime. In the case of the orbifold these degrees of freedom are the twisted states, which are strings "stuck" at the fixed points. When the fields related with these twisted states acquire a non-zero vacuum expectation value, the singularity is deformed, i.e. the metric is changed and becomes regular at this point and around it. An example for a resulting geometry is the Eguchi-Hanson spacetime.

From the point of view of D-branes in the vicinity of the fixed points, the effective theory of the open strings attached to these D-branes is a supersymmetric field theory, whose space of vacua has a singular point, where additional massless degrees of freedom exist. The fields related with the closed string twisted sector couple to the open strings in such a way as to add a Fayet-Iliopoulos term to the supersymmetric field theory Lagrangian, so that when such a field acquires a non-zero vacuum expectation value, the Fayet-Iliopoulos term is non-zero, and thereby deforms the theory (i.e. changes it) so that the singularity no longer exists , .

16.9.1   Calabi–Yau manifolds

Main article: Calabi–Yau manifold

In superstring theory,[12][13] the construction of realistic phenomenological models requires dimensional reduction because the strings naturally propagate in a 10-dimensional space whilst the observed dimension of space-time of the universe is 4. Formal constraints on the theories nevertheless place restrictions on the compactified space in which the extra "hidden" variables live: when looking for realistic 4-dimensional models with supersymmetry, the auxiliary compactified space must be a 6-dimensional Calabi–Yau manifold.[14]

There are a large number of possible Calabi–Yau manifolds (tens of thousands), hence the use of the term "swampland" in the current theoretical physics literature to describe the baffling choice. The general study of Calabi–Yau manifolds is mathematically complex and for a long time examples have been hard to construct explicitly. Orbifolds have therefore proved very useful since they automatically satisfy the constraints imposed by supersymmetry. They provide degenerate examples of Calabi–Yau manifolds due to their singular points,[15] but this is completely acceptable from the point of view of theoretical physics. Such orbifolds are called "supersymmetric": they are technically easier to study than general Calabi–Yau manifolds. It is very often possible to associate a continuous family of non-singular Calabi–Yau manifolds to a singular supersymmetric orbifold. In 4 dimensions this can be illustrated using complex K3 surfaces:

- Every K3 surface admits 16 cycles of dimension 2 that are topologically equivalent to usual 2-spheres. Making the surface of these spheres tend to zero, the K3 surface develops 16 singularities. This limit represents a point on the boundary of the moduli space of K3 surfaces and corresponds to the orbifold $T^4/\mathbb{Z}_2$ obtained by taking the quotient of the torus by the symmetry of inversion.

The study of Calabi–Yau manifolds in string theory and the duality between different models of string theory (type IIA and IIB) led to the idea of mirror symmetry in 1988. The role of orbifolds was first pointed out by Dixon, Harvey, Vafa and Witten around the same time.[16]

16.10 Applications

16.10.1 Music theory

Beyond their manifold and various applications in mathematics and physics, orbifolds have been applied to music theory at least as early as 1985 in the work of Guerino Mazzola[17][18] and later by Dmitri Tymoczko and collaborators (Tymoczko 2006) and (Callender & Tymoczko 2008).[19][20] One of the papers of Tymoczko was the first music theory paper published by the journal *Science*.[21][22][23] Mazzola and Tymoczko have participated in debate regarding their theories documented in a series of commentaries available at their respective web sites.[24][25]

Animated slices of the three-dimensional orbifold T^3/S_3. Slices of cubes standing on end (with their long diagonals perpendicular to the plane of the image) form colored Voronoi regions (colored by chord type) which represent the three note chords at their centers, with augmented triads at the very center, surrounded by major and minor triads (lime green and navy blue). The white regions are degenerate trichords (one-note repeated three times), with the three lines (representing two note chords) connecting their centers forming the walls of the twisted triangular prism, 2D planes perpendicular to plane of the image acting as mirrors.

Tymoczko models musical chords consisting of n notes, not necessarily distinct, as points in the orbifold T^n/S_n – the space of n unordered points (not necessarily distinct) in the circle, realized as the quotient of the n-torus T^n (the space of n ordered points on the circle) by the symmetric group S_n (corresponding from moving from an ordered set to an unordered set).

Musically, this is explained as follows:

- Musical tones depend on the frequency (pitch) of their fundamental, and thus are parametrized by the positive real numbers, $\mathbf{R}^+$.

- Musical tones that differ by an octave (a doubling of frequency) are considered the same tone – this corresponds to taking the logarithm base 2 of frequencies (yielding the real numbers, as $\mathbf{R} = \log_2 \mathbf{R}^+$), then quotienting by the integers (corresponding to differing by some number of octaves), yielding a circle (as $S^1 = \mathbf{R}/\mathbf{Z}$).

- Chords correspond to multiple tones without respect to order – thus t notes (with order) correspond to t ordered points on the circle, or equivalently a single point on the t-torus $T^t := S^1 \times \cdots \times S^1$, and omitting order corresponds to taking the quotient by S_t, yielding an orbifold.

For dyads (two tones), this yields the closed Möbius strip; for triads (three tones), this yields an orbifold that can be described as a triangular prism with the top and bottom triangular faces identified with a 120° twist (a ⅓ twist) – equivalently, as a solid torus in 3 dimensions with a cross-section an equilateral triangle and such a twist.

The resulting orbifold is naturally stratified by repeated tones (properly, by integer partitions of t) – the open set consists of distinct tones (the partition $t = 1 + 1 + \cdots + 1$), while there is a 1-dimensional singular set consisting of all tones being the same (the partition $t = t$), which topologically is a circle, and various intermediate partitions. There is also a notable circle which runs through the center of the open set consisting of equally spaced points. In the case of triads, the three side faces of the prism correspond to two tones being the same and the third different (the partition $3 = 2 + 1$), while the three edges of the prism correspond to the 1-dimensional singular set. The top and bottom faces are part of the open set, and only appear because the orbifold has been cut – if viewed as a triangular torus with a twist, these artifacts disappear.

Tymoczko argues that chords close to the center (with tones equally or almost equally spaced) form the basis of much of traditional Western harmony, and that visualizing them in this way assists in analysis. There are 4 chords on the center (equally spaced under equal temperament – spacing of 4/4/4 between tones), corresponding to the augmented triads (thought of as musical sets) C♯FA, DF♯A♯, D♯GB, and EG♯C (then they cycle: FAC♯ = C♯FA), with the 12 major chords and 12 minor chords being the points next to but not on the center – almost evenly spaced but not quite. Major chords correspond to 4/3/5 (or equivalently, 5/4/3) spacing, while minor chords correspond to 3/4/5 spacing. Key changes then correspond to movement between these points in the orbifold, with smoother changes effected by movement between nearby points.

16.11 See also

- Orbifold notation — a system popularized by the mathematician John Horton Conway for representing types of symmetry groups in two-dimensional spaces of constant curvature

- Orientifold

- Kawasaki's Riemann–Roch formula

- Geometric quotient

16.12 Notes

[1] Satake (1956).

[2] Thurston (1978), Chapter 13.

[3] Haefliger (1990).

[4] Poincaré (1985).

[5] Serre (1970).

[6] Scott (1983).

[7] Bridson and Haefliger (1999).

[8] Di Francesco, Mathieu & Sénéchal (1997)

[9] Bredon (1972).

[10] Theorem of the hyperbolic medians

[11] General introductions to this material can be found in Peter Scott's 1983 notes and the expositions of Boileau, Maillot & Porti and Cooper, Hodgson & Kerckhoff.

[12] M. Green, J. Schwartz and E. Witten, *Superstring theory*, Vol. 1 and 2, Cambridge University Press, 1987, ISBN 0521357527

[13] J. Polchinski, *String theory*, Vol. 2, Cambridge University Press, 1999, ISBN 0-521-63304-4

[14] P. Candelas, *Lectures On Complex Manifolds*, in *Trieste 1987, Proceedings, Superstrings '87* 1-88, 1987

[15] Blumenhagen, Ralph; Lüst, Dieter; Theisen, Stefan (2012), *Basic Concepts of String Theory*, Theoretical and Mathematical Physics, Springer, p. 487, ISBN 9783642294969, Orbifolds can be viewed as singular limits of smooth Calabi–Yau manifolds.

[16] Dixon, Harvey, Vafa and Witten, Nucl.Phys. 1985, B261, 678; 1986, B274, 286.

[17] Guerino Mazzola (1985). *Gruppen und Kategorien in der Musik: Entwurf einer mathematischen Musiktheorie*. Heldermann. ISBN 978-3-88538-210-2. Retrieved 26 February 2012.

[18] Guerino Mazzola; Stefan Müller (2002). *The topos of music: geometric logic of concepts, theory, and performance*. Birkhäuser. ISBN 978-3-7643-5731-3. Retrieved 26 February 2012.

[19] Dmitri Tymoczko, *The Geometry of Music* – links to papers and to visualization software.

[20] *The moduli space of chords: Dmitri Tymoczko on "Geometry and Music", Friday 7 Mar, 2:30pm*, posted 28/Feb/08 – talk abstract and high-level mathematical description.

[21] Michael D. Lemonick, *The Geometry of Music*, Time, 26 January 2007

[22] Elizabeth Gudrais, *Mapping Music*, Harvard Magazine, Jan/Feb 2007

[23] Tony Phillips, *Tony Phillips' Take on Math in the Media*, American Mathematical Society, October 2006

[24] (PDF) http://www.encyclospace.org/special/answer_to_tymoczko.pdf, retrieved 27 February 2012 Missing or empty |title= (help)

[25] (PDF) http://dmitri.tymoczko.com/files/publications/mazzola.pdf, retrieved 27 February 2012 Missing or empty |title= (help)

16.13 References

- Jean-Pierre Serre, *Cours d'arithmétique*, Presse Universitaire de France (1970).

- Glen Bredon, *Introduction to Compact Transformation Groups*, Academic Press (1972). ISBN 0-12-128850-1

- Katsuo Kawakubo, *The Theory of Transformation Groups*, Oxford University Press (1991). ISBN 0-19-853212-1

- Satake, Ichirô (1956). "On a generalization of the notion of manifold". *Proc. Natl. Acad. Sci. U.S.A.* **42**: 359–363. doi:10.1073/pnas.42.6.359.

- William Thurston, *The Geometry and Topology of Three-Manifolds* (Chapter 13), Princeton University lecture notes (1978–1981).

- Thurston, William (1982). "Three-dimensional manifolds, Kleinian groups and hyperbolic geometry". *Bull. Amer. Math. Soc.* **6**: 357–381. doi:10.1090/S0273-0979-1982-15003-0.

- Scott, Peter, *The geometry of 3-manifolds*, Bull. London Math. Soc. **15** (1983), 401–487. (The paper and its errata.)

- Michel Boileau, Geometrizations of 3-manifolds with symmetries

- Michel Boileau, Sylvain Maillot and Joan Porti, *Three-dimensional orbifolds and their geometric structures*. Panoramas and Syntheses **15**. Société Mathématique de France (2003). ISBN 2-85629-152-X.

- Boileau, Michel; Leeb, Bernhard; Porti, Joan (2005). "Geometrization of 3-dimensional orbifolds". *Annals of Mathematics* **162**: 195–290. doi:10.4007/annals.2005.162.195.

- Daryl Cooper, Craig Hodgson and Steven Kerckhoff, *Three-dimensional orbifolds and cone-manifolds*. MSJ Memoirs, **5**. Mathematical Society of Japan, Tokyo (2000). ISBN 4-931469-05-1.

- Matthew Brin, Lecture notes on Seifert fiber spaces.

- Henri Poincaré, *Papers on Fuchsian functions*, translated by John Stillwell, Springer (1985). ISBN 3-540-96215-8.

- Pierre de la Harpe, *An invitation to Coxeter group*, pages 193–253 in "Group theory from a geometrical viewpoint – Trieste 1990", World Scientific (1991). ISBN 981-02-0442-6.

- Werner Ballmann, *Singular spaces of non-positive curvature*, pages 189–201 in "Sur les groupes hyperboliques d'après Mikhael Gromov", Progress in Mathematics **83** (1990), Birkhäuser. ISBN 0-8176-3508-4.

- André Haefliger, *Orbi-espaces*, pages 203–213 in "Sur les groupes hyperboliques d'après Mikhael Gromov", Progress in Mathematics **83** (1990), Birkhäuser. ISBN 0-8176-3508-4.

- John Stallings, *Triangles of groups*, pages 491–503 in "Group theory from a geometrical viewpoint – Trieste 1990", World Scientific (1991). ISBN 981-02-0442-6.

- André Haefliger, *Complexes of groups and orbihedra*, pages 504–540 in "Group theory from a geometrical viewpoint – Trieste 1990", World Scientific (1991). ISBN 981-02-0442-6.

- Martin Bridson and André Haefliger, *Metric Spaces of Non-Positive Curvature*, Grundlehren der math. Wissenschaften **319** (1999), Springer. ISBN 3-540-64324-9.

- Philippe Di Francesco, Pierre Mathieu and David Sénéchal, *Conformal field theory*. Graduate Texts in Contemporary Physics. Springer-Verlag (1997). ISBN 0-387-94785-X.

- Jean-Pierre Serre, *Trees*, Springer (2003) (English translation of "arbres, amalgames, SL_2", 3rd edition, *astérisque* **46** (1983)).

- David Mumford, *An algebraic surface with K ample, $(K^2) = 9$, $pg = q = 0$*, American Journal of Mathematics **101** (1979), 233–244.

- Peter Köhler, Thomas Meixner and Michael Wester, *The 2-adic affine building of type $A_2\tilde{}$ and its finite projections*, J. Combin. Theory **38** (1985), 203–209.

- Donald Cartwright, Anna Maria Mantero, Tim Steger and Anna Zappa, *Groups acting simply transitively on the vertices of a building of type $A_2\tilde{}$*, I, Geometrica Dedicata **47** (1993), 143–166.

- Ballmann, Werner; Brin, Michael (1994). "Polygonal complexes and combinatorial group theory". *Geom. Dedicata* **50**: 165–191. doi:10.1007/BF01265309.

- Świątkowski, Jacek (2001). "A class of automorphism groups of polygonal complexes". *Q. J. Math.* **52**: 231–247. doi:10.1093/qjmath/52.2.231.

- Tymoczko, Dmitri (7 July 2006). "The Geometry of Musical Chords" (PDF). *Science* **313** (5783): 72–74. doi:10.1126/science.1126287. PMID 16825563

- Callender, Clifton; Quinn, Ian; Tymoczko, Dmitri (18 April 2008). "Generalized Voice-Leading Spaces" (PDF). *Science* **320** (5874): 346–348. doi:10.1126/science.1153021

Chapter 17

History of string theory

The **history of string theory** spans several decades of intense research including two superstring revolutions. Through the combined efforts of many different researchers, string theory has developed into a broad and varied subject with connections to quantum gravity, particle and condensed matter physics, cosmology, and pure mathematics.

17.1   1943–1959: S-matrix

String theory is an outgrowth of a research program begun by Werner Heisenberg in 1943, picked up and advocated by many prominent theorists starting in the late 1950s and throughout the 1960s, which was discarded and marginalized in the 1970s to disappear by the 1980s. It was forgotten because a few of the ideas were deeply mistaken, because some of its mathematical methods were alien, and because quantum chromodynamics supplanted it as an approach to the strong interactions.

The program was called the S-matrix theory, and it was a radical rethinking of the foundation of physical law. By the 1940s it was clear that the proton and the neutron were not pointlike particles like the electron. Their magnetic moment differed greatly from that of a pointlike spin-1/2 charged particle, too much to attribute the difference to a small perturbation. Their interactions were so strong that they scattered like a small sphere, not like a point. Heisenberg proposed that the strongly interacting particles were in fact extended objects, and because there are difficulties of principle with extended relativistic particles, he proposed that the notion of a space-time point broke down at nuclear scales.

Without space and time, it is difficult to formulate a physical theory. Heisenberg believed that the solution to this problem is to focus on the observable quantities—those things measurable by experiments. An experiment only sees a microscopic quantity if it can be transferred by a series of events to the classical devices that surround the experimental chamber. The objects that fly to infinity are stable particles, in quantum superpositions of different momentum states.

Heisenberg proposed that even when space and time are unreliable, the notion of momentum state, which is defined far away from the experimental chamber, still works. The physical quantity he proposed as fundamental is the quantum mechanical amplitude for a group of incoming particles to turn into a group of outgoing particles, and he did not admit that there were any steps in between.

The S-matrix is the quantity that describes how a superposition of incoming particles turn into outgoing ones. Heisenberg proposed to study the S-matrix directly, without any assumptions about space-time structure. But when transitions from the far-past to the far-future occur in one step with no intermediate steps, it is difficult to calculate anything. In quantum field theory, the intermediate steps are the fluctuations of fields or equivalently the fluctuations of virtual particles. In this proposed S-matrix theory, there are no local quantities at all.

Heisenberg proposed to use unitarity to determine the S-matrix. In all conceivable situations, the sum of the squares of the amplitudes must be equal to 1. This property can determine the amplitude in a quantum field theory order by order in a perturbation series once the basic interactions are given, and in many quantum field theories the amplitudes grow too fast at high energies to make a unitary S-matrix. But without extra assumptions on the high-energy behavior unitarity is not enough to determine the scattering, and the proposal was ignored for many years.

Heisenberg's proposal was reinvigorated in the late 1950s when several theorists recognized that dispersion relations like those discovered by Hendrik Kramers and Ralph Kronig allow a notion of causality to be formulated, a notion that events in the future would not influence events in the past, even when the microscopic notion of past and future are not clearly defined. The dispersion relations were analytic properties of the S-matrix, and they were more stringent conditions than those that follow from unitarity alone.

Prominent advocates of this approach were Stanley Mandelstam and Geoffrey Chew. Mandelstam had discovered the double-dispersion relations, a new and powerful analytic form, in 1958, and believed that it would be the key to progress in the intractable strong interactions.

17.2 1959–1968: Regge theory and bootstrap models

By this time, many strongly interacting particles of ever higher spins had been discovered, and it became clear that they were not all fundamental. While Japanese physicist Shoichi Sakata proposed that the particles could be understood as bound states of just three of them—the proton, the neutron and the Lambda (see Sakata model), Geoffrey Chew believed that none of these particles are fundamental. Sakata's approach was reworked in the 1960s into the quark model by Murray Gell-Mann and George Zweig by making the charges of the hypothetical constituents fractional and rejecting the idea that they were observed particles. Chew's approach was then considered more mainstream because it did not introduce fractional charges and because it only focused on the experimentally measurable S-matrix elements, not on hypothetical pointlike constituents.

In 1959 Tullio Regge, a young theorist in Italy discovered that bound states in quantum mechanics can be organized into families with different angular momentum called Regge trajectories. This idea was generalized to relativistic quantum mechanics by Mandelstam, Vladimir Gribov and Marcel Froissart, using a mathematical method discovered decades earlier by Arnold Sommerfeld and Kenneth Marshall Watson.

In 1961 Geoffrey Chew and Steven Frautschi recognized that the mesons made Regge trajectories in straight lines, which implied, via Regge theory, that the scattering of these particles would have very strange behavior—it should fall off exponentially quickly at large angles. With this realization, theorists hoped to construct a theory of composite particles on Regge trajectories, whose scattering amplitudes had the asymptotic form demanded by Regge theory. Since the interactions fall off fast at large angles, the scattering theory would have to be somewhat holistic: Scattering off a pointlike constituent leads to large angular deviations at high energies.

17.3 1968–1974: dual resonance model

The first theory of this sort, the dual resonance model, was constructed by Gabriele Veneziano in 1968, who noted that the Euler Beta function could be used to describe 4-particle scattering amplitude data for particles on Regge trajectories. The Veneziano scattering amplitude was quickly generalized to an N-particle amplitude by Ziro Koba and Holger Bech Nielsen, and to what are now recognized as closed strings by Miguel Virasoro and Joel A. Shapiro. Dual resonance models for strong interactions were a popular subject of study 1968-1974.

17.4 1974–1984: superstring theory

In 1969–70, Yoichiro Nambu, Holger Bech Nielsen, and Leonard Susskind presented a physical interpretation of Euler's formula by representing nuclear forces as vibrating, one-dimensional strings. However, this string-based description of the strong force made many predictions that directly contradicted experimental findings. The scientific community lost interest in string theory as a theory of strong interactions in 1974 when quantum chromodynamics became the main focus of theoretical research.

In 1974 John H. Schwarz and Joel Scherk, and independently Tamiaki Yoneya, studied the boson-like patterns of string vibration and found that their properties exactly matched those of the graviton, the gravitational force's hypothetical "messenger" particle. Schwarz and Scherk argued that string theory had failed to catch on because physicists had underestimated its scope. This led to the development of bosonic string theory, which is still the version first taught to many students.

String theory is formulated in terms of the Polyakov action, which describes how strings move through space and time. Like springs, the strings want to contract to minimize their potential energy, but conservation of energy prevents them from disappearing, and instead they oscillate. By applying the ideas of quantum mechanics to strings it is possible to deduce the different vibrational modes of strings, and that each vibrational state appears to be a different particle. The mass of each particle, and the fashion with which it can interact, are determined by the way the string vibrates — in essence, by the "note" the string sounds. The scale of notes, each corresponding to a different kind of particle, is termed the "spectrum" of the theory.

Early models included both *open* strings, which have two distinct endpoints, and *closed* strings, where the endpoints are joined to make a complete loop. The two types of string behave in slightly different ways, yielding two spectra. Not all modern string theories use both types; some incorporate only the closed variety.

The earliest string model, which incorporated only bosons, has problems. Most importantly, the theory has a fundamental instability, believed to result in the decay of spacetime itself. Additionally, as the name implies, the spectrum of particles contains only bosons, particles like the photon that obey particular rules of behavior. While bosons are a critical ingredient of the Universe, they are not its only constituents. Investigating how a string theory may include fermions in its spectrum led to the invention of supersymmetry, a mathematical relation between bosons and fermions. String theories that include fermionic vibrations are now known as superstring theories; several different kinds have been described.

17.5   1984–1994: first superstring revolution

The first superstring revolution is a period of important discoveries roughly between 1984 and 1986. It was realised that string theory was capable of describing all elementary particles as well as the interactions between them. Hundreds of physicists started to work on string theory as the most promising idea to unify physical theories. The revolution was started by a discovery of anomaly cancellation in type I string theory via the Green–Schwarz mechanism in 1984. Several other ground-breaking discoveries, such as the heterotic string, were made in 1985. It was also realised in 1985 that to obtain $N = 1$ supersymmetry, the six small extra dimensions need to be compactified on a Calabi–Yau manifold.

Discover magazine in the November 1986 issue (vol 7, #11) featured a cover story written by Gary Taubes, "Everything's Now Tied to Strings", which explained string theory for a popular audience.

17.6   1994–2003: second superstring revolution

In the early 1990s, Edward Witten and others found strong evidence that the different superstring theories were different limits of a new 11-dimensional theory called M-theory.[1] These discoveries sparked the second superstring revolution that took place approximately between 1994 and 1997.

The different versions of superstring theory were unified, as long hoped, by new equivalences. These are known as S-duality, T-duality, U-duality, mirror symmetry, and conifold transitions. The different theories of strings were also connected to a new 11-dimensional theory called M-theory.

In the mid 1990s, Joseph Polchinski discovered that the theory requires the inclusion of higher-dimensional objects, called D-branes. These added an additional rich mathematical structure to the theory, and opened many possibilities for constructing realistic cosmological models in the theory.

In 1997 Juan Maldacena conjectured a relationship between string theory and a gauge theory called N = 4 supersymmetric Yang–Mills theory. This conjecture, called the AdS/CFT correspondence has generated a great deal of interest in the field and is now well accepted. It is a concrete realization of the holographic principle, which has far-reaching implications for black holes, locality and information in physics, and for the nature of the gravitational interaction.

17.7   2003–present

In 2003 the discovery of the string theory landscape, which suggests that string theory has a large number of inequivalent vacua, led to much discussion of what string theory might eventually be expected to predict, and how cosmology can be incorporated into the theory.[2]

17.8   Notes

[1] When Witten named it M-theory, he did not specify what the "M" stood for, presumably because he did not feel he had the right to name a theory he had not been able to fully describe. The "M" sometimes is said to stand for Mystery, or Magic, or Mother. More serious suggestions include Matrix or Membrane. Sheldon Glashow has noted that the "M" might be an upside down "W", standing for Witten. Others have suggested that the "M" in M-theory should stand for Missing, Monstrous or even Murky. According to Witten himself, as quoted in the PBS documentary based on Brian Greene's *The Elegant Universe*, the "M" in M-theory stands for "magic, mystery, or matrix according to taste."

[2] Rickles 2014, pp. 230–5 and 236 fn. 63.

17.9   References

- Dean Rickles (2014). *A Brief History of String Theory: From Dual Models to M-Theory*. Springer Science & Business Media. ISBN 978-3-642-45128-7.

17.10 Further reading

- Paul Frampton (1974). *Dual Resonance Models.* Frontiers in Physics, W. A. Benjamin. ISBN 978-0-8053-2581-2.

- Shapiro, Joel A. (2007). "Reminiscence on the Birth of String Theory". arXiv:0711.3448.

- Andrea Cappelli; Elena Castellani; Filippo Colomo; Paolo Di Vecchia (2012). *The Birth of String Theory.* Cambridge University Press. ISBN 978-0-521-19790-8.

Chapter 18

String theory

For a more accessible and less technical introduction to this topic, see Introduction to M-theory.

In physics, **string theory** is a theoretical framework in which the point-like particles of particle physics are replaced by one-dimensional objects called strings. It describes how these strings propagate through space and interact with each other. On distance scales larger than the string scale, a string looks just like an ordinary particle, with its mass, charge, and other properties determined by the vibrational state of the string. In string theory, one of the many vibrational states of the string corresponds to the graviton, a quantum mechanical particle that carries gravitational force. Thus string theory is a theory of quantum gravity.

String theory is a broad and varied subject that attempts to address a number of deep questions of fundamental physics. String theory has been applied to a variety of problems in black hole physics, early universe cosmology, nuclear physics, and condensed matter physics, and it has stimulated a number of major developments in pure mathematics. Because string theory potentially provides a unified description of gravity and particle physics, it is a candidate for a theory of everything, a self-contained mathematical model that describes all fundamental forces and forms of matter. Despite much work on these problems, it is not known to what extent string theory describes the real world or how much freedom the theory allows to choose the details.

String theory was first studied in the late 1960s as a theory of the strong nuclear force, before being abandoned in favor of quantum chromodynamics. Subsequently, it was realized that the very properties that made string theory unsuitable as a theory of nuclear physics made it a promising candidate for a quantum theory of gravity. The earliest version of string theory, bosonic string theory, incorporated only the class of particles known as bosons. It later developed into superstring theory, which posits a connection called supersymmetry between bosons and the class of particles called fermions. Five consistent versions of super-

string theory were developed before it was conjectured in the mid-1990s that they were all different limiting cases of a single theory in eleven dimensions known as M-theory. In late 1997, theorists discovered an important relationship called the AdS/CFT correspondence, which relates string theory to another type of physical theory called a quantum field theory.

One of the challenges of string theory is that the full theory does not yet have a satisfactory definition in all circumstances. Another issue is that the theory is thought to describe an enormous landscape of possible universes, and this has complicated efforts to develop theories of particle physics based on string theory. These issues have led some in the community to criticize these approaches to physics and question the value of continued research on string theory unification.

18.1   Fundamentals

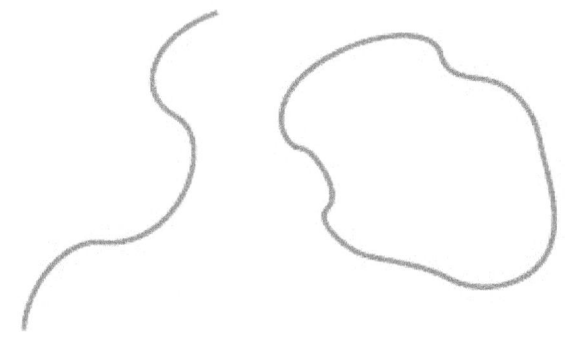

The fundamental objects of string theory are open and closed strings.

In the twentieth century, two theoretical frameworks emerged for formulating the laws of physics. One of these frameworks was Albert Einstein's general theory of relativity, a theory that explains the force of gravity and the structure of space and time. The other was quantum mechanics, a radically different formalism for describing physical phenomena using probability. By the late 1970s, these two frameworks had proven to be sufficient to explain most of the observed features of the universe, from elementary particles to atoms to the evolution of stars and the universe as a whole.[1]

In spite of these successes, there are still many problems that remain to be solved. One of the deepest problems in modern physics is the problem of quantum gravity.[1] The general theory of relativity is formulated within the framework of classical physics, whereas the other fundamental forces are described within the framework of quantum mechanics. A quantum theory of gravity is needed in order to reconcile general relativity with the principles of quantum mechanics, but difficulties arise when one attempts to apply the usual prescriptions of quantum theory to the force of gravity.[2] In addition to the problem of developing a consistent theory of quantum gravity, there are many other fundamental problems in the physics of atomic nuclei, black holes, and the early universe.[lower-alpha 1]

String theory is a theoretical framework that attempts to address these questions and many others. The starting point for string theory is the idea that the point-like particles of particle physics can also be modeled as one-dimensional objects called strings. String theory describes how strings propagate through space and interact with each other. In a given version of string theory, there is only one kind of string, which may look like a small loop or segment of ordinary string, and it can vibrate in different ways. On distance scales larger than the string scale, a string will look just like an ordinary particle, with its mass, charge, and other properties determined by the vibrational state of the string. In this way, all of the different elementary particles may be viewed as vibrating strings. In string theory, one of the vibrational states of the string gives rise to the graviton, a quantum mechanical particle that carries gravitational force. Thus string theory is a theory of quantum gravity.[3]

One of the main developments of the past several decades in string theory was the discovery of certain "dualities", mathematical transformations that identify one physical theory with another. Physicists studying string theory have discovered a number of these dualities between different versions of string theory, and this has led to the conjecture that all consistent versions of string theory are subsumed in a single framework known as M-theory.[4]

Studies of string theory have also yielded a number of results on the nature of black holes and the gravitational interaction. There are certain paradoxes that arise when one attempts to understand the quantum aspects of black holes, and work on string theory has attempted to clarify these issues. In late 1997 this line of work culminated in the discovery of the anti-de Sitter/conformal field theory correspondence or AdS/CFT.[5] This is a theoretical result which relates string theory to other physical theories which are better understood theoretically. The AdS/CFT correspondence has implications for the study of black holes and quantum gravity, and it has been applied to other subjects, including nuclear[6] and condensed matter physics.[7][8]

Since string theory incorporates all of the fundamental interactions, including gravity, many physicists hope that it fully describes our universe, making it a theory of everything. One of the goals of current research in string theory is to find a solution of the theory that reproduces the observed spectrum of elementary particles, with a small cosmological constant, containing dark matter and a plausible mechanism for cosmic inflation. While there has been progress toward these goals, it is not known to what extent string theory describes the real world or how much freedom the theory allows to choose the details.[9]

One of the challenges of string theory is that the full theory does not yet have a satisfactory definition in all circumstances. The scattering of strings is most straightforwardly defined using the techniques of perturbation theory, but it is not known in general how to define string theory nonperturbatively.[10] It is also not clear whether there is any principle by which string theory selects its vacuum state, the physical state that determines the properties of our universe.[11] These problems have led some in the community to criticize these approaches to the unification of physics and question the value of continued research on these problems.[12]

18.1.1 Strings

Main article: String (physics)
The application of quantum mechanics to physical objects

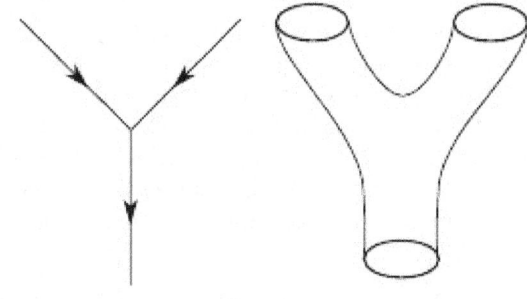

Interaction in the quantum world: worldlines of point-like particles or a worldsheet swept up by closed strings in string theory.

such as the electromagnetic field, which are extended in

space and time, is known as quantum field theory. In particle physics, quantum field theories form the basis for our understanding of elementary particles, which are modeled as excitations in the fundamental fields.[13]

In quantum field theory, one typically computes the probabilities of various physical events using the techniques of perturbation theory. Developed by Richard Feynman and others in the first half of the twentieth century, perturbative quantum field theory uses special diagrams called Feynman diagrams to organize computations. One imagines that these diagrams depict the paths of point-like particles and their interactions.[13]

The starting point for string theory is the idea that the point-like particles of quantum field theory can also be modeled as one-dimensional objects called strings.[14] The interaction of strings is most straightforwardly defined by generalizing the perturbation theory used in ordinary quantum field theory. At the level of Feynman diagrams, this means replacing the one-dimensional diagram representing the path of a point particle by a two-dimensional surface representing the motion of a string.[15] Unlike in quantum field theory, string theory does not yet have a full non-perturbative definition, so many of the theoretical questions that physicists would like to answer remain out of reach.[16]

In theories of particle physics based on string theory, the characteristic length scale of strings is assumed to be on the order of the Planck length, or 10^{-35} meters, the scale at which the effects of quantum gravity are believed to become significant.[15] On much larger length scales, such as the scales visible in physics laboratories, such objects would be indistinguishable from zero-dimensional point particles, and the vibrational state of the string would determine the type of particle. One of the vibrational states of a string corresponds to the graviton, a quantum mechanical particle that carries the gravitational force.[3]

The original version of string theory was bosonic string theory, but this version described only bosons, a class of particles which transmit forces between the matter particles, or fermions. Bosonic string theory was eventually superseded by theories called superstring theories. These theories describe both bosons and fermions, and they incorporate a theoretical idea called supersymmetry. This is a mathematical relation that exists in certain physical theories between the bosons and fermions. In theories with supersymmetry, each boson has a counterpart which is a fermion, and vice versa.[17]

There are several versions of superstring theory: type I, type IIA, type IIB, and two flavors of heterotic string theory ($SO(32)$ and $E_8 \times E_8$). The different theories allow different types of strings, and the particles that arise at low energies exhibit different symmetries. For example, the type I theory includes both open strings (which are segments with

endpoints) and closed strings (which form closed loops), while types IIA and IIB include only closed strings.[18]

18.1.2 Extra dimensions

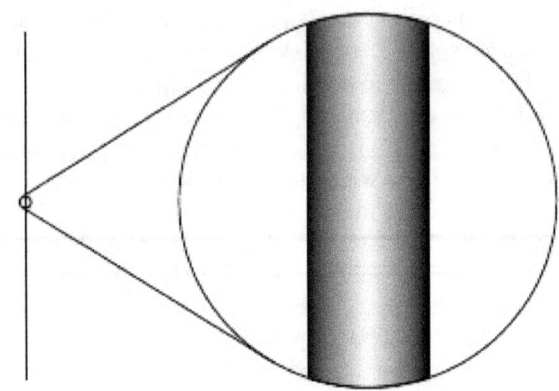

An example of compactification: At large distances, a two dimensional surface with one circular dimension looks one-dimensional.

In everyday life, there are three familiar dimensions of space: height, width and length. Einstein's general theory of relativity treats time as a dimension on par with the three spatial dimensions; in general relativity, space and time are not modeled as separate entities but are instead unified to a four-dimensional spacetime. In this framework, the phenomenon of gravity is viewed as a consequence of the geometry of spacetime.[19]

In spite of the fact that the universe is well described by four-dimensional spacetime, there are several reasons why physicists consider theories in other dimensions. In some cases, by modeling spacetime in a different number of dimensions, a theory becomes more mathematically tractable, and one can perform calculations and gain general insights more easily.[lower-alpha 2] There are also situations where theories in two or three spacetime dimensions are useful for describing phenomena in condensed matter physics.[20] Finally, there exist scenarios in which there could actually be more than four dimensions of spacetime which have nonetheless managed to escape detection.[21]

One notable feature of string theories is that these theories require extra dimensions of spacetime for their mathematical consistency. In bosonic string theory, spacetime is 26-dimensional, while in superstring theory it is ten-dimensional. In order to describe real physical phenomena using string theory, one must therefore imagine scenarios in which these extra dimensions would not be observed in experiments.[22]

Compactification is one way of modifying the number of dimensions in a physical theory. In compactification, some

A cross section of a quintic Calabi–Yau manifold

of the extra dimensions are assumed to "close up" on themselves to form circles.[23] In the limit where these curled up dimensions become very small, one obtains a theory in which spacetime has effectively a lower number of dimensions. A standard analogy for this is to consider a multidimensional object such as a garden hose. If the hose is viewed from a sufficient distance, it appears to have only one dimension, its length. However, as one approaches the hose, one discovers that it contains a second dimension, its circumference. Thus, an ant crawling on the surface of the hose would move in two dimensions.[24]

Compactification can be used to construct models in which spacetime is effectively four-dimensional. However, not every way of compactifying the extra dimensions produces a model with the right properties to describe nature. In a viable model of particle physics, the compact extra dimensions must be shaped like a Calabi–Yau manifold.[23] A Calabi–Yau manifold is a special space which is typically taken to be six-dimensional in applications to string theory. It is named after mathematicians Eugenio Calabi and Shing-Tung Yau.[25]

Another approach to reducing the number of dimensions is the so-called brane-world scenario. In this approach, physicists assume that the observable universe is a four-dimensional subspace of a higher dimensional space. In such models, the force-carrying bosons of particle physics arise from open strings with endpoints attached to the four-dimensional subspace, while gravity arises from closed strings propagating through the larger ambient space. This idea plays an important role in attempts to develop models

of real world physics based on string theory, and it provides a natural explanation for the weakness of gravity compared to the other fundamental forces.[26]

18.1.3 Dualities

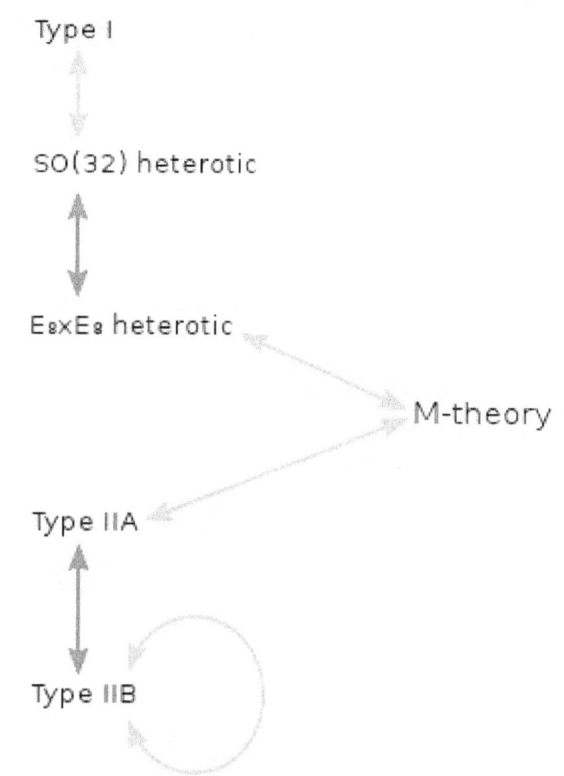

A diagram of string theory dualities. Yellow arrows indicate S-duality. Blue arrows indicate T-duality.

Main articles: S-duality and T-duality

One notable fact about string theory is that the different versions of the theory all turn out to be related in highly nontrivial ways. One of the relationships that can exist between different string theories is called S-duality. This is a relationship which says that a collection of strongly interacting particles in one theory can, in some cases, be viewed as a collection of weakly interacting particles in a completely different theory. Roughly speaking, a collection of particles is said to be strongly interacting if they combine and decay often and weakly interacting if they do so infrequently. Type I string theory turns out to be equivalent by S-duality to the $SO(32)$ heterotic string theory. Similarly, type IIB string theory is related to itself in a nontrivial way by S-duality.[27]

Another relationship between different string theories is T-duality. Here one considers strings propagating around a circular extra dimension. T-duality states that a string propagating around a circle of radius R is equivalent to a string propagating around a circle of radius $1/R$ in the sense that all observable quantities in one description are identified with quantities in the dual description. For example, a string has momentum as it propagates around a circle, and it can also wind around the circle one or more times. The number of times the string winds around a circle is called the winding number. If a string has momentum p and winding number n in one description, it will have momentum n and winding number p in the dual description. For example, type IIA string theory is equivalent to type IIB string theory via T-duality, and the two versions of heterotic string theory are also related by T-duality.[27]

In general, the term *duality* refers to a situation where two seemingly different physical systems turn out to be equivalent in a nontrivial way. Two theories related by a duality need not be string theories. For example, Montonen–Olive duality is example of an S-duality relationship between quantum field theories. The AdS/CFT correspondence is example of a duality which relates string theory to a quantum field theory. If two theories are related by a duality, it means that one theory can be transformed in some way so that it ends up looking just like the other theory. The two theories are then said to be *dual* to one another under the transformation. Put differently, the two theories are mathematically different descriptions of the same phenomena.[28]

18.1.4 Branes

Main article: Brane

In string theory and related theories, a brane is a physi-

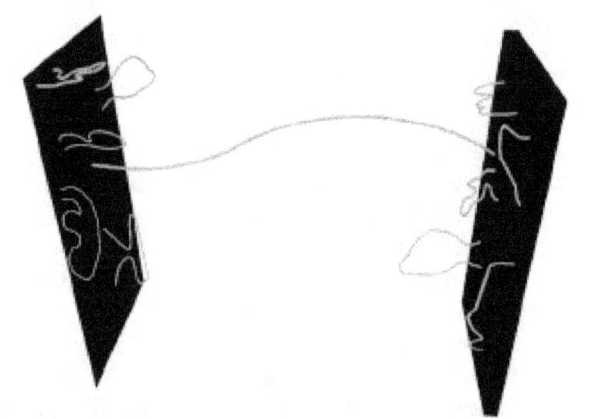

Open strings attached to a pair of D-branes

cal object that generalizes the notion of a point particle to higher dimensions. For example, a point particle can be

viewed as a brane of dimension zero, while a string can be viewed as a brane of dimension one. It is also possible to consider higher-dimensional branes. In dimension p, these are called p-branes. The word brane comes from the word "membrane" which refers to a two-dimensional brane.[29]

Branes are dynamical objects which can propagate through spacetime according to the rules of quantum mechanics. They have mass and can have other attributes such as charge. A p-brane sweeps out a $(p+1)$-dimensional volume in spacetime called its *worldvolume*. Physicists often study fields analogous to the electromagnetic field which live on the worldvolume of a brane.[29]

In string theory, D-branes are an important class of branes that arise when one considers open strings. As an open string propagates through spacetime, its endpoints are required to lie on a D-brane. The letter "D" in D-brane refers to a certain mathematical condition on the system known as the Dirichlet boundary condition. The study of D-branes in string theory has led to important results such as the AdS/CFT correspondence, which has shed light on many problems in quantum field theory.[30]

Branes are also frequently studied from a purely mathematical point of view. Mathematically, branes can be described as objects of certain categories, such as the derived category of coherent sheaves on a complex algebraic variety, or the Fukaya category of a symplectic manifold.[31] The connection between the physical notion of a brane and the mathematical notion of a category has led to important mathematical insights in the fields of algebraic and symplectic geometry[32] and representation theory.[33]

18.2   M-theory

Main article: M-theory

Prior to 1995, theorists believed that there were five consistent versions of superstring theory (type I, type IIA, type IIB, and two versions of heterotic string theory). This understanding changed in 1995 when Edward Witten suggested that the five theories were just special limiting cases of an eleven-dimensional theory called M-theory. Witten's conjecture was based on the work of a number of other physicists, including Ashoke Sen, Chris Hull, Paul Townsend, and Michael Duff. His announcement led to a flurry of research activity now known as the second superstring revolution.[34]

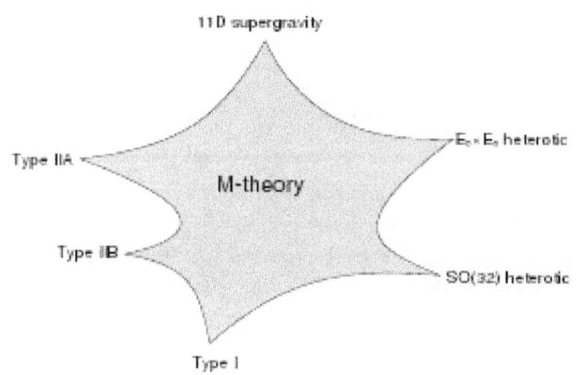

A schematic illustration of the relationship between M-theory, the five superstring theories, and eleven-dimensional supergravity. The shaded region represents a family of different physical scenarios that are possible in M-theory. In certain limiting cases corresponding to the cusps, it is natural to describe the physics using one of the six theories labeled there.

18.2.1 Unification of superstring theories

In the 1970s, many physicists became interested in supergravity theories, which combine general relativity with supersymmetry. Whereas general relativity makes sense in any number of dimensions, supergravity places an upper limit on the number of dimensions.[35] In 1978, work by Werner Nahm showed that the maximum spacetime dimension in which one can formulate a consistent supersymmetric theory is eleven.[36] In the same year, Eugene Cremmer, Bernard Julia, and Joel Scherk of the École Normale Supérieure showed that supergravity not only permits up to eleven dimensions but is in fact most elegant in this maximal number of dimensions.[37][38]

Initially, many physicists hoped that by compactifying eleven-dimensional supergravity, it might be possible to construct realistic models of our four-dimensional world. The hope was that such models would provide a unified description of the four fundamental forces of nature: electromagnetism, the strong and weak nuclear forces, and gravity. Interest in eleven-dimensional supergravity soon waned as various flaws in this scheme were discovered. One of the problems was that the laws of physics appear to distinguish between clockwise and counterclockwise, a phenomenon known as chirality. Edward Witten and others observed this chirality property cannot be readily derived by compactifying from eleven dimensions.[38]

In the first superstring revolution in 1984, many physicists turned to string theory as a unified theory of particle physics and quantum gravity. Unlike supergravity theory, string theory was able to accommodate the chirality of the standard model, and it provided a theory of gravity consistent

with quantum effects.[38] Another feature of string theory that many physicists were drawn to in the 1980s and 1990s was its high degree of uniqueness. In ordinary particle theories, one can consider any collection of elementary particles whose classical behavior is described by an arbitrary Lagrangian. In string theory, the possibilities are much more constrained: by the 1990s, physicists had argued that there were only five consistent supersymmetric versions of the theory.[38]

Although there were only a handful of consistent superstring theories, it remained a mystery why there was not just one consistent formulation.[38] However, as physicists began to examine string theory more closely, they realized that these theories are related in intricate and nontrivial ways. They found that a system of strongly interacting strings can, in some cases, be viewed as a system of weakly interacting strings. This phenomenon is known as S-duality. It was studied by Ashoke Sen in the context of heterotic strings in four dimensions[39][40] and by Chris Hull and Paul Townsend in the context of the type IIB theory.[41] Theorists also found that different string theories may be related by T-duality. This duality implies that strings propagating on completely different spacetime geometries may be physically equivalent.[42]

At around the same time, as many physicists were studying the properties of strings, a small group of physicists was examining the possible applications of higher dimensional objects. In 1987, Eric Bergshoeff, Ergin Sezgin, and Paul Townsend showed that eleven-dimensional supergravity includes two-dimensional branes.[43] Intuitively, these objects look like sheets or membranes propagating through the eleven-dimensional spacetime. Shortly after this discovery, Michael Duff, Paul Howe, Takeo Inami, and Kellogg Stelle considered a particular compactification of eleven-dimensional supergravity with one of the dimensions curled up into a circle.[44] In this setting, one can imagine the membrane wrapping around the circular dimension. If the radius of the circle is sufficiently small, then this membrane looks just like a string in ten-dimensional spacetime. In fact, Duff and his collaborators showed that this construction reproduces exactly the strings appearing in type IIA superstring theory.[45]

Speaking at a string theory conference in 1995, Edward Witten made the surprising suggestion that all five superstring theories were in fact just different limiting cases of a single theory in eleven spacetime dimensions. Witten's announcement drew together all of the previous results on S- and T-duality and the appearance of higher dimensional branes in string theory.[46] In the months following Witten's announcement, hundreds of new papers appeared on the Internet confirming different parts of his proposal.[47] Today this flurry of work is known as the second superstring revolution.[48]

Initially, some physicists suggested that the new theory was a fundamental theory of membranes, but Witten was skeptical of the role of membranes in the theory. In a paper from 1996, Hořava and Witten wrote "As it has been proposed that the eleven-dimensional theory is a supermembrane theory but there are some reasons to doubt that interpretation, we will non-committally call it the M-theory, leaving to the future the relation of M to membranes."[49] In the absence of an understanding of the true meaning and structure of M-theory, Witten has suggested that the M should stand for "magic", "mystery", or "membrane" according to taste, and the true meaning of the title should be decided when a more fundamental formulation of the theory is known.[50]

18.2.2 Matrix theory

Main article: Matrix theory (physics)

In mathematics, a matrix is a rectangular array of numbers or other data. In physics, a matrix model is a particular kind of physical theory whose mathematical formulation involves the notion of a matrix in an important way. A matrix model describes the behavior of a set of matrices within the framework of quantum mechanics.[51]

One important example of a matrix model is the BFSS matrix model proposed by Tom Banks, Willy Fischler, Stephen Shenker, and Leonard Susskind in 1997. This theory describes the behavior of a set of nine large matrices. In their original paper, these authors showed, among other things, that the low energy limit of this matrix model is described by eleven-dimensional supergravity. These calculations led them to propose that the BFSS matrix model is exactly equivalent to M-theory. The BFSS matrix model can therefore be used as a prototype for a correct formulation of M-theory and a tool for investigating the properties of M-theory in a relatively simple setting.[51]

The development of the matrix model formulation of M-theory has led physicists to consider various connections between string theory and a branch of mathematics called noncommutative geometry. This subject is a generalization of ordinary geometry in which mathematicians define new geometric notions using tools from noncommutative algebra.[52] In a paper from 1998, Alain Connes, Michael R. Douglas, and Albert Schwarz showed that some aspects of matrix models and M-theory are described by a noncommutative quantum field theory, a special kind of physical theory in which spacetime is described mathematically using noncommutative geometry.[53] This established a link between matrix models and M-theory on the one hand, and noncommutative geometry on the other hand. It quickly led to the discovery of other important links between noncommutative geometry and various physical theories.[54][55]

18.3 Black holes

In general relativity, a black hole is defined as a region of spacetime in which the gravitational field is so strong that no particle or radiation can escape. In the currently accepted models of stellar evolution, black holes are thought to arise when massive stars undergo gravitational collapse, and many galaxies are thought to contain supermassive black holes at their centers. Black holes are also important for theoretical reasons, as they present profound challenges for theorists attempting to understand the quantum aspects of gravity. String theory has proved to be an important tool for investigating the theoretical properties of black holes because it provides a framework in which theorists can study their thermodynamics.[56]

18.3.1 Bekenstein–Hawking formula

In the branch of physics called statistical mechanics, entropy is a measure of the randomness or disorder of a physical system. This concept was studied in the 1870s by the Austrian physicist Ludwig Boltzmann, who showed that the thermodynamic properties of a gas could be derived from the combined properties of its many constituent molecules. Boltzmann argued that by averaging the behaviors of all the different molecules in a gas, one can understand macroscopic properties such as volume, temperature, and pressure. In addition, this perspective led him to give a precise definition of entropy as the natural logarithm of the number of different states of the molecules (also called *microstates*) that give rise to the same macroscopic features.[57]

In the twentieth century, physicists began to apply the same concepts to black holes. In most systems such as gases, the entropy scales with the volume. In the 1970s, the physicist Jacob Bekenstein suggested that the entropy of a black hole is instead proportional to the *surface area* of its event horizon, the boundary beyond which matter and radiation is lost to its gravitational attraction.[58] When combined with ideas of the physicist Stephen Hawking,[59] Bekenstein's work yielded a precise formula for the entropy of a black hole. The formula expresses the entropy S as

$$S = \frac{c^3 k A}{4 \hbar G}$$

where c is the speed of light, k is Boltzmann's constant, $\hbar$ is the reduced Planck constant, G is Newton's constant, and A is the surface area of the event horizon.[60]

Like any physical system, a black hole has an entropy defined in terms of the number of different microstates that lead to the same macroscopic features. The Bekenstein–Hawking entropy formula gives the expected value of the entropy of a black hole, but by the 1990s, physicists still lacked a derivation of this formula by counting microstates in a theory of quantum gravity. Finding such a derivation of this formula was considered an important test of the viability of any theory of quantum gravity such as string theory.[61]

18.3.2 Derivation within string theory

In a paper from 1996, Andrew Strominger and Cumrun Vafa showed how to derive the Beckenstein–Hawking formula for certain black holes in string theory.[62] Their calculation was based on the observation that D-branes—which look like fluctuating membranes when they are weakly interacting—become dense, massive objects with event horizons when the interactions are strong. In other words, a system of strongly interacting D-branes in string theory is indistinguishable from a black hole. Strominger and Vafa analyzed such D-brane systems and calculated the number of different ways of placing D-branes in spacetime so that their combined mass and charge is equal to a given mass and charge for the resulting black hole. Their calculation reproduced the Bekenstein–Hawking formula exactly, including the factor of 1/4.[63] Subsequent work by Strominger, Vafa, and others refined the original calculations and gave the precise values of the "quantum corrections" needed to describe very small black holes.[64][65]

The black holes that Strominger and Vafa considered in their original work were quite different from real astrophysical black holes. One difference was that Strominger and Vafa considered only extremal black holes in order to make the calculation tractable. These are defined as black holes with the lowest possible mass compatible with a given charge.[66] Strominger and Vafa also restricted attention to black holes in five-dimensional spacetime with unphysical supersymmetry.[67]

Although it was originally developed in this very particular and physically unrealistic context in string theory, the entropy calculation of Strominger and Vafa has led to a qualitative understanding of how black hole entropy can be accounted for in any theory of quantum gravity. Indeed, in 1998, Strominger argued that the original result could be generalized to an arbitrary consistent theory of quantum gravity without relying on strings or supersymmetry.[68] In collaboration with several other authors in 2010, he showed that some results on black hole entropy could be extended to non-extremal astrophysical black holes.[69][70]

18.4 AdS/CFT correspondence

Main article: AdS/CFT correspondence

One approach to formulating string theory and studying its properties is provided by the anti-de Sitter/conformal field theory (AdS/CFT) correspondence. This is a theoretical result which implies that string theory is in some cases equivalent to a quantum field theory. In addition to providing insights into the mathematical structure of string theory, the AdS/CFT correspondence has shed light on many aspects of quantum field theory in regimes where traditional calculational techniques are ineffective.[6] The AdS/CFT correspondence was first proposed by Juan Maldacena in late 1997.[71] Important aspects of the correspondence were elaborated in articles by Steven Gubser, Igor Klebanov, and Alexander Markovich Polyakov,[72] and by Edward Witten.[73] By 2010, Maldacena's article had over 7000 citations, becoming the most highly cited article in the field of high energy physics.[lower-alpha 3]

18.4.1 Overview of the correspondence

In the AdS/CFT correspondence, the geometry of spacetime is described in terms of a certain vacuum solution of Einstein's equation called anti-de Sitter space.[74] In very elementary terms, anti-de Sitter space is a mathematical model of spacetime in which the notion of distance between points (the metric) is different from the notion of distance in ordinary Euclidean geometry. It is closely related to hyperbolic space, which can be viewed as a disk as illustrated on the left.[75] This image shows a tessellation of a disk by triangles and squares. One can define the distance between points of this disk in such a way that all the triangles and squares are the same size and the circular outer boundary is infinitely far from any point in the interior.[76]

One can imagine a stack of hyperbolic disks where each disk represents the state of the universe at a given time. The resulting geometric object is three-dimensional anti-de Sitter space.[75] It looks like a solid cylinder in which any cross section is a copy of the hyperbolic disk. Time runs along the vertical direction in this picture. The surface of this cylinder plays an important role in the AdS/CFT correspondence. As with the hyperbolic plane, anti-de Sitter space is curved in such a way that any point in the interior is actually infinitely far from this boundary surface.[76]

This construction describes a hypothetical universe with only two space dimensions and one time dimension, but it can be generalized to any number of dimensions. Indeed, hyperbolic space can have more than two dimensions and one can "stack up" copies of hyperbolic space to get higher-dimensional models of anti-de Sitter space.[75]

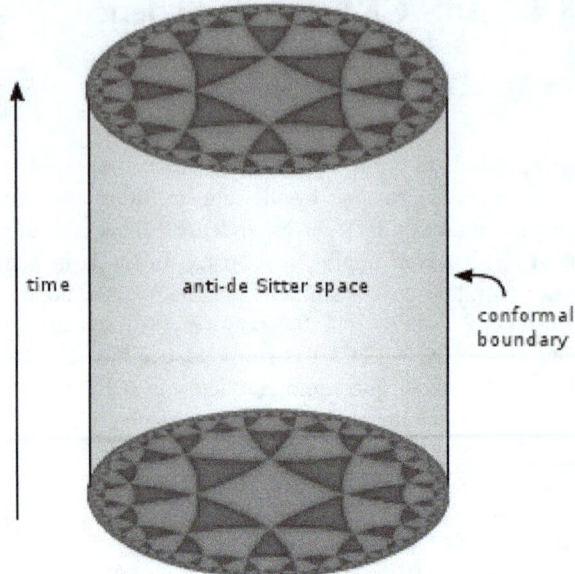

Three-dimensional anti-de Sitter space is like a stack of hyperbolic disks, each one representing the state of the universe at a given time. The resulting spacetime looks like a solid cylinder.

An important feature of anti-de Sitter space is its boundary (which looks like a cylinder in the case of three-dimensional anti-de Sitter space). One property of this boundary is that, within a small region on the surface around any given point, it looks just like Minkowski space, the model of spacetime used in nongravitational physics.[77] One can therefore consider an auxiliary theory in which "spacetime" is given by the boundary of anti-de Sitter space. This observation is the starting point for AdS/CFT correspondence, which states that the boundary of anti-de Sitter space can be regarded as the "spacetime" for a quantum field theory. The claim is that this quantum field theory is equivalent to a gravitational theory, such as string theory, in the bulk anti-de Sitter space in the sense that there is a "dictionary" for translating entities and calculations in one theory into their counterparts in the other theory. For example, a single particle in the gravitational theory might correspond to some collection of particles in the boundary theory. In addition, the predictions in the two theories are quantitatively identical so that if two particles have a 40 percent chance of colliding in the gravitational theory, then the corresponding collections in the boundary theory would also have a 40 percent chance of colliding.[78]

18.4.2 Applications to quantum gravity

The discovery of the AdS/CFT correspondence was a major advance in physicists' understanding of string theory and quantum gravity. One reason for this is that the correspon-

dence provides a formulation of string theory in terms of quantum field theory, which is well understood by comparison. Another reason is that it provides a general framework in which physicists can study and attempt to resolve the paradoxes of black holes.[56]

In 1975, Stephen Hawking published a calculation which suggested that black holes are not completely black but emit a dim radiation due to quantum effects near the event horizon.[59] At first, Hawking's result posed a problem for theorists because it suggested that black holes destroy information. More precisely, Hawking's calculation seemed to conflict with one of the basic postulates of quantum mechanics, which states that physical systems evolve in time according to the Schrödinger equation. This property is usually referred to as unitarity of time evolution. The apparent contradiction between Hawking's calculation and the unitarity postulate of quantum mechanics came to be known as the black hole information paradox.[79]

The AdS/CFT correspondence resolves the black hole information paradox, at least to some extent, because it shows how a black hole can evolve in a manner consistent with quantum mechanics in some contexts. Indeed, one can consider black holes in the context of the AdS/CFT correspondence, and any such black hole corresponds to a configuration of particles on the boundary of anti-de Sitter space.[80] These particles obey the usual rules of quantum mechanics and in particular evolve in a unitary fashion, so the black hole must also evolve in a unitary fashion, respecting the principles of quantum mechanics.[81] In 2005, Hawking announced that the paradox had been settled in favor of information conservation by the AdS/CFT correspondence, and he suggested a concrete mechanism by which black holes might preserve information.[82]

18.4.3 Applications to quantum field theory

Main articles: AdS/QCD correspondence and AdS/CMT correspondence

In addition to its applications to theoretical problems in quantum gravity, the AdS/CFT correspondence has been applied to a variety of problems in quantum field theory. One physical system that has been studied using the AdS/CFT correspondence is the quark–gluon plasma, an exotic state of matter produced in particle accelerators. This state of matter arises for brief instants when heavy ions such as gold or lead nuclei are collided at high energies. Such collisions cause the quarks that make up atomic nuclei to deconfine at temperatures of approximately two trillion kelvins, conditions similar to those present at around 10^{-11} seconds after the Big Bang.[83]

The physics of the quark–gluon plasma is governed by a theory called quantum chromodynamics, but this the-

A magnet levitating above a high-temperature superconductor. Today some physicists are working to understand high-temperature superconductivity using the AdS/CFT correspondence.[7]

ory is mathematically intractable in problems involving the quark–gluon plasma.[lower-alpha 4] In an article appearing in 2005, Đàm Thanh Sơn and his collaborators showed that the AdS/CFT correspondence could be used to understand some aspects of the quark–gluon plasma by describing it in the language of string theory.[84] By applying the AdS/CFT correspondence, Sơn and his collaborators were able to describe the quark gluon plasma in terms of black holes in five-dimensional spacetime. The calculation showed that the ratio of two quantities associated with the quark–gluon plasma, the shear viscosity and volume density of entropy, should be approximately equal to a certain universal constant. In 2008, the predicted value of this ratio for the quark–gluon plasma was confirmed at the Relativistic Heavy Ion Collider at Brookhaven National Laboratory.[85][86]

The AdS/CFT correspondence has also been used to study aspects of condensed matter physics. Over the decades, experimental condensed matter physicists have discovered a number of exotic states of matter, including superconductors and superfluids. These states are described using the formalism of quantum field theory, but some phenomena are difficult to explain using standard field theoretic techniques. Some condensed matter theorists including Subir Sachdev hope that the AdS/CFT correspondence will make it possible to describe these systems in the language of string theory and learn more about their behavior.[85]

So far some success has been achieved in using string theory methods to describe the transition of a superfluid to an insulator. A superfluid is a system of electrically neutral atoms that flows without any friction. Such systems are often produced in the laboratory using liquid helium, but recently experimentalists have developed new ways of producing artificial superfluids by pouring trillions of cold atoms into a lattice of criss-crossing lasers. These atoms

initially behave as a superfluid, but as experimentalists increase the intensity of the lasers, they become less mobile and then suddenly transition to an insulating state. During the transition, the atoms behave in an unusual way. For example, the atoms slow to a halt at a rate that depends on the temperature and on Planck's constant, the fundamental parameter of quantum mechanics, which does not enter into the description of the other phases. This behavior has recently been understood by considering a dual description where properties of the fluid are described in terms of a higher dimensional black hole.[87]

18.5 Phenomenology

Main article: String phenomenology

In addition to being an idea of considerable theoretical interest, string theory provides a framework for constructing models of real world physics that combine general relativity and particle physics. Phenomenology is the branch of theoretical physics in which physicists construct realistic models of nature from more abstract theoretical ideas. String phenomenology is the part of string theory that attempts to construct realistic models based on string theory.

Partly because of theoretical and mathematical difficulties and partly because of the extremely high energies needed to test these theories experimentally, there is so far no experimental evidence that would unambiguously point to any of these models being a correct fundamental description of nature. This has led some in the community to criticize these approaches to unification and question the value of continued research on these problems.[12]

18.5.1 Particle physics

The currently accepted theory describing elementary particles and their interactions is known as the standard model of particle physics. This theory provides a unified description of three of the fundamental forces of nature: electromagnetism and the strong and weak nuclear forces. Despite its remarkable success in explaining a wide range of physical phenomena, the standard model cannot be a complete description of reality. This is because the standard model fails to incorporate the force of gravity and because of problems such as the hierarchy problem and the inability to explain the structure of fermion masses or dark matter.

String theory has been used to construct a variety of models of particle physics going beyond the standard model. Typically, such models are based on the idea of compactification. Starting with the ten- or eleven-dimensional spacetime of string or M-theory, physicists postulate a shape for the extra dimensions. By choosing this shape appropriately, they can construct models roughly similar to the standard model of particle physics, together with additional undiscovered particles.[88] One popular way of deriving realistic physics from string theory is to start with the heterotic theory in ten dimensions and assume that the six extra dimensions of spacetime are shaped like a six-dimensional Calabi–Yau manifold. Such compactifications offer many ways of extracting realistic physics from string theory. Other similar methods can be used to construct realistic models of our four-dimensional world based on M-theory.[89]

18.5.2   Cosmology

Main article: String cosmology

The Big Bang theory is the prevailing cosmological model

A map of the cosmic microwave background produced by the Wilkinson Microwave Anisotropy Probe

for the universe from the earliest known periods through its subsequent large scale evolution. Despite its success in explaining many observed features of the universe including galactic redshifts, the relative abundance of light elements such as hydrogen and helium, and the existence of a cosmic microwave background, there are several questions that remain unanswered. For example, the standard Big Bang model does not explain why the universe appears to be same in all directions, why it appears flat on very large distance scales, or why certain hypothesized particles such as magnetic monopoles are not observed in experiments.[90]

Currently, the leading candidate for a theory going beyond the Big Bang is the theory of cosmic inflation. Developed by Alan Guth and others in the 1980s, inflation postulates a period of extremely rapid accelerated expansion of the universe prior to the expansion described by the standard Big Bang theory. The theory of cosmic inflation preserves the successes of the Big Bang while providing a natural explanation for some of the mysterious features of the universe.[91] The theory has also received striking support from observations of the cosmic microwave background,

the radiation that has filled the sky since around 380,000 years after the Big Bang.[92]

In the theory of inflation, the rapid initial expansion of the universe is caused by a hypothetical particle called the inflaton. The exact properties of this particle are not fixed by the theory but should ultimately be derived from a more fundamental theory such as string theory.[93] Indeed, there have been a number of attempts to identify an inflaton within the spectrum of particles described by string theory and to study inflation using string theory. While these approaches might eventually find support in observational data such as measurements of the cosmic microwave background, the application of string theory to cosmology is still in its early stages.[94]

18.6   Connections to mathematics

In addition to influencing research in theoretical physics, string theory has stimulated a number of major developments in pure mathematics. Like many developing ideas in theoretical physics, string theory does not at present have a mathematically rigorous formulation in which all of its concepts can be defined precisely. As a result, physicists who study string theory are often guided by physical intuition to conjecture relationships between the seemingly different mathematical structures that are used to formalize different parts of the theory. These conjectures are later proved by mathematicians, and in this way, string theory serves as a source of new ideas in pure mathematics.[95]

18.6.1   Mirror symmetry

Main article: Mirror symmetry (string theory)

After Calabi–Yau manifolds had entered physics as a way to compactify extra dimensions in string theory, many physicists began studying these manifolds. In the late 1980s, several physicists noticed that given such a compactification of string theory, it is not possible to reconstruct uniquely a corresponding Calabi–Yau manifold.[96] Instead, two different versions of string theory, type IIA and type IIB, can be compactified on completely different Calabi–Yau manifolds giving rise to the same physics. In this situation, the manifolds are called mirror manifolds, and the relationship between the two physical theories is called mirror symmetry.[97]

Regardless of whether Calabi–Yau compactifications of string theory provide a correct description of nature, the existence of the mirror duality between different string theories has significant mathematical consequences. The Calabi–Yau manifolds used in string theory are of interest in pure mathematics, and mirror symmetry allows math-

The Clebsch cubic is an example of a kind of geometric object called an algebraic variety. A classical result of enumerative geometry states that there are exactly 27 straight lines that lie entirely on this surface.

ematicians to solve problems in enumerative geometry, a branch of mathematics concerned with counting the numbers of solutions to geometric questions.[31][98]

Enumerative geometry studies a class of geometric objects called algebraic varieties which are defined by the vanishing of polynomials. For example, the Clebsch cubic illustrated on the right is an algebraic variety defined using a certain polynomial of degree three in four variables. A celebrated result of nineteenth-century mathematicians Arthur Cayley and George Salmon states that there are exactly 27 straight lines that lie entirely on such a surface.[99]

Generalizing this problem, one can ask how many lines can be drawn on a quintic Calabi–Yau manifold, such as the one illustrated above, which is defined by a polynomial of degree five. This problem was solved by the nineteenth-century German mathematician Hermann Schubert, who found that there are exactly 2,875 such lines. In 1986, geometer Sheldon Katz proved that the number of curves, such as circles, that are defined by polynomials of degree two and lie entirely in the quintic is 609,250.[100]

By the year 1991, most of the classical problems of enumerative geometry had been solved and interest in enumerative geometry had begun to diminish.[101] The field was reinvigorated in May 1991 when physicists Philip Candelas, Xenia de la Ossa, Paul Green, and Linda Parks showed that mirror symmetry could be used to translate difficult mathematical questions about one Calabi–Yau manifold into easier questions about its mirror.[102] In particular, they used mirror symmetry to show that a six-dimensional Calabi–Yau

manifold can contain exactly 317,206,375 curves of degree three.[101] In addition to counting degree-three curves, Candelas and his collaborators obtained a number of more general results for counting rational curves which went far beyond the results obtained by mathematicians.[103]

Originally, these results of Candelas were justified on physical grounds. However, mathematicians generally prefer rigorous proofs that do not require an appeal to physical intuition. Inspired by physicists' work on mirror symmetry, mathematicians have therefore constructed their own arguments proving the enumerative predictions of mirror symmetry.[lower-alpha 5] Today mirror symmetry is an active area of research in mathematics, and mathematicians are working to develop a more complete mathematical understanding of mirror symmetry based on physicists' intuition.[104] Major approaches to mirror symmetry include the homological mirror symmetry program of Maxim Kontsevich[32] and the SYZ conjecture of Andrew Strominger, Shing-Tung Yau, and Eric Zaslow.[105]

18.6.2 Monstrous moonshine

Main article: Monstrous moonshine
Group theory is the branch of mathematics that studies the

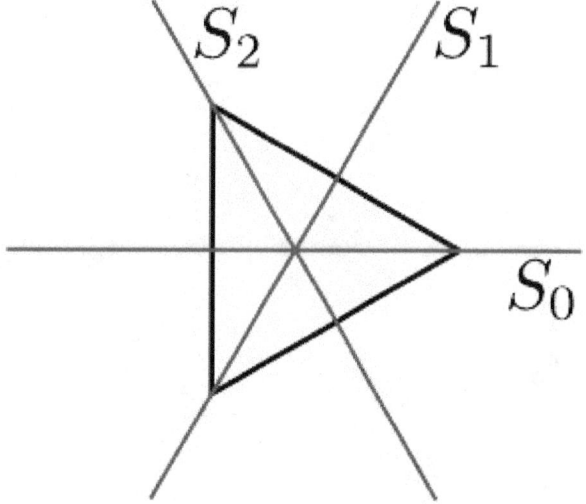

An equilateral triangle can be rotated through 120°, 240°, or 360°, or reflected in any of the three lines pictured without changing its shape.

concept of symmetry. For example, one can consider a geometric shape such as an equilateral triangle. There are various operations that one can perform on this triangle without changing its shape. One can rotate it through 120°, 240°, or 360°, or one can reflect in any of the lines labeled S_0, S_1, or S_2 in the picture. Each of these operations is called a *symmetry*, and the collection of these symmetries satisfies certain technical properties making it into what mathematicians call a group. In this particular example, the group is known as the dihedral group of order 6 because it has six elements. A general group may describe finitely many or infinitely many symmetries; if there are only finitely many symmetries, it is called a finite group.[106]

Mathematicians often strive for a classification (or list) of all mathematical objects of a given type. It is generally believed that finite groups are too diverse to admit a useful classification. A more modest but still challenging problem is to classify all finite *simple* groups. These are finite groups which may be used as building blocks for constructing arbitrary finite groups in the same way that prime numbers can be used to construct arbitrary whole numbers by taking products.[lower-alpha 6] One of the major achievements of contemporary group theory is the classification of finite simple groups, a mathematical theorem which provides a list of all possible finite simple groups.[107]

This classification theorem identifies several infinite families of groups as well as 26 additional groups which do not fit into any family. The latter groups are called the "sporadic" groups, and each one owes its existence to a remarkable combination of circumstances. The largest sporadic group, the so-called monster group, has over 10^{53} elements, more than a thousand times the number of atoms in the Earth.[108]

A graph of the j-*function in the complex plane*

A seemingly unrelated construction is the *j*-function of number theory. This object belongs to a special class of functions called modular functions, whose graphs form a certain kind of repeating pattern.[109] Although this function appears in a branch of mathematics which seems very different from the theory of finite groups, the two subjects turn out to be intimately related. In the late 1970s, mathematicians John McKay and John Thompson noticed that certain numbers arising in the analysis of the monster group (namely, the dimensions of its irreducible representations) are related to numbers that appear in a formula for the *j*-

function (namely, the coefficients of its Fourier series).[110] This relationship was further developed by John Horton Conway and Simon Norton[111] who called it monstrous moonshine because it seemed so far fetched.[112]

In 1992, Richard Borcherds constructed a bridge between the theory of modular functions and finite groups and, in the process, explained the observations of McKay and Thompson.[113][114] Borcherds' work used ideas from string theory in an essential way, extending earlier results of Igor Frenkel, James Lepowsky, and Arne Meurman, who had realized the monster group as the symmetries of a particular version of string theory.[115] In 1998, Borcherds was awarded the Fields medal for his work.[116]

Since the 1990s, the connection between string theory and moonshine has led to further results in mathematics and physics.[108] In 2010, physicists Tohru Eguchi, Hirosi Ooguri, and Yuji Tachikawa discovered connections between a different sporadic group, the Mathieu group M_{24}, and a certain version of string theory.[117] Miranda Cheng, John Duncan, and Jeffrey A. Harvey proposed a generalization of this moonshine phenomenon called umbral moonshine,[118] and their conjecture was proved mathematically by Duncan, Michael Griffin, and Ken Ono.[119] Witten has also speculated that the version of string theory appearing in monstrous moonshine might be related to a certain simplified model of gravity in three spacetime dimensions.[120]

18.7 History

Main article: History of string theory

18.7.1 Early results

Some of the structures reintroduced by string theory arose for the first time much earlier as part of the program of classical unification started by Albert Einstein. The first person to add a fifth dimension to a theory of gravity was Gunnar Nordström in 1914, who noted that gravity in five dimensions describes both gravity and electromagnetism in four. Nordström attempted to unify electromagnetism with his theory of gravitation, which was however superseded by Einstein's general relativity in 1919. Thereafter, German mathematician Theodor Kaluza combined the fifth dimension with general relativity, and only Kaluza is usually credited with the idea. In 1926, the Swedish physicist Oskar Klein gave a physical interpretation of the unobservable extra dimension—it is wrapped into a small circle. Einstein introduced a non-symmetric metric tensor, while much later Brans and Dicke added a scalar component to gravity. These ideas would be revived within string theory, where they are demanded by consistency conditions.

Leonard Susskind

String theory was originally developed during the late 1960s and early 1970s as a never completely successful theory of hadrons, the subatomic particles like the proton and neutron that feel the strong interaction. In the 1960s, Geoffrey Chew and Steven Frautschi discovered that the mesons make families called Regge trajectories with masses related to spins in a way that was later understood by Yoichiro Nambu, Holger Bech Nielsen and Leonard Susskind to be the relationship expected from rotating strings. Chew advocated making a theory for the interactions of these trajectories that did not presume that they were composed of any fundamental particles, but would construct their interactions from self-consistency conditions on the S-matrix. The S-matrix approach was started by Werner Heisenberg in the 1940s as a way of constructing a theory that did not rely on the local notions of space and time, which Heisenberg believed break down at the nuclear scale. While the scale was off by many orders of magnitude, the approach he advocated was ideally suited for a theory of quantum gravity.

Working with experimental data, R. Dolen, D. Horn and C. Schmid developed some sum rules for hadron exchange. When a particle and antiparticle scatter, virtual particles can

be exchanged in two qualitatively different ways. In the s-channel, the two particles annihilate to make temporary intermediate states that fall apart into the final state particles. In the t-channel, the particles exchange intermediate states by emission and absorption. In field theory, the two contributions add together, one giving a continuous background contribution, the other giving peaks at certain energies. In the data, it was clear that the peaks were stealing from the background—the authors interpreted this as saying that the t-channel contribution was dual to the s-channel one, meaning both described the whole amplitude and included the other.

Gabriele Veneziano

The result was widely advertised by Murray Gell-Mann, leading Gabriele Veneziano to construct a scattering amplitude that had the property of Dolen-Horn-Schmid duality, later renamed world-sheet duality. The amplitude needed poles where the particles appear, on straight line trajectories, and there is a special mathematical function whose poles are evenly spaced on half the real line— the Gamma function— which was widely used in Regge theory. By manipulating combinations of Gamma functions, Veneziano was able to find a consistent scattering amplitude with poles on straight lines, with mostly positive residues, which obeyed duality and had the appropriate Regge scaling at high energy. The amplitude could fit near-beam scattering data as well as other Regge type fits, and had a suggestive integral representation that could be used for generalization.

Over the next years, hundreds of physicists worked to complete the bootstrap program for this model, with many surprises. Veneziano himself discovered that for the scattering amplitude to describe the scattering of a particle that appears in the theory, an obvious self-consistency condition, the lightest particle must be a tachyon. Miguel Virasoro and Joel Shapiro found a different amplitude now understood to be that of closed strings, while Ziro Koba and Holger Nielsen generalized Veneziano's integral representation to multiparticle scattering. Veneziano and Sergio Fubini introduced an operator formalism for computing the scattering amplitudes that was a forerunner of world-sheet conformal theory, while Virasoro understood how to remove the poles with wrong-sign residues using a constraint on the states. Claud Lovelace calculated a loop amplitude, and noted that there is an inconsistency unless the dimension of the theory is 26. Charles Thorn, Peter Goddard and Richard Brower went on to prove that there are no wrong-sign propagating states in dimensions less than or equal to 26.

In 1969, Yoichiro Nambu, Holger Bech Nielsen, and Leonard Susskind recognized that the theory could be given a description in space and time in terms of strings. The scattering amplitudes were derived systematically from the action principle by Peter Goddard, Jeffrey Goldstone, Claudio Rebbi, and Charles Thorn, giving a space-time picture to the vertex operators introduced by Veneziano and Fubini and a geometrical interpretation to the Virasoro conditions.

In 1970, Pierre Ramond added fermions to the model, which led him to formulate a two-dimensional supersymmetry to cancel the wrong-sign states. John Schwarz and André Neveu added another sector to the fermi theory a short time later. In the fermion theories, the critical dimension was 10. Stanley Mandelstam formulated a world sheet conformal theory for both the bose and fermi case, giving a two-dimensional field theoretic path-integral to generate the operator formalism. Michio Kaku and Keiji Kikkawa gave a different formulation of the bosonic string, as a string field theory, with infinitely many particle types and with fields taking values not on points, but on loops and curves.

In 1974, Tamiaki Yoneya discovered that all the known string theories included a massless spin-two particle that obeyed the correct Ward identities to be a graviton. John Schwarz and Joel Scherk came to the same conclusion and made the bold leap to suggest that string theory was a theory of gravity, not a theory of hadrons. They reintroduced Kaluza–Klein theory as a way of making sense of the extra dimensions. At the same time, quantum chromodynamics was recognized as the correct theory of hadrons, shifting the attention of physicists and apparently leaving the bootstrap program in the dustbin of history.

String theory eventually made it out of the dustbin, but for the following decade all work on the theory was completely ignored. Still, the theory continued to develop at a steady pace thanks to the work of a handful of devotees. Ferdinando Gliozzi, Joel Scherk, and David Olive realized in 1976 that the original Ramond and Neveu Schwarzstrings were separately inconsistent and needed to be combined. The resulting theory did not have a tachyon, and was proven to have space-time supersymmetry by John Schwarz and Michael Green in 1981. The same year, Alexander Polyakov gave the theory a modern path integral formulation, and went on to develop conformal field theory extensively. In 1979, Daniel Friedan showed that the equations of motions of string theory, which are generalizations of the Einstein equations of General Relativity, emerge from the Renormalization group equations for the two-dimensional field theory. Schwarz and Green discovered T-duality, and constructed two superstring theories—IIA and IIB related by T-duality, and type I theories with open strings. The consistency conditions had been so strong, that the entire theory was nearly uniquely determined, with only a few discrete choices.

18.7.2 First superstring revolution

Edward Witten

In the early 1980s, Edward Witten discovered that most theories of quantum gravity could not accommodate chiral fermions like the neutrino. This led him, in collaboration with Luis Álvarez-Gaumé to study violations of the conservation laws in gravity theories with anomalies, concluding that type I string theories were inconsistent. Green and Schwarz discovered a contribution to the anomaly that Witten and Alvarez-Gaumé had missed, which restricted the gauge group of the type I string theory to be SO(32). In coming to understand this calculation, Edward Witten became convinced that string theory was truly a consistent theory of gravity, and he became a high-profile advocate. Following Witten's lead, between 1984 and 1986, hundreds of physicists started to work in this field, and this is sometimes called the first superstring revolution.

During this period, David Gross, Jeffrey Harvey, Emil Martinec, and Ryan Rohm discovered heterotic strings. The gauge group of these closed strings was two copies of E8, and either copy could easily and naturally include the standard model. Philip Candelas, Gary Horowitz, Andrew Strominger and Edward Witten found that the Calabi–Yau manifolds are the compactifications that preserve a realistic amount of supersymmetry, while Lance Dixon and others worked out the physical properties of orbifolds, distinctive geometrical singularities allowed in string theory. Cumrun Vafa generalized T-duality from circles to arbitrary manifolds, creating the mathematical field of mirror symmetry. Daniel Friedan, Emil Martinec and Stephen Shenker further developed the covariant quantization of the superstring using conformal field theory techniques. David Gross and Vipul Periwal discovered that string perturbation theory was divergent. Stephen Shenker showed it diverged much faster than in field theory suggesting that new nonperturbative objects were missing.

Joseph Polchinski

In the 1990s, Joseph Polchinski discovered that the theory requires higher-dimensional objects, called D-branes

and identified these with the black-hole solutions of supergravity. These were understood to be the new objects suggested by the perturbative divergences, and they opened up a new field with rich mathematical structure. It quickly became clear that D-branes and other p-branes, not just strings, formed the matter content of the string theories, and the physical interpretation of the strings and branes was revealed—they are a type of black hole. Leonard Susskind had incorporated the holographic principle of Gerardus 't Hooft into string theory, identifying the long highly excited string states with ordinary thermal black hole states. As suggested by 't Hooft, the fluctuations of the black hole horizon, the world-sheet or world-volume theory, describes not only the degrees of freedom of the black hole, but all nearby objects too.

18.7.3   Second superstring revolution

In 1995, at the annual conference of string theorists at the University of Southern California (USC), Edward Witten gave a speech on string theory that in essence united the five string theories that existed at the time, and giving birth to a new 11-dimensional theory called M-theory. M-theory was also foreshadowed in the work of Paul Townsend at approximately the same time. The flurry of activity that began at this time is sometimes called the second superstring revolution.[34]

Juan Maldacena

During this period, Tom Banks, Willy Fischler, Stephen Shenker and Leonard Susskind formulated matrix theory, a full holographic description of M-theory using IIA D0 branes.[51] This was the first definition of string theory that was fully non-perturbative and a concrete mathematical realization of the holographic principle. It is an example of a gauge-gravity duality and is now understood to be a special case of the AdS/CFT correspondence. Andrew Strominger and Cumrun Vafa calculated the entropy of certain configurations of D-branes and found agreement with the semi-classical answer for extreme charged black holes.[62] Petr Hořava and Witten found the eleven-dimensional formulation of the heterotic string theories, showing that orbifolds solve the chirality problem. Witten noted that the effective description of the physics of D-branes at low energies is by a supersymmetric gauge theory, and found geometrical interpretations of mathematical structures in gauge theory that he and Nathan Seiberg had earlier discovered in terms of the location of the branes.

In 1997, Juan Maldacena noted that the low energy excitations of a theory near a black hole consist of objects close to the horizon, which for extreme charged black holes looks like an anti-de Sitter space.[71] He noted that in this limit the gauge theory describes the string excitations near the branes. So he hypothesized that string theory on a near-horizon extreme-charged black-hole geometry, an anti-deSitter space times a sphere with flux, is equally well described by the low-energy limiting gauge theory, the $N = 4$ supersymmetric Yang–Mills theory. This hypothesis, which is called the AdS/CFT correspondence, was further developed by Steven Gubser, Igor Klebanov and Alexander Polyakov,[72] and by Edward Witten,[73] and it is now well-accepted. It is a concrete realization of the holographic principle, which has far-reaching implications for black holes, locality and information in physics, as well as the nature of the gravitational interaction.[56] Through this relationship, string theory has been shown to be related to gauge theories like quantum chromodynamics and this has led to more quantitative understanding of the behavior of hadrons, bringing string theory back to its roots.[84]

18.8   Criticism

18.8.1   Number of solutions

Main article: String theory landscape

To construct models of particle physics based on string theory, physicists typically begin by specifying a shape for the extra dimensions of spacetime. Each of these different shapes corresponds to a different possible universe, or

"vacuum state", with a different collection of particles and forces. String theory as it is currently understood has an enormous number of vacuum states, typically estimated to be around 10^{500}, and these might be sufficiently diverse to accommodate almost any phenomena that might be observed at low energies.[121]

Many critics of string theory have expressed concerns about the large number of possible universes described by string theory. In his book *Not Even Wrong*, Peter Woit, a lecturer in the mathematics department at Columbia University, has argued that the large number of different physical scenarios renders string theory vacuous as a framework for constructing models of particle physics. According to Woit,

> The possible existence of, say, 10^{500} consistent different vacuum states for superstring theory probably destroys the hope of using the theory to predict anything. If one picks among this large set just those states whose properties agree with present experimental observations, it is likely there still will be such a large number of these that one can get just about whatever value one wants for the results of any new observation.[122]

Some physicists believe this large number of solutions is actually a virtue because it may allow a natural anthropic explanation of the observed values of physical constants, in particular the small value of the cosmological constant.[122] The anthropic principle is the idea that some of the numbers appearing in the laws of physics are not fixed by any fundamental principle but must be compatible with the evolution of intelligent life. In 1987, Steven Weinberg published an article in which he argued that the cosmological constant could not have been too large, or else galaxies and intelligent life would not have been able to develop.[123] Weinberg suggested that there might be a huge number of possible consistent universes, each with a different value of the cosmological constant, and observations indicate a small value of the cosmological constant only because humans happen to live in a universe that has allowed intelligent life, and hence observers, to exist.[124]

String theorist Leonard Susskind has argued that string theory provides a natural anthropic explanation of the small value of the cosmological constant.[125] According to Susskind, the different vacuum states of string theory might be realized as different universes within a larger multiverse. The fact that the observed universe has a small cosmological constant is just a tautological consequence of the fact that a small value is required for life to exist.[126] Many prominent theorists and critics have disagreed with Susskind's conclusions.[127] According to Woit, "in this case [anthropic reasoning] is nothing more than an excuse for failure. Speculative scientific ideas fail not just when they make incorrect predictions, but also when they turn out to be vacuous and incapable of predicting anything."[128]

18.8.2 Background independence

Main article: Background independence

One of the fundamental properties of Einstein's general theory of relativity is that it is background independent, meaning that the formulation of the theory does not in any way privilege a particular spacetime geometry.[129]

One of the main criticisms of string theory from early on is that it is not manifestly background independent. In string theory, one must typically specify a fixed reference geometry for spacetime, and all other possible geometries are described as perturbations of this fixed one. In his book *The Trouble With Physics*, physicist Lee Smolin of the Perimeter Institute for Theoretical Physics claims that this is the principal weakness of string theory as a theory of quantum gravity, saying that string theory has failed to incorporate this important insight from general relativity.[130]

Others have disagreed with Smolin's characterization of string theory. In a review of Smolin's book, string theorist Joseph Polchinski writes

> [Smolin] is mistaking an aspect of the mathematical language being used for one of the physics being described. New physical theories are often discovered using a mathematical language that is not the most suitable for them... In string theory it has always been clear that the physics is background-independent even if the language being used is not, and the search for more suitable language continues. Indeed, as Smolin belatedly notes, [AdS/CFT] provides a solution to this problem, one that is unexpected and powerful.[131]

Polchinski notes that an important open problem in quantum gravity is to develop holographic descriptions of gravity which do not require the gravitational field to be asymptotically anti-de Sitter.[131]

Smolin responded that the claims about background-independence, which Polchinski presents as "clear", are in fact only an unproven hope for future results, and Smolin is skeptical about them being true at all because of fundamental reasons: "If the strong form of the AdS/CFT conjecture is shown to be correct, then a very weak, and limited form of background will have been achieved. But ... this is still a big if". Smolin points out that current results about

the [AdS/CFT] conjecture rely on global super-symmetry as perturbative physics, "but the whole point of general relativity and quantum gravity is that the generic solutions are governed by no global symmetries because the geometry of spacetime is completely dynamical", which "makes it very non-trivial to show the strong form of the [AdS/CFT] conjecture, because it must extend to solutions of supergravity arbitrarily far from those with global symmetries in the bulk".[132] Smolin summarizes:

> It would be more accurate to say, "Some string theorists believe that the formulations of perturbative string theories and dualities between them that they study concretely are approximations to a deeper, background independent formulation. This missing background independent formulation is not just a different language for the theory, it is hoped to be the statement of the principles and laws that define the theory, from which everything studied so far would be derived as an approximation."[132]

18.8.3 Sociological issues

Since the superstring revolutions of the 1980s and 1990s, string theory has become the dominant paradigm of high energy theoretical physics.[133] Some string theorists have expressed the view that there does not exist an equally successful alternative theory addressing the deep questions of fundamental physics. In an interview from 1987, Nobel laureate David Gross made the following controversial comments about the reasons for the popularity of string theory:

> The most important [reason] is that there are no other good ideas around. That's what gets most people into it. When people started to get interested in string theory they didn't know anything about it. In fact, the first reaction of most people is that the theory is extremely ugly and unpleasant, at least that was the case a few years ago when the understanding of string theory was much less developed. It was difficult for people to learn about it and to be turned on. So I think the real reason why people have got attracted by it is because there is no other game in town. All other approaches of constructing grand unified theories, which were more conservative to begin with, and only gradually became more and more radical, have failed, and this game hasn't failed yet.[134]

Several other high profile theorists and commentators have expressed similar views, suggesting that there are no viable alternatives to string theory.[135]

Many critics of string theory have commented on this state of affairs. In his book criticizing string theory, Peter Woit views the status of string theory research as unhealthy and detrimental to the future of fundamental physics. He argues that the extreme popularity of string theory among theoretical physicists is partly a consequence of the financial structure of academia and the fierce competition for scarce resources.[136] In his book *The Road to Reality*, mathematical physicist Roger Penrose expresses similar views, stating "The often frantic competitiveness that this ease of communication engenders leads to 'bandwagon' effects, where researchers fear to be left behind if they do not join in."[137] Penrose also claims that the technical difficulty of modern physics forces young scientists to rely on the preferences of established researchers, rather than forging new paths of their own.[138] Lee Smolin expresses a slightly different position in his critique, claiming that string theory grew out of a tradition of particle physics which discourages speculation about the foundations of physics, while his preferred approach, loop quantum gravity, encourages more radical thinking. According to Smolin,

> String theory is a powerful, well-motivated idea and deserves much of the work that has been devoted to it. If it has so far failed, the principal reason is that its intrinsic flaws are closely tied to its strengths—and, of course, the story is unfinished, since string theory may well turn out to be part of the truth. The real question is not why we have expended so much energy on string theory but why we haven't expended nearly enough on alternative approaches.[139]

Smolin goes on to offer a number of prescriptions for how scientists might encourage a greater diversity of approaches to quantum gravity research.[140]

18.9 References

18.9.1 Notes

[1] For example, physicists are still working to understand the phenomenon of quark confinement, the paradoxes of black holes, and the origin of dark energy.

[2] For example, in the context of the AdS/CFT correspondence, theorists often formulate and study theories of gravity in unphysical numbers of spacetime dimensions.

[3] "Top Cited Articles during 2010 in hep-th". Retrieved 25 July 2013.

[4] More precisely, one cannot apply the methods of perturbative quantum field theory.

[5] Two independent mathematical proofs of mirror symmetry were given by Givental 1996, 1998 and Lian, Liu, Yau 1997, 1999, 2000.

[6] More precisely, a nontrivial group is called *simple* if its only normal subgroups are the trivial group and the group itself. The Jordan–Hölder theorem exhibits finite simple groups as the building blocks for all finite groups.

18.9.2 Citations

[1] Becker, Becker, and Schwarz 2007, p. 1

[2] Zwiebach 2009, p. 6

[3] Becker, Becker, and Schwarz 2007, pp. 2–3

[4] Becker, Becker, and Schwarz 2007, pp. 9–12

[5] Becker, Becker, and Schwarz 2007, pp. 14–15

[6] Klebanov and Maldacena 2009

[7] Merali 2011

[8] Sachdev 2013

[9] Becker, Becker, and Schwarz 2007, pp. 3, 15–16

[10] Becker, Becker, and Schwarz 2007, p. 8

[11] Becker, Becker, and Schwarz 13–14

[12] Woit 2006

[13] Zee 2010

[14] Becker, Becker, and Schwarz 2007, p. 2

[15] Becker, Becker, and Schwarz 2007, p. 6

[16] Zwiebach 2009, p. 12

[17] Becker, Becker, and Schwarz 2007, p. 4

[18] Zwiebach 2009, p. 324

[19] Wald 1984, p. 4

[20] Zee 2010, Parts V and VI

[21] Zwiebach 2009, p. 9

[22] Zwiebach 2009, p. 8

[23] Yau and Nadis 2010, Ch. 6

[24] Greene 2000, p. 186

[25] Yau and Nadis 2010, p. ix

[26] Randall and Sundrum 1999

[27] Becker, Becker, and Schwarz 2007

[28] Zwiebach 2009, p. 376

[29] Moore 2005, p. 214

[30] Moore 2005, p. 215

[31] Aspinwall et al. 2009

[32] Kontsevich 1995

[33] Kapustin and Witten 2007

[34] Duff 1998

[35] Duff 1998, p. 64

[36] Nahm 1978

[37] Cremmer, Julia, and Scherk 1978

[38] Duff 1998, p. 65

[39] Sen 1994a

[40] Sen 1994b

[41] Hull and Townsend 1995

[42] Duff 1998, p. 67

[43] Bergshoeff, Sezgin, and Townsend 1987

[44] Duff et al. 1987

[45] Duff 1998, p. 66

[46] Witten 1995

[47] Duff 1998, pp. 67–68

[48] Becker, Becker, and Schwarz 2007, p. 296

[49] Hořava and Witten 1996

[50] Duff 1996, sec. 1

[51] Banks et al. 1997

[52] Connes 1994

[53] Connes, Douglas, and Schwarz 1998

[54] Nekrasov and Schwarz 1998

[55] Seiberg and Witten 1999

[56] de Haro et al. 2013, p. 2

[57] Yau and Nadis 2010, p. 187–188

[58] Bekenstein 1973

[59] Hawking 1975

[60] Wald 1984, p. 417

[61] Yau and Nadis 2010, p. 189

[62] Strominger and Vafa 1996

[63] Yau and Nadis 2010, pp. 190–192

[64] Maldacena, Strominger, and Witten 1997

[65] Ooguri, Strominger, and Vafa 2004

[66] Yau and Nadis 2010, pp. 192–193

[67] Yau and Nadis 2010, pp. 194–195

[68] Strominger 1998

[69] Guica et al. 2009

[70] Castro, Maloney, and Strominger 2010

[71] Maldacena 1998

[72] Gubser, Klebanov, and Polyakov 1998

[73] Witten 1998

[74] Klebanov and Maldacena 2009, p. 28

[75] Maldacena 2005, p. 60

[76] Maldacena 2005, p. 61

[77] Zwiebach 2009, p. 552

[78] Maldacena 2005, pp. 61–62

[79] Susskind 2008

[80] Zwiebach 2009, p. 554

[81] Maldacena 2005, p. 63

[82] Hawking 2005

[83] Zwiebach 2009, p. 559

[84] Kovtun, Son, and Starinets 2001

[85] Merali 2011, p. 303

[86] Luzum and Romatschke 2008

[87] Sachdev 2013, p. 51

[88] Candelas et al. 1985

[89] Yau and Nadis 2010, pp. 147–150

[90] Becker, Becker, and Schwarz 2007, pp. 530–531

[91] Becker, Becker, and Schwarz 2007, p. 531

[92] Becker, Becker, and Schwarz 2007, p. 538

[93] Becker, Becker, and Schwarz 2007, p. 533

[94] Becker, Becker, and Schwarz 2007, pp. 539–543

[95] Deligne et al. 1999, p. 1

[96] Hori et al. 2003, p. xvii

[97] Aspinwall et al. 2009, p. 13

[98] Hori et al. 2003

[99] Yau and Nadis 2010, p. 167

[100] Yau and Nadis 2010, p. 166

[101] Yau and Nadis 2010, p. 169

[102] Candelas et al. 1991

[103] Yau and Nadis 2010, p. 171

[104] Hori et al. 2003, p. xix

[105] Strominger, Yau, and Zaslow 1996

[106] Dummit and Foote 2004

[107] Dummit and Foote 2004, pp. 102–103

[108] Klarreich 2015

[109] Gannon 2006, p. 2

[110] Gannon 2006, p. 4

[111] Conway and Norton 1979

[112] Gannon 2006, p. 5

[113] Gannon 2006, p. 8

[114] Borcherds 1992

[115] Frenkel, Lepowsky, and Meurman 1988

[116] Gannon 2006, p. 11

[117] Eguchi, Ooguri, and Tachikawa 2010

[118] Cheng, Duncan, and Harvey 2013

[119] Duncan, Griffin, and Ono 2015

[120] Witten 2007

[121] Woit 2006, pp. 240–242

[122] Woit 2006, p. 242

[123] Weinberg 1987

[124] Woit 2006, p. 243

[125] Susskind 2005

[126] Woit 2006, pp. 242–243

[127] Woit 2006, p. 240

[128] Woit 2006, p. 249

[129] Smolin 2006, p. 81

[130] Smolin 2006, p. 184

[131] Polchinski 2007

[132] Lee Smolin, April 2007:"Archived copy". Archived from the original on November 5, 2015. Retrieved December 31, 2015. Response to review of The Trouble with Physics by Joe Polchinski

[133] Penrose 2004, p. 1017

[134] Woit 2006, pp. 224–225

[135] Woit 2006, Ch. 16

[136] Woit 2006, p. 239

[137] Penrose 2004, p. 1018

[138] Penrose 2004, pp. 1019–1020

[139] Smolin 2006, p. 349

[140] Smolin 2006, Ch. 20

18.9.3   Bibliography

- Aspinwall, Paul; Bridgeland, Tom; Craw, Alastair; Douglas, Michael; Gross, Mark; Kapustin, Anton; Moore, Gregory; Segal, Graeme; Szendröi, Balázs; Wilson, P.M.H., eds. (2009). *Dirichlet Branes and Mirror Symmetry*. American Mathematical Society. ISBN 978-0-8218-3848-8.

- Banks, Tom; Fischler, Willy; Schenker, Stephen; Susskind, Leonard (1997). "M theory as a matrix model: A conjecture". *Physical Review D* **55** (8): 5112–5128. arXiv:hep-th/9610043. Bibcode:1997PhRvD..55.5112B. doi:10.1103/physrevd.55.5112.

- Becker, Katrin; Becker, Melanie; Schwarz, John (2007). *String theory and M-theory: A modern introduction*. Cambridge University Press. ISBN 978-0-521-86069-7.

- Bekenstein, Jacob (1973). "Black holes and entropy". *Physical Review D* **7** (8): 2333–2346. Bibcode:1973PhRvD...7.2333B. doi:10.1103/PhysRevD.7.2333.

- Bergshoeff, Eric; Sezgin, Ergin; Townsend, Paul (1987). "Supermembranes and eleven-dimensional supergravity". *Physics Letters B* **189** (1): 75–78. Bibcode:1987PhLB..189...75B. doi:10.1016/0370-2693(87)91272-X.

- Borcherds, Richard (1992). "Monstrous moonshine and Lie superalgebras". *Inventiones Mathematicae* **109** (1): 405–444. Bibcode:1992InMat.109..405B. doi:10.1007/BF01232032.

- Candelas, Philip; de la Ossa, Xenia; Green, Paul; Parks, Linda (1991). "A pair of Calabi–Yau manifolds as an exactly soluble superconformal field theory". *Nuclear Physics B* **359** (1): 21–74. Bibcode:1991NuPhB.359...21C. doi:10.1016/0550-3213(91)90292-6.

- Candelas, Philip; Horowitz, Gary; Strominger, Andrew; Witten, Edward (1985). "Vacuum configurations for superstrings". *Nuclear Physics B* **258**: 46–74. Bibcode:1985NuPhB.258...46C. doi:10.1016/0550-3213(85)90602-9.

- Castro, Alejandra; Maloney, Alexander; Strominger, Andrew (2010). "Hidden conformal symmetry of the Kerr black hole". *Physical Review D* **82** (2). arXiv:1004.0996. Bibcode:2010PhRvD..82b4008C. doi:10.1103/PhysRevD.82.024008.

- Cheng, Miranda; Duncan, John; Harvey, Jeffrey (2013). "Umbral Moonshine". arXiv:1204.2779.

- Connes, Alain (1994). *Noncommutative Geometry*. Academic Press. ISBN 978-0-12-185860-5.

- Connes, Alain; Douglas, Michael; Schwarz, Albert (1998). "Noncommutative geometry and matrix theory". *Journal of High Energy Physics*. 19981 (2): 003. arXiv:hep-th/9711162. Bibcode:1998JHEP...02..003C. doi:10.1088/1126-6708/1998/02/003.

- Conway, John; Norton, Simon (1979). "Monstrous moonshine". *Bull. London Math. Soc.* **11** (3): 308–339. doi:10.1112/blms/11.3.308.

- Cremmer, Eugene; Julia, Bernard; Scherk, Joel (1978). "Supergravity theory in eleven dimensions". *Physics Letters B* **76** (4): 409–412. Bibcode:1978PhLB...76..409C. doi:10.1016/0370-2693(78)90894-8.

- de Haro, Sebastian; Dieks, Dennis; 't Hooft, Gerard; Verlinde, Erik (2013). "Forty Years of String Theory Reflecting on the Foundations". *Foundations of Physics* **43** (1): 1–7. Bibcode:2013FoPh...43....1D. doi:10.1007/s10701-012-9691-3.

- Deligne, Pierre; Etingof, Pavel; Freed, Daniel; Jeffery, Lisa; Kazhdan, David; Morgan, John; Morrison, David; Witten, Edward, eds. (1999). *Quantum Fields and Strings: A Course for Mathematicians* **1**. American Mathematical Society. ISBN 978-0821820124.

- Duff, Michael (1996). "M-theory (the theory formerly known as strings)". *International Journal of Modern Physics A* **11** (32): 6523–41. arXiv:hep-th/9608117. Bibcode:1996IJMPA..11.5623D. doi:10.1142/S0217751X96002583.

- Duff, Michael (1998). "The theory formerly known as strings". *Scientific American* **278** (2): 64–9. doi:10.1038/scientificamerican0298-64.

- Duff, Michael; Howe, Paul; Inami, Takeo; Stelle, Kellogg (1987). "Superstrings in $D=10$ from supermembranes in $D=11$". *Nuclear Physics B* **191** (1): 70–74. Bibcode:1987PhLB..191...70D. doi:10.1016/0370-2693(87)91323-2.

- Dummit, David; Foote, Richard (2004). *Abstract Algebra*. Wiley. ISBN 978-0-471-43334-7.

- Duncan, John; Griffin, Michael; Ono, Ken (2015). "Proof of the Umbral Moonshine Conjecture". arXiv:1503.01472.

- Eguchi, Tohru; Ooguri, Hirosi; Tachikawa, Yuji (2011). "Notes on the K3 surface and the Mathieu group M_{24}". *Experimental Mathematics* **20** (1): 91–96. doi:10.1080/10586458.2011.544585.

- Frenkel, Igor; Lepowsky, James; Meurman, Arne (1988). *Vertex Operator Algebras and the Monster*. Pure and Applied Mathematics **134**. Academic Press. ISBN 0-12-267065-5.

- Gannon, Terry. *Moonshine Beyond the Monster: The Bridge Connecting Algebra, Modular Forms, and Physics*. Cambridge University Press.

- Givental, Alexander (1996). "Equivariant Gromov-Witten invariants". *International Mathematics Research Notices* **1996** (13): 613–663. doi:10.1155/S1073792896000414.

- Givental, Alexander (1998). "A mirror theorem for toric complete intersections". *Topological field theory, primitive forms and related topics*: 141–175. doi:10.1007/978-1-4612-0705-4_5. ISBN 978-1-4612-6874-1.

- Gubser, Steven; Klebanov, Igor; Polyakov, Alexander (1998). "Gauge theory correlators from non-critical string theory". *Physics Letters B* **428**: 105–114. arXiv:hep-th/9802109. Bibcode:1998PhLB..428..105G. doi:10.1016/S0370-2693(98)00377-3.

- Guica, Monica; Hartman, Thomas; Song, Wei; Strominger, Andrew (2009). "The Kerr/CFT Correspondence". *Physical Review D* **80** (12). arXiv:0809.4266. Bibcode:2009PhRvD..80l4008G. doi:10.1103/PhysRevD.80.124008.

- Hawking, Stephen (1975). "Particle creation by black holes". *Communications in Mathematical Physics* **43** (3): 199–220. Bibcode:1975CMaPh..43..199H. doi:10.1007/BF02345020.

- Hawking, Stephen (2005). "Information loss in black holes". *Physical Review D* **72** (8). arXiv:hep-th/0507171. Bibcode:2005PhRvD..72h4013H. doi:10.1103/PhysRevD.72.084013.

- Hořava, Petr; Witten, Edward (1996). "Heterotic and Type I string dynamics from eleven dimensions". *Nuclear Physics B* **460** (3): 506–524. arXiv:hep-th/9510209. Bibcode:1996NuPhB.460..506H. doi:10.1016/0550-3213(95)00621-4.

- Hori, Kentaro; Katz, Sheldon; Klemm, Albrecht; Pandharipande, Rahul; Thomas, Richard; Vafa, Cumrun; Vakil, Ravi; Zaslow, Eric, eds. (2003). *Mirror Symmetry* (PDF). American Mathematical Society. ISBN 0-8218-2955-6.

- Hull, Chris; Townsend, Paul (1995). "Unity of superstring dualities". *Nuclear Physics B* **4381** (1): 109–137. arXiv:hep-th/9410167. Bibcode:1995NuPhB.438..109H. doi:10.1016/0550-3213(94)00559-W.

- Kapustin, Anton; Witten, Edward (2007). "Electric-magnetic duality and the geometric Langlands program". *Communications in Number Theory and Physics* **1** (1): 1–236. arXiv:hep-th/0604151. Bibcode:2007CNTP....1....1K. doi:10.4310/cntp.2007.v1.n1.a1.

- Klarreich, Erica. "Mathematicians chase moonshine's shadow". *Quanta Magazine*. Retrieved March 2015.

- Klebanov, Igor; Maldacena, Juan (2009). "Solving Quantum Field Theories via Curved Spacetimes" (PDF). *Physics Today* **62**: 28–33. Bibcode:2009PhT....62a..28K. doi:10.1063/1.3074260. Archived from the original (PDF) on July 2, 2013. Retrieved May 2013.

- Kontsevich, Maxim (1995). "Homological algebra of mirror symmetry". *Proceedings of the International Congress of Mathematicians*: 120–139. arXiv:alg-geom/9411018. Bibcode:1994alg.geom.11018K.

- Kovtun, P. K.; Son, Dam T.; Starinets, A. O. (2001). "Viscosity in strongly interacting quantum field theories from black hole physics". *Physical Review Letters* **94** (11): 111601. arXiv:hep-th/0405231. Bibcode:2005PhRvL..94k1601K. doi:10.1103/PhysRevLett.94.111601. PMID 15903845.

- Lian, Bong; Liu, Kefeng; Yau, Shing-Tung (1997). "Mirror principle, I". *Asian Journal of Mathematics* **1**: 729–763. arXiv:alg-geom/9712011. Bibcode:1997alg.geom.12011L.

- Lian, Bong; Liu, Kefeng; Yau, Shing-Tung (1999a). "Mirror principle, II". *Asian Journal of Mathematics* **3**: 109–146. arXiv:math/9905006. Bibcode:1999math......5006L.

- Lian, Bong; Liu, Kefeng; Yau, Shing-Tung (1999b). "Mirror principle, III". *Asian Journal of Mathematics* **3**: 771–800. arXiv:math/9912038. Bibcode:1999math.....12038L.

- Lian, Bong; Liu, Kefeng; Yau, Shing-Tung (2000). "Mirror principle, IV". *Surveys in Differential Geometry* **7**: 475–496. arXiv:math/0007104. Bibcode:2000math......7104L. doi:10.4310/sdg.2002.v7.n1.a15.

- Luzum, Matthew; Romatschke, Paul (2008). "Conformal relativistic viscous hydrodynamics: Applications to RHIC results at $\sqrt{s_{NN}}$=200 GeV". *Physical Review C* **78** (3). arXiv:0804.4015. doi:10.1103/PhysRevC.78.034915.

- Maldacena, Juan (1998). "The Large N limit of superconformal field theories and supergravity". *Advances in Theoretical and Mathematical Physics* **2**: 231–252. arXiv:hep-th/9711200. Bibcode:1998AdTMP...2..231M. doi:10.1063/1.59653.

- Maldacena, Juan (2005). "The Illusion of Gravity" (PDF). *Scientific American* **293** (5): 56–63. Bibcode:2005SciAm.293e..56M. doi:10.1038/scientificamerican1105-56. PMID 16318027. Archived from the original (PDF) on November 1, 2014. Retrieved July 2013.

- Maldacena, Juan; Strominger, Andrew; Witten, Edward (1997). "Black hole entropy in M-theory". *Journal of High Energy Physics* **1997** (12). doi:10.1088/1126-6708/1997/12/002.

- Merali, Zeeya (2011). "Collaborative physics: string theory finds a bench mate". *Nature* **478** (7369): 302–304. Bibcode:2011Natur.478..302M. doi:10.1038/478302a. PMID 22012369.

- Moore, Gregory (2005). "What is ... a Brane?" (PDF). *Notices of the AMS* **52**: 214. Retrieved June 2013.

- Nahm, Walter (1978). "Supersymmetries and their representations". *Nuclear Physics B* **135** (1): 149–166. Bibcode:1978NuPhB.135..149N. doi:10.1016/0550-3213(78)90218-3.

- Nekrasov, Nikita; Schwarz, Albert (1998). "Instantons on noncommutative $\mathbf{R}^4$ and (2,0) superconformal six dimensional theory". *Communications in*

Mathematical Physics **198** (3): 689–703. arXiv:hep-th/9802068. Bibcode:1998CMaPh.198..689N. doi:10.1007/s002200050490.

- Ooguri, Hirosi; Strominger, Andrew; Vafa, Cumrun (2004). "Black hole attractors and the topological string". *Physical Review D* **70** (10). doi:10.1103/physrevd.70.106007.

- Polchinski, Joseph (2007). "All Strung Out?". *American Scientist*. Retrieved April 2015.

- Penrose, Roger (2005). *The Road to Reality: A Complete Guide to the Laws of the Universe*. Knopf. ISBN 0-679-45443-8.

- Randall, Lisa; Sundrum, Raman (1999). "An alternative to compactification". *Physical Review Letters* **83** (23): 4690–4693. arXiv:hep-th/9906064. Bibcode:1999PhRvL..83.4690R. doi:10.1103/PhysRevLett.83.4690.

- Sachdev, Subir (2013). "Strange and stringy". *Scientific American* **308** (44): 44–51. Bibcode:2012SciAm.308a..44S. doi:10.1038/scientificamerican0113-44.

- Seiberg, Nathan; Witten, Edward (1999). "String Theory and Noncommutative Geometry". *Journal of High Energy Physics* **1999** (9): 032. arXiv:hep-th/9908142. Bibcode:1999JHEP...09..032S. doi:10.1088/1126-6708/1999/09/032.

- Sen, Ashoke (1994a). "Strong-weak coupling duality in four-dimensional string theory". *International Journal of Modern Physics A* **9** (21): 3707–3750. arXiv:hep-th/9402002. Bibcode:1994IJMPA...9.3707S. doi:10.1142/S0217751X94001497.

- Sen, Ashoke (1994b). "Dyon-monopole bound states, self-dual harmonic forms on the multi-monopole moduli space, and $SL(2,\mathbf{Z})$ invariance in string theory". *Physics Letters B* **329** (2): 217–221. arXiv:hep-th/9402032. Bibcode:1994PhLB..329..217S. doi:10.1016/0370-2693(94)90763-3.

- Smolin, Lee (2006). *The Trouble with Physics: The Rise of String Theory, the Fall of a Science, and What Comes Next*. New York: Houghton Mifflin Co. ISBN 0-618-55105-0.

- Strominger, Andrew (1998). "Black hole entropy from near-horizon microstates". *Journal of High Energy Physics* **1998** (2): 009. arXiv:hep-th/9712251. Bibcode:1998JHEP...02..009S. doi:10.1088/1126-6708/1998/02/009.

- Strominger, Andrew; Vafa, Cumrun (1996). "Microscopic origin of the Bekenstein–Hawking entropy". *Physics Letters B* **379** (1): 99–104. arXiv:hep-th/9601029. Bibcode:1996PhLB..379...99S. doi:10.1016/0370-2693(96)00345-0.

- Strominger, Andrew; Yau, Shing-Tung; Zaslow, Eric (1996). "Mirror symmetry is T-duality". *Nuclear Physics B* **479** (1): 243–259. arXiv:hep-th/9606040. Bibcode:1996NuPhB.479..243S. doi:10.1016/0550-3213(96)00434-8.

- Susskind, Leonard (2005). *The Cosmic Landscape: String Theory and the Illusion of Intelligent Design*. Back Bay Books. ISBN 978-0316013338.

- Susskind, Leonard (2008). *The Black Hole War: My Battle with Stephen Hawking to Make the World Safe for Quantum Mechanics*. Little, Brown and Company. ISBN 978-0-316-01641-4.

- Wald, Robert (1984). *General Relativity*. University of Chicago Press. ISBN 978-0-226-87033-5.

- Weinberg, Steven (1987). *Anthropic bound on the cosmological constant* **59**. Physical Review Letters. p. 2607.

- Witten, Edward (1995). "String theory dynamics in various dimensions". *Nuclear Physics B* **443** (1): 85–126. arXiv:hep-th/9503124. Bibcode:1995NuPhB.443...85W. doi:10.1016/0550-3213(95)00158-O.

- Witten, Edward (1998). "Anti-de Sitter space and holography". *Advances in Theoretical and Mathematical Physics* **2**: 253–291. arXiv:hep-th/9802150. Bibcode:1998AdTMP...2..253W.

- Witten, Edward (2007). "Three-dimensional gravity revisited". arXiv:0706.3359 [hep-th].

- Woit, Peter (2006). *Not Even Wrong: The Failure of String Theory and the Search for Unity in Physical Law*. Basic Books. p. 105. ISBN 0-465-09275-6.

- Yau, Shing-Tung; Nadis, Steve (2010). *The Shape of Inner Space: String Theory and the Geometry of the Universe's Hidden Dimensions*. Basic Books. ISBN 978-0-465-02023-2.

- Zee, Anthony (2010). *Quantum Field Theory in a Nutshell* (2nd ed.). Princeton University Press. ISBN 978-0-691-14034-6.

- Zwiebach, Barton (2009). *A First Course in String Theory*. Cambridge University Press. ISBN 978-0-521-88032-9.

18.10 Further reading

18.10.1 Popularizations

General

- Greene, Brian (2003). *The Elegant Universe: Superstrings, Hidden Dimensions, and the Quest for the Ultimate Theory*. New York: W.W. Norton & Company. ISBN 0-393-05858-1.

- Greene, Brian (2004). *The Fabric of the Cosmos: Space, Time, and the Texture of Reality*. New York: Alfred A. Knopf. ISBN 0-375-41288-3.

Critical

- Penrose, Roger (2005). *The Road to Reality: A Complete Guide to the Laws of the Universe*. Knopf. ISBN 0-679-45443-8.

- Smolin, Lee (2006). *The Trouble with Physics: The Rise of String Theory, the Fall of a Science, and What Comes Next*. New York: Houghton Mifflin Co. ISBN 0-618-55105-0.

- Woit, Peter (2006). *Not Even Wrong: The Failure of String Theory And the Search for Unity in Physical Law*. London: Jonathan Cape &: New York: Basic Books. ISBN 978-0-465-09275-8.

18.10.2 Textbooks

For physicists

- Becker, Katrin; Becker, Melanie; Schwarz, John (2007). *String Theory and M-theory: A Modern Introduction*. Cambridge University Press. ISBN 978-0-521-86069-7.

- Green, Michael; Schwarz, John; Witten, Edward (2012). *Superstring theory. Vol. 1: Introduction*. Cambridge University Press. ISBN 978-1107029118.

- Green, Michael; Schwarz, John; Witten, Edward (2012). *Superstring theory. Vol. 2: Loop amplitudes, anomalies and phenomenology*. Cambridge University Press. ISBN 978-1107029132.

- Polchinski, Joseph (1998). *String Theory Vol. 1: An Introduction to the Bosonic String*. Cambridge University Press. ISBN 0-521-63303-6.

- Polchinski, Joseph (1998). *String Theory Vol. 2: Superstring Theory and Beyond*. Cambridge University Press. ISBN 0-521-63304-4.

- Zwiebach, Barton (2009). *A First Course in String Theory*. Cambridge University Press. ISBN 978-0-521-88032-9.

For mathematicians

- Deligne, Pierre; Etingof, Pavel; Freed, Daniel; Jeffery, Lisa; Kazhdan, David; Morgan, John; Morrison, David; Witten, Edward, eds. (1999). *Quantum Fields and Strings: A Course for Mathematicians, Vol. 2*. American Mathematical Society. ISBN 978-0821819883.

18.11 External links

- *The Elegant Universe*—A three-hour miniseries with Brian Greene by *NOVA* (original PBS Broadcast Dates: October 28, 8–10 p.m. and November 4, 8–9 p.m., 2003). Various images, texts, videos and animations explaining string theory.

- Not Even Wrong—A blog critical of string theory

- The Official String Theory Web Site

- Why String Theory—An introduction to string theory.

Chapter 19

String duality

This article is about string duality. For other forms of duality, see Duality (disambiguation).

String duality is a class of symmetries in physics that link different string theories, theories which assume that the fundamental building blocks of the universe are strings instead of point particles.

19.1 Overview

Before the so-called "duality revolution" there were believed to be five distinct versions of string theory, plus the (unstable) bosonic and gluonic theories.

Note that in the type IIA and type IIB string theories closed strings are allowed to move everywhere throughout the ten-dimensional space-time (called the *bulk*), while open strings have their ends attached to D-branes, which are membranes of lower dimensionality (their dimension is odd - 1,3,5,7 or 9 - in type IIA and even - 0,2,4,6 or 8 - in type IIB, including the time direction).

Before the 1990s, string theorists believed there were five distinct superstring theories: type I, types IIA and IIB, and the two heterotic string theories (SO(32) and $E_8 \times E_8$). The thinking was that out of these five candidate theories, only one was the actual theory of everything, and that theory was the theory whose low energy limit, with ten dimensions spacetime compactified down to four, matched the physics observed in our world today. It is now known that the five superstring theories are not fundamental, but are instead different limits of a more fundamental theory, dubbed M-theory. These theories are related by transformations called dualities. If two theories are related by a duality transformation, each observable of the first theory can be mapped in some way to the second theory to yield equivalent predictions. The two theories are then said to be dual to one another under that transformation. Put differently, the two theories are two mathematically different descriptions of the same phenomena. A simple example of a duality is

the equivalence of particle physics upon replacing matter with antimatter; describing our universe in terms of anti-particles would yield identical predictions for any possible experiment.

String dualities often link quantities that appear to be separate: Large and small distance scales, strong and weak coupling strengths. These quantities have always marked very distinct limits of behavior of a physical system, in both classical field theory and quantum particle physics. But strings can obscure the difference between large and small, strong and weak, and this is how these five very different theories end up being related.

19.2 T-duality

Main article: T-duality

Suppose we are in ten spacetime dimensions, which means we have nine space dimensions and one time. Take one of those nine space dimensions and make it a circle of radius R, so that traveling in that direction for a distance L = 2πR takes you around the circle and brings you back to where you started. A particle traveling around this circle will have a quantized momentum around the circle, because its momentum is linked to its wavelength (see Wave-particle duality), and 2πR must be a multiple of that. In fact, the particle momentum around the circle - and the contribution to its energy - is of the form n/R (in standard units, for an integer n), so that at large R there will be many more states compared to small R (for a given maximum energy). A string, in addition to traveling around the circle, may also wrap around it. The number of times the string winds around the circle is called the winding number, and that is also quantized (as it must be an integer). Winding around the circle requires energy, because the string must be stretched against its tension, so it contributes an amount of energy of the form wR/L_{st}^2, where L_{st} is a constant called the *string length* and w is the winding number (an integer). Now (for a given

151

maximum energy) there will be many different states (with different momenta) at large R, but there will also be many different states (with different windings) at small R. In fact, a theory with large R and a theory with small R are equivalent, where the role of momentum in the first is played by the winding in the second, and vice versa. Mathematically, taking R to L_{st}^2/R and switching n and w will yield the same equations. So exchanging momentum and winding modes of the string exchanges a large distance scale with a small distance scale.

This type of duality is called T-duality. T-duality relates type IIA superstring theory to type IIB superstring theory. That means if we take type IIA and Type IIB theory and compactify them both on a circle (one with a large radius and the other with a small radius) then switching the momentum and winding modes, and switching the distance scale, changes one theory into the other. The same is also true for the two heterotic theories. T-duality also relates type I superstring theory to both type IIA and type IIB superstring theories with certain boundary conditions (termed orientifold).

Formally, the location of the string on the circle is described by two fields living on it, one which is left-moving and another which is right-moving. The movement of the string center (and hence its momentum) is related to the sum of the fields, while the string stretch (and hence its winding number) is related to their difference. T-duality can be formally described by taking the left-moving field to minus itself, so that the sum and the difference are interchanged, leading to switching of momentum and winding.

19.3   S-duality

Main articles: S-duality and M-theory

Every force has a coupling constant, which is a measure of its strength, and determines the chances of one particle to emit or absorb another particle. For electromagnetism, the coupling constant is proportional to the square of the electric charge. When physicists study the quantum behavior of electromagnetism, they can't solve the whole theory exactly, because every particle may emit and absorb many other particles, which may also do the same, endlessly. So events of emission and absorption are considered as perturbations and are dealt with by a series of approximations, first assuming there is only one such event, then correcting the result for allowing two such events, etc. (this method is called Perturbation theory). This is a reasonable approximation only if the coupling constant is small, which is the case for electromagnetism. But if the coupling constant gets large, that method of calculation breaks down, and the lit-

tle pieces become worthless as an approximation to the real physics.

This also can happen in string theory. String theories have a coupling constant. But unlike in particle theories, the string coupling constant is not just a number, but depends on one of the oscillation modes of the string, called the dilaton. Exchanging the dilaton field with minus itself exchanges a very large coupling constant with a very small one. This symmetry is called S-duality. If two string theories are related by S-duality, then one theory with a strong coupling constant is the same as the other theory with weak coupling constant. The theory with strong coupling cannot be understood by means of perturbation theory, but the theory with weak coupling can. So if the two theories are related by S-duality, then we just need to understand the weak theory, and that is equivalent to understanding the strong theory.

Superstring theories related by S-duality are: type I superstring theory with heterotic SO(32) superstring theory, and type IIB theory with itself.

Furthermore, type IIA theory in strong coupling behaves like an 11-dimensional theory, with the dilaton field playing the role of an eleventh dimension. This 11-dimensional theory is known as M-theory.

Unlike the T-duality, however, S-duality has not been proven to even a physics level of rigor for any of the aforementioned cases. It remains, strictly speaking, a conjecture, although most string theorists believe in its validity.

19.4   See also

- U-duality

- Mirror Symmetry

Chapter 20

D-brane

In string theory, **D-branes** are a class of extended objects upon which open strings can end with Dirichlet boundary conditions, after which they are named. D-branes were discovered by Dai, Leigh and Polchinski, and independently by Hořava in 1989. In 1995, Polchinski identified D-branes with black p-brane solutions of supergravity, a discovery that triggered the Second Superstring Revolution and led to both holographic and M-theory dualities.

D-branes are typically classified by their spatial dimension, which is indicated by a number written after the D. A D0-brane is a single point, a D1-brane is a line (sometimes called a "D-string"), a D2-brane is a plane, and a D25-brane fills the highest-dimensional space considered in bosonic string theory. There are also instantonic $D(-1)$-branes, which are localized in both space and time.

20.1 Theoretical background

The equations of motion of string theory require that the endpoints of an open string (a string with endpoints) satisfy one of two types of boundary conditions: The Neumann boundary condition, corresponding to free endpoints moving through spacetime at the speed of light, or the Dirichlet boundary conditions, which pin the string endpoint. Each coordinate of the string must satisfy one or the other of these conditions. There can also exist strings with mixed boundary conditions, where the two endpoints satisfy NN, DD, ND and DN boundary conditions. If p spatial dimensions satisfy the Neumann boundary condition, then the string endpoint is confined to move within a p-dimensional hyperplane. This hyperplane provides one description of a Dp-brane.

Although rigid in the limit of zero coupling, the spectrum of open strings ending on a D-brane contains modes associated with its fluctuations, implying that D-branes are dynamical objects. When N D-branes are nearly coincident, the spectrum of strings stretching between them becomes very rich. One set of modes produce a non-abelian gauge theory on the world-volume. Another set of modes is an $N \times N$ dimensional matrix for each transverse dimension of the brane. If these matrices commute, they may be diagonalized, and the eigenvalues define the position of the N D-branes in space. More generally, the branes are described by non-commutative geometry, which allows exotic behavior such as the Myers effect, in which a collection of Dp-branes expand into a D(p+2)-brane.

Tachyon condensation is a central concept in this field. Ashoke Sen has argued that in Type IIB string theory, tachyon condensation allows (in the absence of Neveu-Schwarz 3-form flux) an arbitrary D-brane configuration to be obtained from a stack of D9 and anti D9-branes. Edward Witten has shown that such configurations will be classified by the K-theory of the spacetime. Tachyon condensation is still very poorly understood. This is due to the lack of an exact string field theory that would describe the off-shell evolution of the tachyon.

20.2 Braneworld cosmology

This has implications for physical cosmology. Because string theory implies that the Universe has more dimensions than we expect—26 for bosonic string theories and 10 for superstring theories—we have to find a reason why the extra dimensions are not apparent. One possibility would be that the visible Universe is in fact a very large D-brane extending over three spatial dimensions. Material objects, made of open strings, are bound to the D-brane, and cannot move "at right angles to reality" to explore the Universe outside the brane. This scenario is called a brane cosmology. The force of gravity is *not* due to open strings; the gravitons which carry gravitational forces are vibrational states of *closed* strings. Because closed strings do not have to be attached to D-branes, gravitational effects could depend upon the extra dimensions orthogonal to the brane.

20.3   D-brane scattering

When two D-branes approach each other the interaction is captured by the one loop annulus amplitude of strings between the two branes. The scenario of two parallel branes approaching each other at a constant velocity can be mapped to the problem of two stationary branes that are rotated relative to each other by some angle. The annulus amplitude yields singularities that correspond to the on-shell production of open strings stretched between the two branes. This is true irrespective of the charge of the D-branes. At non-relativistic scattering velocities the open strings may be described by a low-energy effective action that contains two complex scalar fields that are coupled via a term $\phi^2\chi^2$. Thus, as the field ϕ (separation of the branes) changes, the mass of the field χ changes. This induces open string production and as a result the two scattering branes will be trapped.

20.4   Gauge theories

The arrangement of D-branes constricts the types of string states which can exist in a system. For example, if we have two parallel D2-branes, we can easily imagine strings stretching from brane 1 to brane 2 or vice versa. (In most theories, strings are *oriented* objects: each one carries an "arrow" defining a direction along its length.) The open strings permissible in this situation then fall into two categories, or "sectors": those originating on brane 1 and terminating on brane 2, and those originating on brane 2 and terminating on brane 1. Symbolically, we say we have the [1 2] and the [2 1] sectors. In addition, a string may begin and end on the same brane, giving [1 1] and [2 2] sectors. (The numbers inside the brackets are called *Chan-Paton indices*, but they are really just labels identifying the branes.) A string in either the [1 2] or the [2 1] sector has a minimum length: it cannot be shorter than the separation between the branes. All strings have some tension, against which one must pull to lengthen the object; this pull does work on the string, adding to its energy. Because string theories are by nature relativistic, adding energy to a string is equivalent to adding mass, by Einstein's relation $E = mc^2$. Therefore, the separation between D-branes controls the minimum mass open strings may have.

Furthermore, affixing a string's endpoint to a brane influences the way the string can move and vibrate. Because particle states "emerge" from the string theory as the different vibrational states the string can experience, the arrangement of D-branes controls the types of particles present in the theory. The simplest case is the [1 1] sector for a Dp-brane, that is to say the strings which begin and end on any particular D-brane of p dimensions. Examining the conse-

quences of the Nambu-Goto action (and applying the rules of quantum mechanics to quantize the string), one finds that among the spectrum of particles is one resembling the photon, the fundamental quantum of the electromagnetic field. The resemblance is precise: a p-dimensional version of the electromagnetic field, obeying a p-dimensional analogue of Maxwell's equations, exists on every Dp-brane.

In this sense, then, one can say that string theory "predicts" electromagnetism: D-branes are a necessary part of the theory if we permit open strings to exist, and all D-branes carry an electromagnetic field on their volume.

Other particle states originate from strings beginning and ending on the same D-brane. Some correspond to massless particles like the photon; also in this group are a set of massless scalar particles. If a Dp-brane is embedded in a spacetime of d spatial dimensions, the brane carries (in addition to its Maxwell field) a set of d - p massless scalars (particles which do not have polarizations like the photons making up light). Intriguingly, there are just as many massless scalars as there are directions perpendicular to the brane; the *geometry* of the brane arrangement is closely related to the *quantum field theory* of the particles existing on it. In fact, these massless scalars are Goldstone excitations of the brane, corresponding to the different ways the symmetry of empty space can be broken. Placing a D-brane in a universe breaks the symmetry among locations, because it defines a particular place, assigning a special meaning to a particular location along each of the d - p directions perpendicular to the brane.

The quantum version of Maxwell's electromagnetism is only one kind of gauge theory, a $U(1)$ gauge theory where the gauge group is made of unitary matrices of order 1. D-branes can be used to generate gauge theories of higher order, in the following way:

Consider a group of N separate Dp-branes, arranged in parallel for simplicity. The branes are labeled 1,2,...,N for convenience. Open strings in this system exist in one of many sectors: the strings beginning and ending on some brane i give that brane a Maxwell field and some massless scalar fields on its volume. The strings stretching from brane i to another brane j have more intriguing properties. For starters, it is worthwhile to ask which sectors of strings can interact with one another. One straightforward mechanism for a string interaction is for two strings to join endpoints (or, conversely, for one string to "split down the middle" and make two "daughter" strings). Since endpoints are restricted to lie on D-branes, it is evident that a [1 2] string may interact with a [2 3] string, but not with a [3 4] or a [4 17] one. The masses of these strings will be influenced by the separation between the branes, as discussed above, so for simplicity's sake we can imagine the branes squeezed closer and closer together, until they lie atop one another. If we regard two overlapping branes as distinct objects, then we still have all the sectors we had before, but without the effects due to the brane separations.

The zero-mass states in the open-string particle spectrum for a system of N coincident D-branes yields a set of interacting quantum fields which is exactly a $U(N)$ gauge theory. (The string theory does contain other interactions, but they are only detectable at very high energies.) Gauge theories were not invented starting with bosonic or fermionic strings; they originated from a different area of physics, and have become quite useful in their own right. If nothing else, the relation between D-brane geometry and gauge theory offers a useful pedagogical tool for explaining gauge interactions, even if string theory fails to be the "theory of everything".

20.5   Black holes

Another important use of D-branes has been in the study of black holes. Since the 1970s, scientists have debated the problem of black holes having entropy. Consider, as a thought experiment, dropping an amount of hot gas into a black hole. Since the gas cannot escape from the hole's gravitational pull, its entropy would seem to have vanished from the universe. In order to maintain the second law of thermodynamics, one must postulate that the black hole gained whatever entropy the infalling gas originally had. Attempting to apply quantum mechanics to the study of black holes, Stephen Hawking discovered that a hole should emit energy with the characteristic spectrum of thermal radiation. The characteristic temperature of this Hawking radiation is given by

$$T_{\mathrm{H}} = \frac{\hbar c^3}{8\pi G M k_B} \quad \left(\approx \frac{1.227 \times 10^{23}\ kg}{M}\ K\right)$$

where G is Newton's gravitational constant, M is the black hole's mass and kB is Boltzmann's constant.

Using this expression for the Hawking temperature, and assuming that a zero-mass black hole has zero entropy, one can use thermodynamic arguments to derive the "Bekenstein entropy":

$$S_{\mathrm{B}} = \frac{k_B 4\pi G}{\hbar c} M^2.$$

The Bekenstein entropy is proportional to the black hole mass squared; because the Schwarzschild radius is proportional to the mass, the Bekenstein entropy is proportional to the black hole's *surface area*. In fact,

$$S_{\mathrm{B}} = \frac{A k_B}{4 l_{\mathrm{P}}^2},$$

where l_{P} is the Planck length.

The concept of black hole entropy poses some interesting conundra. In an ordinary situation, a system has entropy when a large number of different "microstates" can satisfy the same macroscopic condition. For example, given a box full of gas, many different arrangements of the gas atoms can have the same total energy. However, a black hole was believed to be a featureless object (in John Wheeler's catchphrase, "Black holes have no hair"). What, then, are the "degrees of freedom" which can give rise to black hole entropy?

String theorists have constructed models in which a black hole is a very long (and hence very massive) string. This model gives rough agreement with the expected entropy of a Schwarzschild black hole, but an exact proof has yet to be found one way or the other. The chief difficulty is that it is relatively easy to count the degrees of freedom quantum strings possess *if they do not interact with one another*. This is analogous to the ideal gas studied in introductory thermodynamics: the easiest situation to model is when the gas atoms do not have interactions among themselves. Developing the kinetic theory of gases in the case where the gas atoms or molecules experience inter-particle forces (like the van der Waals force) is more difficult. However, a world without interactions is an uninteresting place: most significantly for the black hole problem, gravity is an interaction, and so if the "string coupling" is turned off, no black hole could ever arise. Therefore, calculating black hole entropy requires working in a regime where string interactions exist.

Extending the simpler case of non-interacting strings to the regime where a black hole could exist requires supersymmetry. In certain cases, the entropy calculation done for zero string coupling remains valid when the strings interact. The challenge for a string theorist is to devise a situation in which a black hole can exist which does not "break" supersymmetry. In recent years, this has been done by building black holes out of D-branes. Calculating the entropies of these hypothetical holes gives results which agree with the expected Bekenstein entropy. Unfortunately, the cases studied so far all involve higher-dimensional spaces — D5-branes in nine-dimensional space, for example. They do not directly apply to the familiar case, the Schwarzschild black holes observed in our own universe.

20.6   History

Dirichlet boundary conditions and D-branes had a long "pre-history" before their full significance was recognized. Mixed Dirichlet/Neumann boundary conditions were first considered by Warren Siegel in 1976 as a means of lowering the critical dimension of open string theory from 26 or

10 to 4 (Siegel also cites unpublished work by Halpern, and a 1974 paper by Chodos and Thorn, but a reading of the latter paper shows that it is actually concerned with linear dilation backgrounds, not Dirichlet boundary conditions). This paper, though prescient, was little-noted in its time (a 1985 parody by Siegel, "The Super-g String," contains an almost dead-on description of braneworlds). Dirichlet conditions for all coordinates including Euclidean time (defining what are now known as D-instantons) were introduced by Michael Green in 1977 as a means of introducing pointlike structure into string theory, in an attempt to construct a string theory of the strong interaction. String compactifications studied by Harvey and Minahan, Ishibashi and Onogi, and Pradisi and Sagnotti in 1987-89 also employed Dirichlet boundary conditions.

The fact that T-duality interchanges the usual Neumann boundary conditions with Dirichlet boundary conditions was discovered independently by Horava and by Dai, Leigh, and Polchinski in 1989; this result implies that such boundary conditions must necessarily appear in regions of the moduli space of any open string theory. The Dai et al. paper also notes that the locus of the Dirichlet boundary conditions is dynamical, and coins the term Dirichlet-brane (D-brane) for the resulting object (this paper also coins orientifold for another object that arises under string T-duality). A 1989 paper by Leigh showed that D-brane dynamics are governed by the Dirac-Born-Infeld action. D-instantons were extensively studied by Green in the early 1990s, and were shown by Polchinski in 1994 to produce the $e^{-1/g}$ nonperturbative string effects anticipated by Shenker. In 1995 Polchinski showed that D-branes are the sources of electric and magnetic Ramond–Ramond fields that are required by string duality, leading to rapid progress in the nonperturbative understanding of string theory.

20.7   See also

- Bogomol'nyi–Prasad–Sommerfield bound

- M-theory

20.8   References

- Bachas, C. P. "Lectures on D-branes" (1998). arXiv:hep-th/9806199.

- Giveon, A. and Kutasov, D. "Brane dynamics and gauge theory", *Rev. Mod. Phys.* **71**, 983 (1999). arXiv:hep-th/9802067.

- Hashimoto, Koji, *D-Brane: Superstrings and New Perspective of Our World.* Springer (2012). ISBN 978-3-642-23573-3

- Johnson, Clifford (2003). *D-branes.* Cambridge: Cambridge University Press. ISBN 0-521-80912-6.

- Polchinski, Joseph, *TASI Lectures on D-branes*, arXiv:hep-th/9611050. Lectures given at TASI '96.

- Polchinski, Joseph, *Phys. Rev. Lett.* **75**, 4724 (1995). An article which established D-branes' significance in string theory.

- Zwiebach, Barton. *A First Course in String Theory.* Cambridge University Press (2004). ISBN 0-521-83143-1.

Chapter 21

Brane

For other uses, see Brane (disambiguation).

In string theory and related theories such as supergravity theories, a **brane** is a physical object that generalizes the notion of a point particle to higher dimensions. For example, a point particle can be viewed as a brane of dimension zero, while a string can be viewed as a brane of dimension one. It is also possible to consider higher-dimensional branes. In dimension p, these are called p-branes. The word "brane" comes from the word "membrane" which refers to a two-dimensional brane.[1]

Branes are dynamical objects which can propagate through spacetime according to the rules of quantum mechanics. They have mass and can have other attributes such as charge. A p-brane sweeps out a $(p+1)$-dimensional volume in spacetime called its *worldvolume*. Physicists often study fields analogous to the electromagnetic field, which live on the worldvolume of a brane.[2]

In string theory, D-branes are an important class of branes that arise when one considers open strings. As an open string propagates through spacetime, its endpoints are required to lie on a D-brane. The letter "D" in D-brane refers to a certain mathematical condition on the system known as the Dirichlet boundary condition. The study of D-branes in string theory has led to important results such as the AdS/CFT correspondence, which has shed light on many problems in quantum field theory.

Branes are also frequently studied from a purely mathematical point of view since they are related to subjects such as homological mirror symmetry and noncommutative geometry. Mathematically, branes may be represented as objects of certain categories, such as the derived category of coherent sheaves on a Calabi–Yau manifold, or the Fukaya category.

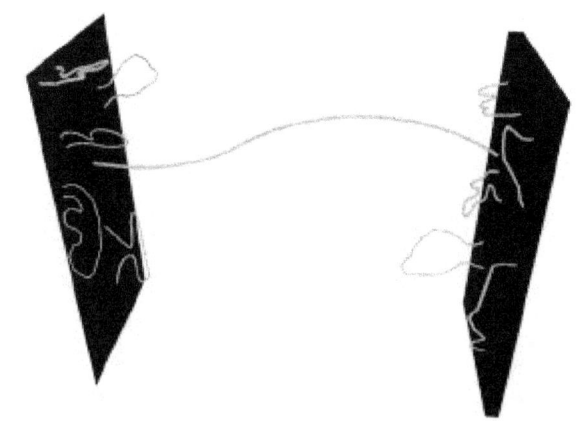

Open strings attached to a pair of D-branes

21.1 D-branes

Main article: D-brane

In string theory, a string may be open (forming a segment with two endpoints) or closed (forming a closed loop). D-branes are an important class of branes that arise when one considers open strings. As an open string propagates through spacetime, its endpoints are required to lie on a D-brane. The letter "D" in D-brane refers to a condition that it satisfies, the Dirichlet boundary condition.[3]

One crucial point about D-branes is that the dynamics on the D-brane worldvolume is described by a gauge theory, a kind of highly symmetric physical theory which is also used to describe the behavior of elementary particles in the standard model of particle physics. This connection has led to many important insights into gauge theory. For example, it led to the discovery of the AdS/CFT correspondence, a theoretical tool that physicists use to translate difficult problems in gauge theory into more mathematically tractable problems in string theory.[4]

21.2 Mathematical viewpoint

Mathematically, branes can be described using the notion of a category.[5] This is a mathematical structure consisting of *objects*, and for any pair of objects, a set of *morphisms* between them. In most examples, the objects are mathematical structures (such as sets, vector spaces, or topological spaces) and the morphisms are functions between these structures.[6] One can also consider categories where the objects are D-branes and the morphisms between two branes α and β are states of open strings stretched between α and β.[7]

A cross section of a Calabi–Yau manifold

In one version of string theory known as the topological B-model, the D-branes are complex submanifolds of certain six-dimensional shapes called Calabi–Yau manifolds, together with additional data that arise physically from having charges at the endpoints of strings.[8] Intuitively, one can think of a submanifold as a surface embedded inside of a Calabi–Yau manifold, although submanifolds can also exist in dimensions different from two.[9] In mathematical language, the category having these branes as its objects is known as the derived category of coherent sheaves on the Calabi–Yau.[10] In another version of string theory called the topological A-model, the D-branes can again be viewed as submanifolds of a Calabi–Yau manifold. Roughly speaking, they are what mathematicians call special Lagrangian submanifolds.[11] This means among other things that they have half the dimension of the space in which they sit, and they are length-, area-, or volume-minimizing.[12] The category having these branes as its objects is called the Fukaya category.[13]

The derived category of coherent sheaves is constructed using tools from complex geometry, a branch of mathematics that describes geometric curves in algebraic terms and solves geometric problems using algebraic equations.[14] On the other hand, the Fukaya category is constructed using symplectic geometry, a branch of mathematics that arose from studies of classical physics. Symplectic geometry studies spaces equipped with a symplectic form, a mathematical tool that can be used to compute area in two-dimensional examples.[15]

The homological mirror symmetry conjecture of Maxim Kontsevich states that the derived category of coherent sheaves on one Calabi–Yau manifold is equivalent in a certain sense to the Fukaya category of a completely different Calabi–Yau manifold.[16] This equivalence provides an unexpected bridge between two branches of geometry, namely complex and symplectic geometry.[17]

21.3 See also

- Black brane
- Brane cosmology
- Dirac membrane
- M2-brane
- M5-brane
- NS5-brane

21.4 Notes

[1] Moore 2005, p. 214

[2] Moore 2005, p. 214

[3] Moore 2005, p. 215

[4] Moore 2005, p. 215

[5] Aspinwall et al. 2009

[6] A basic reference on category theory is Mac Lane 1998.

[7] Zaslow 2008, p. 536

[8] Zaslow 2008, p. 536

[9] Yau and Nadis 2010, p. 165

[10] Aspinwal et al. 2009, p. 575

[11] Aspinwal et al. 2009, p. 575

[12] Yau and Nadis 2010, p. 175

[13] Aspinwal et al. 2009, p. 575

[14] Yau and Nadis 2010, pp. 180–1

[15] Zaslow 2008, p. 531

[16] Aspinwall et al. 2009, p. 616

[17] Yau and Nadis 2010, p. 181

21.5   References

- Aspinwall, Paul; Bridgeland, Tom; Craw, Alastair; Douglas, Michael; Gross, Mark; Kapustin, Anton; Moore, Gregory; Segal, Graeme; Szendröi, Balázs; Wilson, P.M.H., eds. (2009). *Dirichlet Branes and Mirror Symmetry*. American Mathematical Society. ISBN 978-0-8218-3848-8.

- Mac Lane, Saunders (1998). *Categories for the Working Mathematician*. ISBN 978-0-387-98403-2.

- Moore, Gregory (2005). "What is ... a Brane?" (PDF). *Notices of the AMS* 52: 214. Retrieved June 2013.

- Yau, Shing-Tung; Nadis, Steve (2010). *The Shape of Inner Space: String Theory and the Geometry of the Universe's Hidden Dimensions*. Basic Books. ISBN 978-0-465-02023-2.

- Zaslow, Eric (2008). "Mirror Symmetry". In Gowers, Timothy. *The Princeton Companion to Mathematics*. ISBN 978-0-691-11880-2.

Chapter 22

Perturbation theory

This article is about perturbation theory as a general mathematical method. For perturbation theory as applied to quantum mechanics, see Perturbation theory (quantum mechanics).

Perturbation theory comprises mathematical methods for finding an approximate solution to a problem, by starting from the exact solution of a related, simpler problem. A critical feature of the technique is a middle step that breaks the problem into "solvable" and "perturbation" parts.[1] Perturbation theory is applicable if the problem at hand cannot be solved exactly, but can be formulated by adding a "small" term to the mathematical description of the exactly solvable problem.

Perturbation theory leads to an expression for the desired solution in terms of a formal power series in some "small" parameter – known as a **perturbation series** – that quantifies the deviation from the exactly solvable problem. The leading term in this power series is the solution of the exactly solvable problem, while further terms describe the deviation in the solution, due to the deviation from the initial problem. Formally, we have for the approximation to the full solution A, a series in the small parameter (here called ε), like the following:

$$A = A_0 + \varepsilon^1 A_1 + \varepsilon^2 A_2 + \cdots$$

In this example, A_0 would be the known solution to the exactly solvable initial problem and A_1, A_2, ... represent the **higher-order terms** which may be found iteratively by some systematic procedure. For small ε these higher-order terms in the series become successively smaller.

An approximate "perturbation solution" is obtained by truncating the series, usually by keeping only the first two terms, the initial solution and the "first-order" perturbation correction

$$A \approx A_0 + \varepsilon A_1 .$$

22.1 General description

Perturbation theory is closely related to methods used in numerical analysis. The earliest use of what would now be called *perturbation theory* was to deal with the otherwise unsolvable mathematical problems of celestial mechanics: for example the orbit of the Moon, which moves noticeably differently from a simple Keplerian ellipse because of the competing gravitation of the Earth and the Sun.

Perturbation methods start with a simplified form of the original problem, which is *simple enough* to be solved exactly. In celestial mechanics, this is usually a Keplerian ellipse. Under non-relativistic gravity, an ellipse is exactly correct when there are only two gravitating bodies (say, the Earth and the Moon) but not quite correct when there are three or more objects (say, the Earth, Moon, Sun, and the rest of the solar system) and not quite correct when the gravitational interaction is stated using formulas from General relativity.

The solved, but simplified problem is then *"perturbed"* to make the conditions that the perturbed solution actually satisfies closer to the real problem, such as including the gravitational attraction of a third body (the Sun). The "conditions" are a formula (or several) that represent reality, often something arising from a physical law like Newton's second law, the force-acceleration equation,

$$\mathbf{F} = m\mathbf{a} .$$

In the case of the example, the force **F** is calculated based on the number of gravitationally relevant bodies; the acceleration a is obtained, using calculus, from the path of the Moon in its orbit. Both of these come in two forms: approximate values for force and acceleration, which result from simplifications, and hypothetical exact values for force and acceleration, which would require the complete answer to calculate.

The slight changes that result from accommodating the perturbation, which themselves may have been simplified yet again, are used as corrections to the approximate solution. Because of simplifications introduced along every step of the way, the corrections are never perfect, and the conditions met by the corrected solution do not perfectly match the equation demanded by reality. However, even only one cycle of corrections often provides an excellent approximate answer to what the real solution should be.

There is no requirement to stop at only one cycle of corrections. A partially corrected solution can be re-used as the new starting point for yet another cycle of perturbations and corrections. In principle, cycles of finding increasingly better corrections could go on indefinitely. In practice, one typically stops at one or two cycles of corrections. The usual difficulty with the method is that the corrections progressively make the new solutions very much more complicated, so each cycle is much more difficult to manage than the previous cycle of corrections. Isaac Newton is reported to have said, regarding the problem of the Moon's orbit, that *"It causeth my head to ache."*[2]

This general procedure is a widely used mathematical tool in advanced sciences and engineering: start with a simplified problem and gradually add corrections that make the formula that the corrected problem matches closer and closer to the formula that represents reality. It is the natural extension to mathematical functions of the "guess, check, and fix" method used by older civilisations to compute certain numbers, such as square roots.

22.2 Examples

Examples for the "mathematical description" are: an algebraic equation, a differential equation (e.g., the equations of motion in celestial mechanics or a wave equation), a free energy (in statistical mechanics), a Hamiltonian operator (in quantum mechanics).

Examples for the kind of solution to be found perturbatively: the solution of the equation (e.g., the trajectory of a particle), the statistical average of some physical quantity (e.g., average magnetization), the ground state energy of a quantum mechanical problem.

Examples for the exactly solvable problems to start with: linear equations, including linear equations of motion (harmonic oscillator, linear wave equation), statistical or quantum-mechanical systems of non-interacting particles (or in general, Hamiltonians or free energies containing only terms quadratic in all degrees of freedom).

Examples of "perturbations" to deal with: Nonlinear contributions to the equations of motion, interactions between particles, terms of higher powers in the Hamiltonian/Free Energy.

For physical problems involving interactions between particles, the terms of the perturbation series may be displayed (and manipulated) using Feynman diagrams.

22.3 History

Perturbation theory was first devised to solve otherwise intractable problems in the calculation of the motions of planets in the solar system. The gradually increasing accuracy of astronomical observations led to incremental demands in the accuracy of solutions to Newton's gravitational equations, which led several notable 18th and 19th century mathematicians to extend and generalize the methods of perturbation theory. These well-developed perturbation methods were adopted and adapted to solve new problems arising during the development of Quantum Mechanics in 20th century atomic and subatomic physics.

22.3.1 Beginnings in the study of planetary motion

Since the planets are very remote from each other, and since their mass is small as compared to the mass of the Sun, the gravitational forces between the planets can be neglected, and the planetary motion is considered, to a first approximation, as taking place along Kepler's orbits, which are defined by the equations of the two-body problem, the two bodies being the planet and the Sun.[3]

Since astronomic data came to be known with much greater accuracy, it became necessary to consider how the motion of a planet around the Sun is affected by other planets. This was the origin of the three-body problem; thus, in studying the system Moon–Earth–Sun the mass ratio between the Moon and the Earth was chosen as the small parameter. Lagrange and Laplace were the first to advance the view that the constants which describe the motion of a planet around the Sun are "perturbed", as it were, by the motion of other planets and vary as a function of time; hence the name "perturbation theory".[3]

Perturbation theory was investigated by the classical scholars — Laplace, Poisson, Gauss — as a result of which the computations could be performed with a very high accuracy. The discovery of the planet Neptune in 1848 by Urbain Le Verrier, based on the deviations in motion of the planet Uranus (he sent the coordinates to Johann Gottfried Galle who successfully observed Neptune through his telescope), represented a triumph of perturbation theory.[3]

22.3.2 Rise of understanding of chaotic systems

The development of basic perturbation theory for differential equations was fairly complete by the middle of the 19th century. It was at that time that Charles-Eugène Delaunay was studying the perturbative expansion for the Earth-Moon-Sun system, and discovered the so-called "problem of small denominators". Here, the denominator appearing in the n-th term of the perturbative expansion could become arbitrarily small, causing the n-th correction to be as large or larger than the first-order correction.

At the turn of the 20th century, this problem led Henri Poincaré to make one of the first deductions of the existence of chaos, and what is poetically called the "butterfly effect": that even a very small perturbation can ultimately have a very large effect on non-dissipative or "friction-free" dynamic systems.

A partial resolution of the small-divisor problem was given by the statement of the KAM theorem in 1954. Developed by Andrey Kolmogorov, Vladimir Arnold and Jürgen Moser, this theorem stated the conditions under which a system of partial differential equations will have only mildly chaotic behaviour under small perturbations.

22.3.3 Application to new problems in 20th century physics

Perturbation theory saw a particularly dramatic expansion and evolution with the arrival of quantum mechanics. Although perturbation theory was used in the semi-classical theory of the Bohr atom, the calculations were monstrously complicated, and subject to somewhat ambiguous interpretation. The discovery of Heisenberg's matrix mechanics allowed a vast simplification of the application of perturbation theory. Notable examples are the Stark effect and the Zeeman effect, which have a simple enough theory to be included in standard undergraduate textbooks in quantum mechanics. Other early applications include the fine structure and the hyperfine structure in the hydrogen atom.

In modern times, perturbation theory underlies much of quantum chemistry and quantum field theory. In chemistry, perturbation theory was used to obtain the first solutions for the helium atom.

In the middle of the 20th century, Richard Feynman realized that the perturbative expansion could be given a dramatic and beautiful graphical representation in terms of what are now called Feynman diagrams. Although originally applied only in quantum field theory, such diagrams now find increasing use in any area where perturbative expansions are studied.

22.3.4 Search for better methods for quantum mechanics

In the late 20th century, broad dissatisfaction with perturbation theory in the quantum physics community, including not only the difficulty of going beyond second order in the expansion, but also questions about whether the perturbative expansion is even convergent, has led to a strong interest in the area of non-perturbative analysis, that is, the study of exactly solvable models.

Much of the theoretical work in non-perturbative analysis goes under the name of quantum groups and non-commutative geometry. The prototypical model is the Korteweg–de Vries equation, a highly non-linear equation for which the interesting solutions, the solitons, cannot be reached by perturbation theory, even if the perturbations were carried out to infinite order.

22.4 Perturbation orders

The standard exposition of perturbation theory is given in terms of the *order* to which the perturbation is carried out: **first-order perturbation theory** or **second-order perturbation theory**, and whether the perturbed states are degenerate, which requires **singular perturbation**. In the singular case extra care must be taken, and the theory is slightly more elaborate.

22.4.1 First-order, non-singular perturbation theory

This section develops, in simple terms,[4] the general theory for the perturbative solution to a differential equation to the first order. To keep the exposition simple, a crucial assumption is made: that the solutions to the unperturbed system are not *degenerate*, so that the perturbation series can be inverted. There are ways of dealing with the degenerate (or *singular*) case; these require extra care.

Suppose one wants to solve a differential equation of the form

$$Dg(x) = \lambda g(x) \,,$$

where D is some specific differential operator, and λ is an eigenvalue. Many problems involving ordinary or partial differential equations can be cast in this form.

It is presumed that the differential operator can be written in the form

$$D = D^{(0)} + \varepsilon D^{(1)}$$

where ε is presumed to be small, and that, furthermore, the complete set of solutions for $D^{(0)}$ are known.

That is, one has a set of solutions $f_n^{(0)}(x)$, labelled by some arbitrary index n, such that

$$D^{(0)} f_n^{(0)}(x) = \lambda_n^{(0)} f_n^{(0)}(x).$$

Furthermore, one assumes that the set of solutions $\{f_n^{(0)}(x)\}$ form an orthonormal set,

$$\int f_m^{(0)}(x) f_n^{(0)}(x)\, dx = \delta_{mn}$$

with δ_{mn} the Kronecker delta function.

To zeroth order, one expects that the solutions $g(x)$ are then somehow "close" to one of the unperturbed solutions $f_n^{(0)}(x)$. That is,

$$g(x) = f_n^{(0)}(x) + \mathcal{O}(\varepsilon)$$

and

$$\lambda = \lambda_n^{(0)} + \mathcal{O}(\varepsilon).$$

where $\mathcal{O}$ denotes the relative size, in big-O notation, of the perturbation.

To solve this problem, one assumes that the solution $g(x)$ can be written as a linear combination of the $f_n^{(0)}(x)$,

$$g(x) = \sum_m c_m f_m^{(0)}(x)$$

with all of the constants $c_m = \mathcal{O}(\varepsilon)$ except for n, where $c_n = \mathcal{O}(1)$.

Substituting this last expansion into the differential equation, taking the inner product of the result with $f_n^{(0)}(x)$, and making use of orthogonality, one obtains

$$c_n \lambda_n^{(0)} + \varepsilon \sum_m c_m \int f_n^{(0)}(x) D^{(1)} f_m^{(0)}(x)\, dx = \lambda c_n .$$

This can be trivially rewritten as a simple linear algebra problem of finding the eigenvalue of a matrix, where

$$\sum_m A_{nm} c_m = \lambda c_n$$

where the matrix elements A_{nm} are given by

$$A_{nm} = \delta_{nm} \lambda_n^{(0)} + \varepsilon \int f_n^{(0)}(x) D^{(1)} f_m^{(0)}(x)\, dx .$$

Rather than solving this full matrix equation, one notes that, of all the c_m in the linear equation, only one, namely c_n, is not small. Thus, to the *first order* in ε, the linear equation may be solved trivially as

$$\lambda = \lambda_n^{(0)} + \varepsilon \int f_n^{(0)}(x) D^{(1)} f_n^{(0)}(x)\, dx$$

since all of the other terms in the linear equation are of order $\mathcal{O}(\varepsilon^2)$. The above gives the solution of the eigenvalue to first order in perturbation theory.

The function $g(x)$ to first order is obtained through similar reasoning. Substituting

$$g(x) = f_n^{(0)}(x) + \varepsilon f_n^{(1)}(x)$$

so that

$$\left(D^{(0)} + \varepsilon D^{(1)}\right)\left(f_n^{(0)}(x) + \varepsilon f_n^{(1)}(x)\right) = \left(\lambda_n^{(0)} + \varepsilon \lambda_n^{(1)}\right)\left(f_n^{(0)}(x) + \varepsilon f_n^{(1)}\right.$$

gives an equation for $f_n^{(1)}(x)$.

It may be solved integrating with the partition of unity

$$\delta(x - y) = \sum_n f_n^{(0)}(x) f_n^{(0)}(y)$$

to give

$$f_n^{(1)}(x) = \sum_{m\,(\neq n)} \frac{f_m^{(0)}(x)}{\lambda_n^{(0)} - \lambda_m^{(0)}} \int f_m^{(0)}(y) D^{(1)} f_n^{(0)}(y)\, dy$$

which finally gives the exact solution to the perturbed differential equation to first order in the perturbation ε.

Several observations may be made about the form of this solution. First, the sum over functions with differences of eigenvalues in the denominator evokes the resolvent in Fredholm theory. This is no accident; the resolvent acts essentially as a kind of Green's function or propagator, passing the perturbation along. Higher-order perturbations resemble this form, with an additional sum over a resolvent appearing at each order.

The form of this solution also illustrates the idea behind the small-divisor problem. If, for whatever reason, two eigenvalues are close, so that the difference $\lambda_n^{(0)} - \lambda_m^{(0)}$ becomes small, the corresponding term in the above sum will become disproportionately large. In particular, if this happens in higher-order terms, the higher-order perturbation may become as large or larger in magnitude than the first-order perturbation. Such a situation calls into question the validity of utilizing a perturbative analysis to begin with, which can be understood to be a fairly catastrophic situation; it is frequently encountered in chaotic dynamical systems, and requires the development of techniques other than perturbation theory to solve the problem.

Curiously, the situation is not at all bad if two or more eigenvalues are *exactly equal*. This case is referred to as singular or degenerate perturbation theory, addressed below. The degeneracy of eigenvalues indicates that the unperturbed system has some sort of symmetry, and that the generators of that symmetry commute with the unperturbed differential operator. Typically, the perturbing term does not possess the symmetry, and so the full solutions do not, either; one says that the perturbation *lifts* or *breaks* the degeneracy. In this case, the perturbation can still be performed, as in following sections; however, care must be taken to work in a basis for the unperturbed states, so that these map one-to-one to the perturbed states, rather than being a mixture.

22.4.2   Perturbation theory of degenerate states

One may note that a problem occurs in the above first order perturbation theory when two or more eigenfunctions of the unperturbed system correspond to the same eigenvalue, i.e. when the eigenvalue equation becomes

$$D^{(0)} f_{n,i}^{(0)}(x) = \lambda_n^{(0)} f_{n,i}^{(0)}(x) \, ,$$

and the index i labels *several states with the same eigenvalue* $\lambda_n^{(0)}$. The expression for the eigenfunctions which has energy differences in the denominators becomes infinite. In that case, degenerate perturbation theory must be applied.

The degeneracy must first be removed for higher order perturbation theory. First, consider the eigenfunction which is a linear combination of eigenfunctions with the same eigenvalue only,

$$g(x) = \sum_k c_{n,k} f_{n,k}^{(0)}(x) \, ,$$

which, again from the orthogonality of $f_{n,k}^{(0)}$, leads to the following equation,

$$c_{n,i}\lambda_n^{(0)} + \varepsilon \sum_k c_{n,k} \int f_{n,i}^{(0)}(x) D^{(1)} f_{n,k}^{(0)}(x)\, dx = \lambda c_{n,i}$$

for each n.

As for the majority of low quantum numbers n, i changes over a *small range of integers*, so often the later equation can be solved analytically as an at most 4×4 matrix equation. Once the degeneracy is removed, the first and any order of the above perturbation theory may be further applied relying on the new eigenfunctions.

22.4.3   An example of second-order singular perturbation theory

Main article: Singular perturbation

Consider the following equation for the unknown variable x:

$$x = 1 + \varepsilon x^5 .$$

For the initial problem with $\varepsilon = 0$, the solution is $x_0 = 1$. For small ε the lowest-order approximation may be found by inserting the ansatz

$$x = x_0 + \varepsilon x_1 + \cdots$$

into the equation and demanding the equation to be fulfilled up to terms that involve powers of ε higher than the first. This yields $x_1 = 1$. In the same way, the higher orders may be found. However, even in this simple example it may be observed that for (arbitrarily) small positive ε there are four other solutions to the equation (with very large magnitude). The reason we don't find these solutions in the above perturbation method is because these solutions diverge when $\varepsilon \to 0$ while the ansatz assumes regular behavior in this limit.

The four additional solutions can be found using the methods of singular perturbation theory. In this case this works as follows. Since the four solutions diverge at $\varepsilon = 0$, it makes sense to rescale x. We put

$$x = y\varepsilon^{-\nu}$$

such that in terms of y the solutions stay finite. This means that we need to choose the exponent ν to match the rate at which the solutions diverge. In terms of y the equation reads:

$$\varepsilon^{-\nu} y = 1 + \varepsilon^{1-5\nu} y^5$$

The 'right' value for ν is obtained when the exponent of ε in the prefactor of the term proportional to y is equal to the exponent of ε in the prefactor of the term proportional to y^5, i.e. when $\nu = 1/4$. This is called 'significant degeneration'. If we choose ν larger, then the four solutions will collapse to zero in terms of y and they will become degenerate with the solution we found above. If we choose ν smaller, then the four solutions will still diverge to infinity.

Putting $\nu = 1/4$ in the above equation yields:

$$y = \varepsilon^{\frac{1}{4}} + y^5$$

This equation can be solved using ordinary perturbation theory in the same way as regular expansion for x was obtained. Since the expansion parameter is now $\varepsilon^{1/4}$ we put:

$$y = y_0 + \varepsilon^{\frac{1}{4}} y_1 + \varepsilon^{\frac{1}{2}} y_2 + \cdots$$

There are five solutions for y_0: $\{0, \pm 1, \pm i\}$. We must disregard the solution $y = 0$ since it corresponds to the original regular solution which appears to be at zero for $\varepsilon = 0$, because in the limit $\varepsilon \to 0$ we are rescaling by an infinite amount. The next term is $y_1 = -1/4$. In terms of x the four solutions are thus given as:

$$x = \varepsilon^{-\frac{1}{4}} \left[y_0 - \frac{1}{4}\varepsilon^{\frac{1}{4}} + \cdots \right]$$

22.4.4 Example of degenerate perturbation theory – Stark effect in resonant rotating wave

Let us consider a hydrogen atom rotating with a constant angular frequency ω in an electric field. The Hamiltonian is given by:

$$H = H_0 + \varepsilon x$$

where the unperturbed Hamiltonian is

$$H_0 = \frac{p^2}{2} - \frac{1}{r} - \omega L_z,$$

and Lz is the operator for the z-component of angular momentum: $Lz = i\partial/\partial\varphi$. The perturbation εx can be seen as the strength of the applied electric field multiplied by one of the space coordinates (This calculation is in atomic units, so that every quantity is dimensionless).

The eigenvalues of H_0 are

$$E_{n,m} = -\frac{1}{2}n^2 - m\omega$$

For the lowest energy eigenstates of Hydrogen $|n, l, m\rangle$, $|1, 0, 0\rangle$ and $|2, 1, 1\rangle$ in the resonance $E_{2,1} - E_{1,0} = 0$ their energies are therefore equal $E_{1,0} = E_{2,1} = -1/2$, while the eigenstates are different.

The eigenvalue equation for the Hamiltonian takes the form

$$\begin{bmatrix} E_{1,0} & \varepsilon d \\ \varepsilon d & E_{1,0} \end{bmatrix} \begin{bmatrix} a \\ b \end{bmatrix} = E \begin{bmatrix} a \\ b \end{bmatrix}$$

where

$$d = \frac{128}{243} a_0$$

which leads to the quadratic equation which can be readily solved

$$(E_{1,0} - E)^2 - d^2 \varepsilon^2 = 0$$

with the solution

$$|\chi 1\rangle = \frac{1}{\sqrt{2}} (|1, 0, 0\rangle + |2, 1, 1\rangle)$$
$$E(1) = E_{1,0} + d\varepsilon$$

$$|\chi 2\rangle = \frac{1}{\sqrt{2}} (|1, 0, 0\rangle - |2, 1, 1\rangle)$$
$$E(2) = E_{1,0} - d\varepsilon$$

These states are the Stark states in the rotating frame, they are Trojan (higher eigenvalue) and anti-Trojan wavepackets.

22.5 Some modern applications and limitations

Both regular and singular perturbation theory are frequently used in physics and engineering. Regular perturbation theory may only be used to find those solutions of a problem that evolve smoothly out of the initial solution when changing the parameter (that are "adiabatically connected" to the initial solution).

A well-known example from physics where regular perturbation theory fails is in fluid dynamics when one treats the viscosity as a small parameter. Close to a boundary, the fluid velocity goes to zero, even for very small viscosity (the no-slip condition). For zero viscosity, it is not possible to impose this boundary condition and a regular perturbative expansion amounts to an expansion about an unrealistic physical solution. Singular perturbation theory can, however, be applied here and this amounts to 'zooming in' at the boundaries (using the method of matched asymptotic expansions).

Perturbation theory can fail when the system can transition to a different "phase" of matter, with a qualitatively different behaviour, that cannot be modelled by the physical formulas put into the perturbation theory (e.g., a solid crystal melting into a liquid). In some cases, this failure manifests itself by divergent behavior of the perturbation series. Such divergent series can sometimes be resummed using techniques such as Borel resummation.

Perturbation techniques can be also used to find approximate solutions to non-linear differential equations. Examples of techniques used to find approximate solutions to these types of problems are the Lindstedt–Poincaré technique and the method of multiple time scales.

There is absolutely no guarantee that perturbative methods result in a convergent solution. In fact, asymptotic series are the norm.

22.6   Perturbation theory in chemistry

Many of the ab initio quantum chemistry methods use perturbation theory directly or are closely related methods. Implicit perturbation theory[5] works with the complete Hamiltonian from the very beginning and never specifies a perturbation operator as such. Møller–Plesset perturbation theory uses the difference between the Hartree–Fock Hamiltonian and the exact non-relativistic Hamiltonian as the perturbation. The zero-order energy is the sum of orbital energies. The first-order energy is the Hartree–Fock energy and electron correlation is included at second-order or higher. Calculations to second, third or fourth order are very common and the code is included in most ab initio quantum chemistry programs. A related but more accurate method is the coupled cluster method.

22.7   See also

- Cosmological perturbation theory
- Dynamic nuclear polarisation

- Alternative approach to perturbation theory[6]
- Eigenvalue perturbation
- Interval FEM
- Orders of approximation
- Structural stability
- Lyapunov stability
- Homotopy perturbation method

22.8   References

[1] William E. Wiesel (2010). *Modern Astrodynamics*. Ohio: Aphelion Press. p. 107. ISBN 978-145378-1470.

[2] Cropper, William H. (2004), *Great Physicists: The Life and Times of Leading Physicists from Galileo to Hawking*, Oxford University Press, p. 34, ISBN 978-0-19-517324-6.

[3] Perturbation theory. N. N. Bogolyubov, jr. (originator), Encyclopedia of Mathematics. URL: http://www.encyclopediaofmath.org/index.php?title=Perturbation_theory&oldid=11676

[4] • Sakurai, J.J., and Napolitano, J. (1964,2011). *Modern quantum mechanics* (2nd ed.), Addison Wesley ISBN 978-0-8053-8291-4 . Chapter 5

[5] King, Matcha (1976). "Theory of the Chemical Bond". *JACS* **98** (12): 3415–3420. doi:10.1021/ja00428a004.

[6] Martínez-Carranza, J.; Soto-Eguibar, F.; Moya-Cessa, H. (2012). "Alternative analysis to perturbation theory in quantum mechanics". *The European Physical Journal D* **66**. doi:10.1140/epjd/e2011-20654-5.

22.9   External links

- Introduction to regular perturbation theory by Eric Vanden-Eijnden (PDF)
- Perturbation Method of Multiple Scales

Chapter 23

UV completion

In theoretical physics, **ultraviolet completion**, or **UV completion**, of a quantum field theory is the passing from a lower energy quantum field theory to a more general quantum field theory above a threshold value known as the cut-off. In particular, the more general high energy theory must be well-defined at arbitrarily high energies, the so-called ultraviolet (UV) regime.

The ultraviolet theory must be renormalizable; it can have no Landau poles; and most typically, it enjoys asymptotic freedom in the case that it is a quantum field theory (or at least has a nontrivial fixed point). However, it may also be a background of string theory whose ultraviolet behavior is at least as good as that of renormalizable quantum field theories. Besides these two known examples (QFT and string theory), it could be a completely different theory than string theory that behaves well at very high energies.

23.1   See also

- Ultraviolet limit

- Fermi's interaction

- Quantum mechanics

- String theory

Chapter 24

N=8 Supergravity

N=8 Supergravity is the most symmetric quantum field theory which involves gravity and a finite number of fields. It can be found from a dimensional reduction of 11D supergravity by making the size of 7 of the dimensions go to zero. It has 8 supersymmetries which is the most any gravitational theory can have since there are 8 half-steps between spin 2 and spin -2. (The spin 2 graviton is the particle with the highest spin in this theory). More supersymmetries would mean the particles would have superpartners with spins higher than 2. The only theories with spins higher than 2 which are consistent involve an infinite number of particles (such as String Theory and Higher-Spin Theories). Stephen Hawking in his Brief History of Time speculated that this theory could be the Theory of Everything. However, in later years this was abandoned in favour of String Theory. There has been renewed interest in the 21st century with the possibility that this theory may be finite.

24.1 Calculations

It has been found recently that the expansion of N=8 Supergravity in terms of Feynman diagrams has shown that N=8 Supergravity is in some ways a product of two $N = 4$ super Yang–Mills theories. This is written schematically as:

N = 8 Supergravity = (N = 4 Super Yang–Mills) × (N = 4 Super Yang–Mills)

This is not so surprising as N=8 supergravity contains 6 independent representations of N=4 Super Yang–Mills.

24.2 Particle content

The theory contains 1 graviton (spin 2), 8 gravitinos (spin 3/2), 28 vector bosons (spin 1), 56 fermions (spin 1/2), 70 scalar fields (spin 0) where we don't distinguish particles with negative spin. These numbers are simple combinatorial numbers that come from Pascal's Triangle.

One reason why the theory was abandoned was that the 28 vector bosons which form an O(8) gauge group is too small to contain the standard model U(1)xSU(2)xSU(3) gauge group which can only fit within the orthogonal group O(10).

For model building, it has been assumed that almost all the supersymmetries would be broken in nature leaving just 1 supersymmetry (N=1) although nowadays because of the lack of evidence for N=1 supersymmetry higher supersymmetries are now being considered such as N=2.

24.3 Connection with superstring theory

N=8 Supergravity can be viewed as the low energy approximation of 11D M-Theory with 7 of its dimensions on a certain compact surface.

24.4 Extensions

When derived from Heterotic Superstring Theory compactified on higher-dimensional tori, there are two extensions to N=8 Supergravity. They are the coupling of additional fields which form E_8xE_8 N=4 Super-Yang Mills and SO(32) N=4 Super Yang–Mills. When adding those fields the number of particles increases to 256 + 496x16 = 8192 = 2^{13} particles.

24.5 Global symmetries

Some surprising global symmetries have been found in this theory. For example, it has been shown that there is an E_7 global symmetry but in order for the theory to be finite it is thought that there may be other symmetries not yet found.

24.6   References

Chapter 25

Higher-dimensional supergravity

Higher-dimensional supergravity is the supersymmetric generalization of general relativity in higher dimensions. Supergravity can be formulated in any number of dimensions up to eleven. This article focuses upon supergravity (SUGRA) in greater than four dimensions.

25.1 Supermultiplets

Fields related by supersymmetry transformations form a supermultiplet; the one that contains a graviton is called the supergravity multiplet.

The name of a supergravity theory generally includes the number of dimensions of spacetime that it inhabits, and also the number $\mathcal{N}$ of gravitinos that it has. Sometimes one also includes the choices of supermultiplets in the name of theory. For example, an $\mathcal{N} = 2$, $(9 + 1)$-dimensional supergravity enjoys 9 spatial dimensions, one time and 2 gravitinos. While the field content of different supergravity theories varies considerably, all supergravity theories contain at least one gravitino and they all contain a single graviton. Thus every supergravity theory contains a single supergravity supermultiplet. It is still not known whether one can construct theories with multiple gravitons that are not equivalent to multiple decoupled theories with a single graviton in each. In maximal supergravity theories (see below), all fields are related by supersymmetry transformations so that there is only one supermultiplet: the supergravity multiplet.

25.2 Gauged supergravity versus Yang–Mills supergravity

Often an abuse of nomenclature is used when "gauge supergravity" refers to a supergravity theory in which fields in the theory are charged with respect to vector fields in the theory. However, when the distinction is important, the following is the correct nomenclature. If a global (i.e. rigid) R-symmetry is gauged, the gravitino is charged with respect to some vector fields, and the theory is called gauged supergravity. When other global (rigid) symmetries (e.g., if the theory is a non-linear sigma model) of the theory are gauged such that some (non-gravitino) fields are charged with respect to vectors, it is known as a Yang–Mills–Einstein supergravity theory. Of course, one can imagine having a "gauged Yang–Mills–Einstein" theory using a combination of the above gaugings.

25.3 Counting gravitinos

Gravitinos are fermions, which means that according to the spin-statistics theorem they must have an odd number of spinorial indices. In fact the gravitino field has one spinor and one vector index, which means that gravitinos transform as a tensor product of a spinorial representation and the vector representation of the Lorentz group. This is a Rarita-Schwinger spinor.

While there is only one vector representation for each Lorentz group, in general there are several different spinorial representations. Technically these are really representations of the double cover of the Lorentz group called a spin group.

The canonical example of a spinorial representation is the Dirac spinor, which exists in every number of space-time dimensions. However the Dirac spinor representation is not always irreducible. When calculating the number $\mathcal{N}$, one always counts the number of *real* irreducible representations. The spinors with spins less than 3/2 that exist in each number of dimensions will be classified in the following subsection.

25.4 A classification of spinors

The available spinor representations depends on k; The maximal compact subgroup of the little group of the

Lorentz group that preserves the momentum of a massless particle is $\text{Spin}(d-1) \times \text{Spin}(d-k-1)$, where k is equal to the number d of spatial dimensions minus the number $d - k$ of time dimensions. (See helicity (particle physics)) For example, in our world, this is $3 - 1 = 2$. Due to the mod 8 Bott periodicity of the homotopy groups of the Lorentz group, really we only need to consider k modulo 8.

For any value of k there is a Dirac representation, which is always of real dimension $2^{1+\lfloor \frac{2d-k}{2} \rfloor}$ where $\lfloor x \rfloor$ is the greatest integer less than or equal to x. When $-2 \leq k \leq 2 \pmod 8$ there is a real Majorana spinor representation, whose dimension is half that of the Dirac representation. When k is even there is a Weyl spinor representation, whose real dimension is again half that of the Dirac spinor. Finally when k is divisible by eight, that is, when k is zero modulo eight, there is a Majorana-Weyl spinor, whose real dimension is one quarter that of the Dirac spinor.

Occasionally one also considers symplectic Majorana spinor which exist when $3 \leq k \leq 5$, which have half has many components as Dirac spinors. When $k=4$ these may also be Weyl, yielding Weyl symplectic Majorana spinors which have one quarter as many components as Dirac spinors.

25.5 Choosing chiralities

Spinors in n-dimensions are representations (really modules) not only of the n-dimensional Lorentz group, but also of a Lie algebra called the n-dimensional Clifford algebra. The most commonly used basis of the complex $2^{\lfloor n \rfloor}$-dimensional representation of the Clifford algebra, the representation that acts on the Dirac spinors, consists of the gamma matrices.

When n is even the product of all of the gamma matrices, which is often referred to as Γ_5 as it was first considered in the case $n = 4$, is not itself a member of the Clifford algebra. However, being a product of elements of the Clifford algebra, it is in the algebra's universal cover and so has an action on the Dirac spinors.

In particular, the Dirac spinors may be decomposed into eigenspaces of Γ_5 with eigenvalues equal to $\pm(-1)^{-k/2}$, where k is the number of spatial minus temporal dimensions in the spacetime. The spinors in these two eigenspaces each form projective representations of the Lorentz group, known as Weyl spinors. The eigenvalue under Γ_5 is known as the chirality of the spinor, which can be left or right-handed.

A particle that transforms as a single Weyl spinor is said to be chiral. The CPT theorem, which is required by Lorentz invariance in Minkowski space, implies that when there is a single time direction such particles have antiparticles of the opposite chirality.

Recall that the eigenvalues of Γ_5, whose eigenspaces are the two chiralities, are $\pm(-1)^{-k/2}$. In particular, when k is equal to two modulo four the two eigenvalues are complex conjugate and so the two chiralities of Weyl representations are complex conjugate representations.

Complex conjugation in quantum theories corresponds to time inversion. Therefore, the CPT theorem implies that when the number of Minkowski dimensions is divisible by four (so that k is equal to 2 modulo 4) there be an equal number of left-handed and right-handed supercharges. On the other hand, if the dimension is equal to 2 modulo 4, there can be different numbers of left and right-handed supercharges, and so often one labels the theory by a doublet $\mathcal{N} = (\mathcal{N}_L, \mathcal{N}_R)$ where $\mathcal{N}_L$ and $\mathcal{N}_R$ are the number of left-handed and right-handed supercharges respectively.

25.6 Counting supersymmetries

All supergravity theories are invariant under transformations in the super-Poincaré algebra, although individual configurations are not in general invariant under every transformation in this group. The super-Poincaré group is generated by the Super-Poincaré algebra, which is a Lie superalgebra. A Lie superalgebra is a $\mathbf{Z}_2$ graded algebra in which the elements of degree zero are called bosonic and those of degree one are called fermionic. A commutator, that is an antisymmetric bracket satisfying the Jacobi identity is defined between each pair of generators of fixed degree except for pairs of fermionic generators, for which instead one defines a symmetric bracket called an anticommutator.

The fermionic generators are also called supercharges. Any configuration which is invariant under any of the supercharges is said to be BPS, and often nonrenormalization theorems demonstrate that such states are particularly easily treated because they are unaffected by many quantum corrections.

The supercharges transform as spinors, and the number of irreducible spinors of these fermionic generators is equal to the number of gravitinos $\mathcal{N}$ defined above. Often $\mathcal{N}$ is defined to be the number of fermionic generators, instead of the number of gravitinos, because this definition extends to supersymmetric theories without gravity.

Sometimes it is convenient to characterize theories not by the number $\mathcal{N}$ of irreducible representations of gravitinos or supercharges, but instead by the total Q of their dimensions. This is because some features of the theory have the same Q-dependence in any number of dimensions. For example, one is often only interested in theories in which all particles have spin less than or equal to two. This requires that Q not exceed 32, except possibly in special cases in which the supersymmetry is realized in an unconventional, nonlinear fashion with products of bosonic generators in the anticommutators of the fermionic generators.

25.7 Examples

25.7.1 Why less than 32 SUSYs?

The supergravity theories that have attracted the most interest contain no spins higher than two. This means, in particular, that they do not contain any fields that transform as symmetric tensors of rank higher than two under Lorentz transformations. The consistency of interacting higher spin field theories is, however, presently a field of very active interest.

The supercharges in every super-Poincaré algebra are generated by a multiplicative basis of m fundamental supercharges, and an additive basis of the supercharges (this definition of supercharges is a bit more broad than that given above) is given by a product of any subset of these m fundamental supercharges. The number of subsets of m elements is 2^m, thus the space of supercharges is 2^m-dimensional.

The fields in a supersymmetric theory form representations of the super-Poincaré algebra. It can be shown that when m is greater than 5 there are no representations that contain only fields of spin less than or equal to two. Thus we are interested in the case in which m is less than or equal to 5, which means that the maximal number of supercharges is 32. A supergravity theory with precisely 32 supersymmetries is known as a maximal supergravity.

Above we saw that the number of supercharges in a spinor depends on the dimension and the signature of spacetime. The supercharges occur in spinors. Thus the above limit on the number of supercharges cannot be satisfied in a spacetime of arbitrary dimension. Below we will describe some of the cases in which it is satisfied.

25.7.2 A 12-dimensional two-time theory

The highest dimension in which spinors exist with only 32 supercharges is 12. If there are 11 spatial directions and 1 time direction then there will be Weyl and Majorana spinors which both are of dimension 64, and so are too large. However, some authors have considered nonlinear actions of the supersymmetry in which higher spin fields may not appear.

If instead one considers 10 spatial direction and a second temporal dimension then there is a Majorana-Weyl spinor, which as desired has only 32 components. For an overview

of two-time theories by one of their main proponents, Itzhak Bars, see his paper Two-Time Physics and Two-Time Physics on arxiv.org. He considered 12-dimensional supergravity in Supergravity, p-brane duality and hidden space and time dimensions.

It was widely, but not universally, thought that two-time theories may have problems. For example, there could be causality problems (disconnect between cause and effect) and unitarity problems (negative probability, ghosts). Also, the Hamiltonian-based approach to quantum mechanics may have to be modified in the presence of a second Hamiltonian for the other time. However, in Two-Time Physics it was demonstrated that such potential problems are solved with an appropriate gauge symmetry.

Some other two time theories describe low-energy behavior, such as Cumrun Vafa's F-theory that is also formulated with the help of 12 dimensions. F-theory itself however is not a two-time theory. One can understand 2 of the 12-dimensions of F-theory as a bookkeeping device; they should not be confused with the other 10 spacetime coordinates. These two dimensions are somehow dual to each other and should not be treated independently.

25.7.3 11-dimensional maximal SUGRA

This maximal supergravity is the classical limit of M-theory. There is, classically, only one 11-dimensional supergravity theory. Like all maximal supergravities, it contains a single supermultiplet, the supergravity supermultiplet. This supermultiplet contains the graviton, a Majorana gravitino and a 3-form gauge field often called the C-field.

It contains two p-brane solutions, a 2-brane and a 5-brane, which are electrically and magnetically charged, respectively, with respect to the C-field. This means that 2-brane and 5-brane charge are the violations of the Bianchi identities for the dual C-field and original C-field respectively. The supergravity 2-brane and 5-brane are the long-wavelength limits (see also the historical survey above) of the M2-brane and M5-brane in M-theory.

25.7.4 10d SUGRA theories

Type IIA SUGRA: $N = (1, 1)$

This maximal supergravity is the classical limit of type IIA string theory. The field content of the supergravity supermultiplet consists of a graviton, a Majorana gravitino, a Kalb-Ramond field, odd-dimensional Ramond-Ramond gauge potentials, a dilaton and a dilatino.

The Bianchi identities of the Ramond-Ramond gauge potentials C_{2k-1} can be violated by adding sources ρ, which are called D$(8 - 2k)$-branes

$$ddC_{2k-1} = \rho.$$

In the democratic formulation of type IIA supergravity there exist Ramond-Ramond gauge potentials for $0 < k < 6$, which leads to D0-branes (also called D-particles), D2-branes, D4-branes, D6-branes and, if one includes the case $k = -1$, D8-branes. In addition there are fundamental strings and their electromagnetic duals, which are called NS5-branes.

Although obviously there are no -1-form gauge connections, the corresponding 0-form field strength, G_0 may exist. This field strength is called the **Romans mass** and when it is not equal to zero the supergravity theory is called massive IIA supergravity or Romans IIA supergravity. From the above Bianchi identity we see that a D8-brane is a domain wall between zones of differing G_0, thus in the presence of a D8-brane at least part of the spacetime will be described by the Romans theory.

IIA SUGRA from 11d SUGRA

IIA SUGRA is the dimensional reduction of 11-dimensional supergravity on a circle. This means that 11d supergravity on the spacetime $M^{10} \times S^1$ is equivalent to IIA supergravity on the 10-manifold M^{10} where one eliminates modes with masses proportional to the inverse radius of the circle S^1.

In particular the field and brane content of IIA supergravity can be derived via this dimensional reduction procedure. The field G_0 however does not arise from the dimensional reduction, massive IIA is not known to be the dimensional reduction of any higher-dimensional theory. The 1-form Ramond-Ramond potential C_1 is the usual 1-form connection that arises from the Kaluza–Klein procedure, it arises from the components of the 11-d metric that contain one index along the compactified circle. The IIA 3-form gauge potential C_3 is the reduction of the 11d 3-form gauge potential components with indices that do not lie along the circle, while the IIA Kalb-Ramond 2-form B-field consists of those components of the 11-dimensional 3-form with one index along the circle. The higher forms in IIA are not independent degrees of freedom, but are obtained from the lower forms using Hodge duality.

Similarly the IIA branes descend from the 11-dimension branes and geometry. The IIA D0-brane is a Kaluza–Klein momentum mode along the compactified circle. The IIA fundamental string is an 11-dimensional membrane which wraps the compactified circle. The IIA D2-brane is an 11-dimensional membrane that does not wrap the compactified circle. The IIA D4-brane is an 11-dimensional 5-brane that wraps the compactified circle. The IIA NS5-brane is an 11-dimensional 5-brane that does not wrap the compactified circle. The IIA D6-brane is a Kaluza–Klein monopole, that is, a topological defect in the compact circle fibration. The lift of the IIA D8-brane to 11-dimensions is not known, as one side of the IIA geometry as a nontrivial Romans mass, and an 11-dimensional original of the Romans mass is unknown.

Type IIB SUGRA: $N = (2, 0)$

This maximal supergravity is the classical limit of type IIB string theory. The field content of the supergravity supermultiplet consists of a graviton, a Weyl gravitino, a Kalb-Ramond field, even-dimensional Ramond-Ramond gauge potentials, a dilaton and a dilatino.

The Ramond-Ramond fields are sourced by odd-dimensional D$(2k + 1)$-branes, which host supersymmetric $U(1)$ gauge theories. As in IIA supergravity, the fundamental string is an electric source for the Kalb-Ramond B-field and the NS5-brane is a magnetic source. Unlike that of the IIA theory, the NS5-brane hosts a worldvolume $U(1)$ supersymmetric gauge theory with $\mathcal{N} = (1,1)$ supersymmetry, although some of this supersymmetry may be broken depending on the geometry of the spacetime and the other branes that are present.

This theory enjoys an SL(2, **R**) symmetry known as S-duality that interchanges the Kalb-Ramond field and the RR 2-form and also mixes the dilaton and the RR 0-form axion.

Type I gauged SUGRA: $N = (1, 0)$

These are the classical limits of type I string theory and the two heterotic string theories. There is a single Majorana-Weyl spinor of supercharges, which in 10 dimensions contains 16 supercharges. As 16 is less than 32, the maximal number of supercharges, type I is not a maximal supergravity theory.

In particular this implies that there is more than one variety of supermultiplet. In fact, there are two. As usual, there is a supergravity supermultiplet. This is smaller than the supergravity supermultiplet in type II, it contains only the graviton, a Majorana-Weyl gravitino, a 2-form gauge potential, the dilaton and a dilatino. Whether this 2-form is considered to be a Kalb-Ramond field or Ramond-Ramond field depends on whether one considers the supergravity theory to be a classical limit of a heterotic string theory or type I string theory. There is also a vector supermultiplet, which contains a one-form gauge potential called a gluon and also a Majorana-Weyl gluino.

Unlike type IIA and IIB supergravities, for which the classical theory is unique, as a classical theory $\mathcal{N} = 1$ supergravity is consistent with a single supergravity supermultiplet and any number of vector multiplets. It is also consistent without the supergravity supermultiplet, but then it would contain no graviton and so would not be a supergravity theory. While one may add multiple supergravity supermultiplets, it is not known if they may consistently interact. One is free not only to determine the number, if any, of vector supermultiplets, but also there is some freedom in determining their couplings. They must describe a classical super Yang–Mills gauge theory, but the choice of gauge group is arbitrary. In addition one is free to make some choices of gravitational couplings in the classical theory.

While there are many varieties of classical $\mathcal{N} = 1$ supergravities, not all of these varieties are the classical limits of quantum theories. Generically the quantum versions of these theories suffer from various anomalies, as can be seen already at 1-loop in the hexagon Feynman diagrams. In 1984 and 1985 Michael Green and John H. Schwarz have shown that if one includes precisely 496 vector supermultiplets and chooses certain couplings of the 2-form and the metric then the gravitational anomalies cancel. This is called the Green-Schwarz anomaly cancellation mechanism.

In addition, anomaly cancellation requires one to cancel the gauge anomalies. This fixes the gauge symmetry algebra to be either $\mathfrak{so}(32)$, $\mathfrak{e}_8 \oplus \mathfrak{e}_8$, $\mathfrak{e}_8 \oplus 248\mathfrak{u}(1)$ or $496\mathfrak{u}(1)$. However, only the first two Lie algebras can be gotten from superstring theory. Quantum theories with at least 8 supercharges tend to have continuous moduli spaces of vacua. In compactifications of these theories, which have 16 supercharges, there exist degenerate vacua with different values of various Wilson loops. Such Wilson loops may be used to break the gauge symmetries to various subgroups. In particular the above gauge symmetries may be broken to obtain not only the standard model gauge symmetry but also symmetry groups such as SO(10) and SU(5) that are popular in GUT theories.

25.7.5 9d SUGRA theories

In 9-dimensional Minkowski space the only irreducible spinor representation is the Majorana spinor, which has 16 components. Thus supercharges inhabit Majorana spinors of which there are at most two.

Maximal 9d SUGRA from 10d

In particular, if there are two Majorana spinors then one obtains the 9-dimensional maximal supergravity theory. Recall that in 10 dimensions there were two inequivalent maximal supergravity theories, IIA and IIB. The dimensional reduction of either IIA or IIB on a circle is the unique 9-dimensional supergravity. In other words, IIA or IIB on the product of a 9-dimensional space M^9 and a circle is equivalent to the 9-dimension theory on M^9, with Kaluza–Klein modes if one does not take the limit in which the circle shrinks to zero.

T-duality

More generally one could consider the 10-dimensional theory on a nontrivial circle bundle over M^9. Dimensional reduction still leads to a 9-dimensional theory on M^9, but with a 1-form gauge potential equal to the connection of the circle bundle and a 2-form field strength which is equal to the Chern class of the old circle bundle. One may then lift this theory to the other 10-dimensional theory, in which case one finds that the 1-form gauge potential lifts to the Kalb-Ramond field. Similarly, the connection of the fibration of the circle in the second 10-dimensional theory is the integral of the Kalb-Ramond field of the original theory over the compactified circle.

This transformation between the two 10-dimensional theories is known as T-duality. While T-duality in supergravity involves dimensional reduction and so loses information, in the full quantum string theory the extra information is stored in string winding modes and so T-duality is a duality between the two 10-dimensional theories. The above construction can be used to obtain the relation between the circle bundle's connection and dual Kalb-Ramond field even in the full quantum theory.

$N = 1$ Gauged SUGRA

Main article: Gauged supergravity

As was the case in the parent 10-dimensional theory, 9-dimensional N=1 supergravity contains a single supergravity multiplet and an arbitrary number of vector multiplets. These vector multiplets may be coupled so as to admit arbitrary gauge theories, although not all possibilities have quantum completions. Unlike the 10-dimensional theory, as was described in the previous subsection, the supergravity multiplet itself contains a vector and so there will always be at least a U(1) gauge symmetry, even in the N=2 case.

25.8 The mathematics

The Lagrangian for 11D supergravity found by brute force by Cremmer, Julia and Scherk[1] is:

$$
\begin{aligned}
L \;=\; & +\tfrac{1}{2\kappa^2}eR - \tfrac{1}{2}e\overline{\psi}_M \Gamma^{MNP} D_N[\tfrac{1}{2}(\omega-\varpi)]\psi_P \\
& +\tfrac{1}{48}eF^2_{MNPQ} + \tfrac{\sqrt{2}\kappa}{384}e(\overline{\psi}_M \Gamma^{MNPQRS}\psi_S \\
& +12\overline{\psi}^N \Gamma^{PQ}\psi^R)(F+\mathcal{F})_{NPQR} + \tfrac{\sqrt{2}\kappa}{3456}\varepsilon^{M_1\ldots M_{11}} F_{M_1\ldots M_4} F_{M_5\ldots M_8} A_{M_9 M_{10} M_{11}}
\end{aligned}
$$

which contains the three types of field:

$$
e^A_M, \psi_M, A_{MNP}
$$

The symmetry of this supergravity theory is given by the supergroup OSp(1l32) which gives the subgroups O(1) for the bosonic symmetry and Sp(32) for the fermion symmetry. This is because spinors need 32 components in 11 dimensions. 11D supergravity can be compactified down to 4 dimensions which then has OSp(8l4) symemtry. (We still have $8 \times 4 = 32$ so there are still the same number of components.) Spinors need 4 components in 4 dimensions. This gives O(8) for the gauge group which is too small to contain the Standard Model gauge group U(1) × SU(2) × SU(3) which would need at least O(10).

25.9 Notes and references

[1] E. Cremmer, B. Julia, J. Scherk: Phys. Lett. 76B, 409 (1978) .

25.10   Text and image sources, contributors, and licenses

25.10.1   Text

- **Supergravity** *Source:* https://en.wikipedia.org/wiki/Supergravity?oldid=709480809 *Contributors:* AxelBoldt, Michael Hardy, TakuyaMurata, Angela, Charles Matthews, Phys, Bevo, Robbot, Gandalf61, Giftlite, Herbee, LeYaYa, Fropuff, Moyogo, Jeremy Henty, Leonard G., Urvabara, Arivero, Masudr, Dmr2, Srbauer, Markryherd, Physicistjedi, Axl, Wtmitchell, Japanese Searobin, Linas, Kzollman, Mpatel, GregorB, Canderson7, Marasama, Gurch, LeCire~enwiki, Chobot, Roboto de Ajvol, Hillman, Conscious, E. Menay, Wimt, Smoggyrob, QmunkE, Ilmari Karonen, Caco de vidro, SmackBot, Melchoir, FlashSheridan, Vald, Chris the speller, Colonies Chris, QFT, BWDuncan, TheST, Kuru, Jim.belk, JarahE, Michaelbusch, Zero sharp, CapitalR, Jorbesch, Crichigno, CmdrObot, Myasuda, Equendil, Phatom87, Pyro95819, Mbell, Marek69, WVhybrid, West Brom 4ever, Icep, Shlomi Hillel, Yill577, David Eppstein, N.Nahber, Andre.holzner, Mschel, EdBever, Freeboson, Wesino, WJBscribe, Fuenfundachtzig, Signalhead, Cuzkatzimhut, Jickle, Robdunst, WereSpielChequers, Caltas, Wing gundam, Paolo.dL, Oxymoron83, Lightmouse, JL-Bot, EmanWilm, RS1900, ClueBot, ArdClose, Mild Bill Hiccup, JavierReynaldo, Vivio Testarossa, Mastertek, Pqnelson, AnonyScientist, Goulu, Truthnlove, Addbot, Some jerk on the Internet, Wentuq, Luckas-bot, Yobot, Bility, AnomieBOT, ArthurBot, Omnipaedista, Gsard, Hep thinker, FrescoBot, Paine Ellsworth, Pxpt, Tom.Reding, Casimir9999, Wornsear, EmausBot, Slightsmile, Wikipelli, HiW-Bot, ZéroBot, Cogiati, Arbnos, Quantumor, Terraflorin, Bbeehvh, ClueBot NG, Joefromrandb, Helpful Pixie Bot, Bibcode Bot, BG19bot, Altair, BattyBot, Jeremy112233, M0532062613, Jamesx12345, Mamzypig99, Bitprior, Monkbot, AHusain3141, KasparBot, Profusionex, JohnStarling and Anonymous: 75

- **Field (physics)** *Source:* https://en.wikipedia.org/wiki/Field_(physics)?oldid=718332797 *Contributors:* Patrick, Michael Hardy, Dcljr, Angela, Andres, Wooster, Charles Matthews, Dino, Reddi, Phys, Cncs wikipedia, Robbot, Ojigiri~enwiki, Wikibot, Fuelbottle, Ancheta Wis, Giftlite, Waltpohl, LucasVB, Antandrus, Karol Langner, AmarChandra, Mschlindwein, Starfoxy, Lucidish, CALR, Laoma, Masudr, YUL89YYZ, Kbh3rd, El C, Laurascudder, Army1987, Guiltyspark, Varuna, Bobrg~enwiki, MIT Trekkie, Linas, Natcase, Polyparadigm, Dodiad, Mpatel, Ketiltrout, Tbone, Nihiltres, Borgx, RobotE, Bambaiah, Stephenb, NawlinWiki, Albedo, Beanyk, Epipelagic, Arthur Rubin, Sbyrnes321, SmackBot, TheLeopard, Sbharris, Colonies Chris, Hongooi, Sergio.ballestrero, Vegard, John, JHunterJ, Mets501, Treyp, Trevor.tombe, Chmee2, Cydebot, WISo, Soetermans, Skittleys, Thijs!bot, Epbr123, Headbomb, Dalahäst, MichaelMaggs, JBouwman, JAnDbot, Husond, Jpod2, Ed!, Hdt83, R'n'B, Fconaway, HEL, Kimse, Metamusing, Maurice Carbonaro, Lamp90, BernardZ, Squids and Chips, VolkovBot, TXiKiBoT, Reibot, Thomas.schick, SieBot, Malcolmxl5, OKBot, Laurentseries, Anchor Link Bot, ClueBot, Mild Bill Hiccup, Djr32, Brews ohare, Schreiber-Bike, PCHS-NJROTC, TimothyRias, Heinsaar, Mhsb, Truthnlove, Addbot, Fgnievinski, Bte99, Xgambler, Luckas-bot, Yobot, AnomieBOT, Palpher, Ulric1313, Citation bot, ArthurBot, Xqbot, Calcio33, J04n, Topherwhelan, Gsard, WaysToEscape, LucienBOT, Citation bot 1, Rapsar, Pinethicket, Loudubewe, Tom.Reding, TobeBot, Jfmantis, RjwilmsiBot, Ripchip Bot, EmausBot, Jjspinorfield1, Chrisman62, Maschen, RockMagnetist, ClueBot NG, Gilderien, Admock, Helpful Pixie Bot, Shivsagardharam, BG19bot, Jcderico, Uioplk, Ema--or, Hmainsbot1, Katterjohn, DavidLeighEllis, Noyster, BHBrunt, Peterfreed, MCarsten, Mpcalkins, Isambard Kingdom, KasparBot, Crosleybendix and Anonymous: 87

- **Supersymmetry** *Source:* https://en.wikipedia.org/wiki/Supersymmetry?oldid=716142716 *Contributors:* Bryan Derksen, Taw, Andre Engels, Roadrunner, Maury Markowitz, Ewen, Stevertigo, Edward, Michael Hardy, Arpingstone, Theresa knott, IMSoP, Jeandré du Toit, Samw, Smack, Charles Matthews, Maximus Rex, Phys, Raul654, BenRG, Rursus, Mor~enwiki, Ancheta Wis, Giftlite, Mporter, Ferkelparade, Monedula, Fropuff, Xerxes314, Anville, Gus Polly, Moyogo, Unconcerned, DO'Neil, Maarten van Vliet, Pharotic, LiDaobing, Sam Hocevar, Lumidek, Deglr6328, Arivero, Rich Farmbrough, Roybb95~enwiki, Bender235, El C, Nornagon~enwiki, Duk, Tweet Tweet, Russ3Z, LostLeviathan, Pearle, Gary, Francescog~enwiki, Wtmitchell, RJFJR, Reaverdrop, Blaxthos, Killing Vector, Jordan14, Ted BJ, MONGO, Mpatel, MFH, SeventyThree, Bodera, VermillionBird, Drbogdan, Rjwilmsi, Josiah Rowe, R.e.b., Bubba73, Maxim Razin, Drrngrvy, FlaBot, Cless Alvein, Nowhither, Itinerant1, Gparker, KFP, Lmatt, Chobot, Vyroglyph, YurikBot, Wavelength, RussBot, Ohwilleke, Bhny, Epolk, Sasuke Sarutobi, Maxim Leyenson, Chaos, Romanc19s, Bota47, Mgnbar, Closedmouth, Arthur Rubin, RG2, That Guy, From That Show!, A bit iffy, SmackBot, Mira, Kurochka, Wangjiaji, Gilliam, Bluebot, Cadmasteradam, Complexica, Bazonka, Colonies Chris, Can't sleep, clown will eat me, QFT, Ruff ilb, Wen D House, Solarapex, Radagast83, Jgwacker, TheMaster42, Martijn Hoekstra, Ligulembot, Acjohnson55, Yevgeny Kats, Charleswestbrook, TriTertButoxy, Lambiam, Tktktk, Xiaphias, JarahE, Mdanziger, Dan Gluck, Newone, Marysunshine, Tawkerbot2, Banedon, Cydebot, Hydraton31, Bazzargh, David edwards, Michael C Price, Crum375, Koeplinger, Headbomb, J.christianson, Escarbot, Salgueiro~enwiki, Kborland, Jpod2, Cgingold, Maliz, TimidGuy, C9, Kostisl, R'n'B, Zentropa77, Natsirtguy, Maurice Carbonaro, Kevin Hickerson, Shawn in Montreal, Idioma-bot, Sheliak, Cuzkatzimhut, Nxavar, Kawakameha, Cuboidal, Ptrslv72, PhysPhD, Kbrose, SieBot, Nn123645, ClueBot, Jcpilman, Chessmaster7m, Kitsunegami, Rhododendrites, Mastertek, Mishas42, Scrabby~enwiki, TimothyRias, WikHead, MystBot, Addbot, DOI bot, Zahd, Barak Sh, F Notebook, Lightbot, Windward1, Luckas-bot, Yobot, Ibayn, TaBOT-zerem, Amirobot, Nonnormalizable, AnomieBOT, Girl Scout cookie, Materialscientist, Citation bot, ArthurBot, Plumpurple, Tomwsulcer, Omnipaedista, Gsard, CES1596, FrescoBot, HaloStereo1, Paine Ellsworth, Xmikywayx, Citation bot 1, Gil987, Kikeku, Jonesey95, Eddie Nixon, MondalorBot, Aknochel, Tom1661, Gagoga ju, TobeBot, Puzl bustr, Andraas, EmausBot, Djloststylez, Klbrain, Ddimensões, Arbnos, Susy is it, ChuispastonBot, Isocliff, ClueBot NG, KagakuKyouju, IJVin, Frietjes, Helpful Pixie Bot, Bibcode Bot, BG19bot, Teika kazura, JayBeeEye, Ninmacer20, ChrisGualtieri, Dexbot, Logosun, AHusain314, NA48, Rfassbind, Katherine Pendleton, Lioinnisfree, Laplacemat, Liquidityinsta, TaiSakuma, Stamptrader, Kdmeaney, Qxxxxxq, Almaionescu, Monkbot, Janhaithabu, Mammoth2011, Jwill530, Stacie Croquet, Cuttlas1, AHusain3141, Wave system, Archaon593 and Anonymous: 178

- **General relativity** *Source:* https://en.wikipedia.org/wiki/General_relativity?oldid=716703634 *Contributors:* AxelBoldt, Mav, Bryan Derksen, The Anome, AstroNomer, Ap, RK, Andre Engels, XJaM, Chrislintott, JeLuF, Christian List, William Avery, Roadrunner, Ktsquare, B4hand, Stevertigo, Frecklefoot, Patrick, Boud, Michael Hardy, Menchi, Ixfd64, Bcrowell, Nimrod~enwiki, TakuyaMurata, Mcarling, Minesweeper, Alfio, Looxix~enwiki, ArnoLagrange, Ellywa, Ahoerstemeier, Stevenj, William M. Connolley, Snoyes, Angela, Mark Foskey, Julesd, Salsa Shark, AugPi, Andres, Evercat, Hectorthebat, Hick ninja, A.Tigges~enwiki, Gingekerr, Jitse Niesen, Gutza, Rednblu, Doradus, Wik, Dragons flight, Tero~enwiki, Phys, Shizhao, Elwoz, Jerzy, BenRG, Banno, Northgrove, Phil Boswell, Robbot, Craig Stuntz, Sdedeo, Bvc2000, Goethean, Altenmann, Romanm, Lowellian, Mayooranathan, Gandalf61, Blainster, Diderot, DHN, Hadal, Alba, Johnstone, Fuelbottle, Isopropyl, Xanzzibar, Carnildo, Tea2min, Enochlau, Ancheta Wis, Tosha, Giftlite, JamesMLane, Graeme Bartlett, Mikez, BenFrantzDale, Lethe, Tom harrison, Fropuff, Everyking, Physman, Curps, Michael Devore, Jason Quinn, Alvestrand, SWAdair, Glengarry, Bobblewik, Edcolins, DefLog~enwiki, Pgan002, Knutux, GeneralPatton, HorsePunchKid, Robert Brockway, Kaldari, MadIce, Karol Langner, Rjpetti, Rdsmith4, JimWae, Anythingyouwant, Martin Wisse, Thincat, Euphoria, Icairns, Zfr, AmarChandra, Zondor, Econrad, JimJast, Discospinster, Rich Farmbrough,

Guanabot, Pak21, ThomasK, Masudr, Pjacobi, Vsmith, Cdyson37, Jowr, Paul August, SpookyMulder, Dmr2, Bender235, Dcabrilo, Ground, Ben Standeven, Nabla, Livajo, El C, Worldtraveller, Shanes, Etimbo, Causa sui, Bobo192, Robotje, Smalljim, Rbj, JW1805, ParticleMan, I9Q79oL78KiL0QTFHgyc, Mr2001, Matt McIrvin, PWilkinson, Haham hanuka, Schnolle, Varuna, Jumbuck, Jérôme, Alansohn, Hackwrench, Cctoide, Crebbin, Wikidea, SlimVirgin, Benefros, Alexwg, Wtmitchell, Orionix, CloudNine, Bsadowski1, DV8 2XL, LordLoki, HenryLi, Oleg Alexandrov, Kelly Martin, Linas, FeanorStar7, Sabejias, Moneky, Kzollman, Cleonis, Mpatel, Jok2000, Schzmo, Pdn~enwiki, GregorB, Plrk, Wayward, Joke137, Christopher Thomas, Mandarax, Colodia, Canderson7, Rjwilmsi, WCFrancis, MarSch, Eyu100, JoshuacUK, JHMM13, Mike Peel, SanitysEdge, R.e.b., Ems57fcva, Bubba73, Gringo300, Ian Pitchford, RobertG, Mishuletz, Amero, Mathbot, Nihiltres, Vsion, Perfect Tommy~enwiki, Itinerant1, Alfred Centauri, Gparker, Slant, Carrionluggage, Srleffler, Chobot, DVdm, Bgwhite, Dresdnhope, Manscher, PointedEars, Roboto de Ajvol, YurikBot, Wavelength, Bcarm1185, Splintercellguy, Hillman, EDG, MattWright, RussBot, Loom91, AVM, KSmrq, DanMS, SpuriousQ, Shawn81, Eleassar, Shanel, Syth, Madcoverboy, Tailpig, Schlafly, Dputig07, Beanyk, Tony1, Dna-webmaster, Enormousdude, 2over0, KGasso, Petri Krohn, GraemeL, Rlove, Sambc, LeonardoRob0t, Geoffrey.landis, HereToHelp, Willtron, Caballero1967, Meegs, Bsod2, Finell, Luk, Sardanaphalus, SmackBot, Kurochka, Hydrogen Iodide, Pavlovič, Gnangarra, Unyoyega, Nickst, Delldot, Motorneuron, Cessator, Harald88, Edgar181, Shai-kun, Sectryan, Gilliam, Skizzik, Dauto, Saros136, Silly rabbit, Complexica, Colonies Chris, Zven, Abyssal, RProgrammer, Hve, RedHillian, BentSm, Phaedriel, Khoikhoi, Cybercobra, Downwards, Coolbho3000, Nakon, Peterwhy, SkyWriter, DMacks, Nairebis, Henning Makholm, UncleFester, Bidabadi~enwiki, Byelf2007, SashatoBot, Lambiam, Lapaz, Cronholm144, Gizzakk, CPMcE, JorisvS, Goodnightmush, Ckatz, Frokor, Garthbarber, SirFozzie, SandyGeorgia, Midnightblueowl, RichardF, Novangelis, Peter Horn, MTSbot~enwiki, Kvng, JarahE, Licorne, Quaeler, Fan-1967, Editor.singapore, MFago, JoeBot, ShyK, MOBle, RekishiEJ, CapitalR, MD:astronomer, Courcelles, Tawkerbot2, JRSpriggs, Kurtan~enwiki, Harold f, JForget, Sakurambo, Thermochap, Avanu, NickW557, MarsRover, Harrigan, Ian Beynon, Cydebot, Jasperdoomen, WillowW, Fl, MC10, Mato, Pascal.Tesson, Michael C Price, Christian75, Dumb-BOT, Biblbroks, Omicronpersei8, Crum375, N. Macchiavelli, Epbr123, Fisherjs, Markus Pössel, Martin Hogbin, MrXow, Oliver202, Headbomb, Pjvpjv, Tom Barlow, Davidhorman, D.H, AntiVandalBot, Abu-Fool Danyal ibn Amir al-Makhiri, Tkirkman, Gnixon, VectorPosse, TimVickers, Scepia, Dawz, Billevans~enwiki, Tim Shuba, Rico402, Archmagusrm, Jaredroberts, JAnDbot, Vorpal blade, Hut 8.5, YK Times, Acroterion, Pervect, Magioladitis, Connormah, RogierBrussee, WolfmanSF, JamesBWatson, Swpb, Ling.Nut, Soulbot, Pixel ;-), KConWiki, WhatamIdoing, BatteryIncluded, Eldumpo, Allstarecho, User A1, Mollwollfumble, Chris G, Archen~enwiki, Thompson.matthew, STBot, Mermaid from the Baltic Sea, Shentino, Mschel, CommonsDelinker, Pbroks13, J.delanoy, DrKay, R. Baley, Numbo3, Leafsfan85, Aveh8, Lantonov, M C Y 1008, Mathlabster, Zedmelon, Aboutmovies, C quest000, Tcisco, Marrilpet, Nwbeeson, Aatomic1, Potatoswatter, Kolja21, Lseixas, Rémih, Caracalocelot, DemonicInfluence, Sheliak, Deor, Part Deux, JohnBlackburne, Philip Trueman, TXiKiBoT, Oshwah, Coder Dan, GimmeBot, Gombo, Hqb, Rei-bot, IPSOS, Qxz, T doffing, Molinogi, Fizzackerly, JhsBot, Leafyplant, Geometry guy, Ilyushka88, Thebigbendizzle, SwordSmurf, Andy Dingley, Gabrielsleitao, Lamro, Antixt, Vector Potential, James-Chin, Arcfrk, Ccheese4, StevenJohnston, Katzmik, YohanN7, Dnarby, SieBot, Tiddly Tom, Work permit, Yintan, RadicalOne, Wizzard2k, SteakNShake, Arbor to SJ, Babareddeer, JSpung, Phil Bridger, Wmpearl, Oxymoron83, Henry Delforn (old), Csloomis, Thehotelambush, Lightmouse, BrightRoundCircle, OpTioNiGhT, The-G-Unit-Boss, Emgg, AWeishaupt, Divinestuff, Coldcreation, Adam Cuerden, Duae Quartunciae, Heptarchy of teh Anglo-Saxons, baby, Randomblue, TFCforever, Danthewhale, Martarius, Sfan00 IMG, ClueBot, The Thing That Should Not Be, Rjd0060, Metaprimer, Wwheaton, Der Golem, JTBX, TheAmigo42, CounterVandalismBot, Viran, Blanchardb, Rotational, Agge1000, Itzguru, Tanketz, CohesionBot, Eeekster, Stealth500, Brews ohare, NuclearWarfare, PhySusie, SockPuppetForTomruen, SchreiberBike, Another Believer, RubenGarciaHernandez, AC+79 3888, MasterOfHisOwnDomain, He6kd, TimothyRias, Lazyrussian, PseudoOne, Skarebo, NellieBly, JinJian, Truthnlove, Everydayidiot, Tayste, Balungifrancis, Addbot, Mortense, Some jerk on the Internet, Fizzycyst, DOI bot, Mistyocean3, Metagraph, Stariki, Fluffernutter, Schmoolik, MrOllie, Download, EconoPhysicist, Delaszk, Favonian, LinkFA-Bot, Tuition, Tassedethe, Nnedass, Tide rolls, Lightbot, Knutls, Luckas-bot, Ptbotgourou, Legobot II, Julia W, Trickyboarder93, Superamoeba, AnomieBOT, Kristen Eriksen, Giordano.ferdinandi, Jim1138, Jo3sampl, Materialscientist, Wandering Courier, The High Fin Sperm Whale, Citation bot, Xqbot, Stlwebs, Sionus, Amareto2, Unigfjkl, Nickkid5, Stsang, Coretheapple, GrouchoBot, Collin21594, RibotBOT, Rucko123, GhalyBot, Acannas, LucienBOT, Paine Ellsworth, Lagelspeil, Steve Quinn, Knowandgive, Pokyrek, Citation bot 1, Citation bot 4, Electrozity8, Pinethicket, LittleWink, Jonesey95, A412, Tom.Reding, Yougeeaw, Barras, Jauhienij, Meier99, Citator, Comet Tuttle, Hughston, Defender of torch, Duoduoduo, Aribashka, Iibbmm, Diannaa, Earthandmoon, Tbhotch, Brambleclawx, Marie Poise, RjwilmsiBot, Aznhero3793, Ripchip Bot, EmausBot, WikitanvirBot, Immunize, Zhaskey, Fly by Night, DuKu, GoingBatty, Jmencisom, Slightsmile, Dcirovic, Hhhippo, JSquish, ZéroBot, Cogiati, Stanford96, Empty Buffer, Sanford123456, H3llBot, Quondum, REkaxkjdsc, Monterey Bay, Mr little irish, TonyMath, Brandmeister, Maschen, Puffin, Carmichael, Newstv11, RockMagnetist, Sona11235, WizardofCalculus, Milk Coffee, Whoop whoop pull up, Mjbmrbot, Helpsome, ClueBot NG, Manubot, Hagenfeldt, This lousy T-shirt, SusikMkr, Ggonzalm, Jj1236, Mgvongoeden, Snotbot, Widr, Jamester234, Pluma, Ginger.spice14, Bibcode Bot, Jeraphine Gryphon, Lowercase sigmabot, BG19bot, Quarkgluonsoup, Bolatbek, Marsambe, Amp71, Mark Arsten, Lovepool1220, Marsambe1, Benzband, ENG.F.Younis, 123matt123, DeviantFrog, IrishDevil2, F=q(E+v^B), Egbertus2, Harizotoh9, Doctor Lipschitz, Snow Blizzard, Physicsch, Zoldyick, Roozitaa, BattyBot, Reed07, Vanobamo, JoshuSasori, Stigmatella aurantiaca, Cyberbot II, Abhay ravi, ChrisGualtieri, Maestro814, Deathlasersonline, Plokijnu, Billyshiverstick, Read Blooded, Theeditor6079, Flyer1997, Dexbot, Suffian Akhtar, Irondome, Kryomaxim, Twhitguy14, CuriousMind01, J0437-4715, Jamesx12345, Among Men, Leprof 7272, WorldWideJuan, Devinray1991, 1888software, EvergreenFir, Enchantedscience, Mohamed F. El-Hewie, Vai ra'a toa Taina, NeapleBerlina, Jwratner1, Gigantmozg, Ginsuloft, SirKesuma, Anrnusna, JaconaFrere, Osamabin7, Juenni32, Filedelinkerbot, SantiLak, Aryabhatt 21, Willbh15, S11027158, Cjsmith.us, ChamithN, Cris Cyborg, PeterShawhan, Evgeniy E., Sweeeeeeeed, Tetra quark, Absolutelypuremilk, Praveece, JLT2045, LL221W, Jf2839, GeneralizationsAreBad, KasparBot, Jmc76, Sir Cumference, Lemonberry622, Pizzaman62, Dgray101, Amrespi2007, Narasimha Kanduri, רוקמורגלי, J1738, Soopdish and Anonymous: 732

- **Local symmetry** *Source:* https://en.wikipedia.org/wiki/Local_symmetry?oldid=674751475 *Contributors:* Physicistjedi, Mpatel, BD2412, Brad-Beattie, SmackBot, Maksim-e~enwiki, Myasuda, Michael C Price, Headbomb, Dougher, R'n'B, Jeepday, Cuzkatzimhut, Niceguyedc, JavierReynaldo, Jaime Saldarriaga, SchreiberBike, AnomieBOT, FrescoBot, Citation bot 1, AlexUT, GoingBatty, Maschen, MerlIwBot, Helpful Pixie Bot, Brad7777 and Anonymous: 10

- **Minimal Supersymmetric Standard Model** *Source:* https://en.wikipedia.org/wiki/Minimal_Supersymmetric_Standard_Model?oldid= 715985338 *Contributors:* Phys, Dmytro, Gandalf61, Rursus, Connelly, Marcika, Waltpohl, Pharotic, Carandol~enwiki, HorsePunchKid, Grunt, Pjacobi, Jensbn, El C, Jag123, JohnyDog, RJFJR, DV8 2XL, Woohookitty, Mpatel, VermillionBird, Rjwilmsi, Goudzovski, Bhny, Jabber-Wok, Shawn81, SCZenz, Closedmouth, Caco de vidro, Tom Lougheed, Stepa, Dauto, Chris the speller, Bluebot, Colonies Chris, S11982, QFT, MBlume, Jgwacker, Pulu, CenozoicEra, NNemec, Waggers, Dan Gluck, Iridescent, Antonio Prates, Lottamiata, CmdrObot, Myasuda, Michael C Price, Dchristle, RoadMap, Headbomb, CannedhamX, Knotwork, Yill577, Paulnilsson, Maliz, Dr. Morbius, Andre.holzner, Freeboson, Wilsonge, Red Act, Pjoef, StewartMH, PipepBot, ArdClose, Mastertek, Chaosdruid, Rreagan007, SkyLined, Addbot, DOI bot, Mjamja, Tokikake, Luckas-bot, Yobot, Wireader, AnomieBOT, Archon 2488, Citation bot, GenQuest, GrouchoBot, Omnipaedista, Ernsts, Paine Ellsworth, Identitaamore, Citation bot 1, PigFlu Oink, Puzl bustr, RjwilmsiBot, Akrose, EmausBot, WCEngineer, Arbnos, Suslindisambiguator, AManWithNoPlan, Isocliff, Zukertort, Bibcode Bot, BG19bot, ElphiBot, Physlad, ChrisGualtieri, Cinaro, Stamptrader and Anonymous: 50

- **Poincaré group** *Source:* https://en.wikipedia.org/wiki/Poincar%C3%A9_group?oldid=717756936 *Contributors:* AxelBoldt, Zundark, The Anome, XJaM, Mbecker, Stevertigo, Patrick, Michael Hardy, Marco Krohn, AugPi, Stupidmoron, Charles Matthews, Phys, Anupamsr, Giftlite, Lethe, Fropuff, DefLog~enwiki, Jossi, Rich Farmbrough, Bender235, Rgdboer, Aronbeekman, Danski14, Keenan Pepper, Gene Nygaard, Oleg Alexandrov, JFG, Mpatel, Allen3, Rjwilmsi, DVdm, YurikBot, That Guy, From That Show!, SmackBot, Incnis Mrsi, Nbarth, Tsca.bot, Kcordina, Cybercobra, JRSpriggs, Cydebot, Headbomb, Nearyan, JAnDbot, Fetchcomms, Yill577, SHCarter, Sullivan.t.j, Cuzkatzimhut, XCelam, Drschawrz, YohanN7, VVVBot, Phe-bot, Addbot, Luckas-bot, Ptbotgourou, AnomieBOT, Omnipaedista, Paine Ellsworth, Thinking of England, Meaghan, Jowa fan, Skater00, ZéroBot, Quondum, Git2010, Maschen, JFB80, Dexbot, CsDix, Prokaryotes, Kfitzell29, Ryanexler, Unknown111111 and Anonymous: 34

- **Super-Poincaré algebra** *Source:* https://en.wikipedia.org/wiki/Super-Poincar%C3%A9_algebra?oldid=716694774 *Contributors:* Michael Hardy, Charles Matthews, David Shay, Phys, Centrx, Anville, HorsePunchKid, D6, BD2412, R.e.b., Nowhither, Wavelength, Gaius Cornelius, That Guy, From That Show!, SmackBot, PKT, Headbomb, David Eppstein, Freeboson, Geometry guy, Rzs8750, Nilradical, AnonyScientist, Addbot, Omnipaedista, Charvest, Erik9bot, FrescoBot, Molitorppd22, PigFlu Oink, ZéroBot, ClueBot NG, Snotbot, BattyBot, AHusain3141 and Anonymous: 11

- **Cartan connection** *Source:* https://en.wikipedia.org/wiki/Cartan_connection?oldid=708585202 *Contributors:* SebastianHelm, Charles Matthews, Dysprosia, Phys, Tosha, Giftlite, Fropuff, Jason Quinn, Neilc, Pjacobi, Gauge, Rgdboer, Jholland, Mdd, Phils, Oleg Alexandrov, Isnow, BD2412, Rjwilmsi, Salix alba, Juan Marquez, Mathbot, Chobot, YurikBot, Wavelength, SmackBot, Paxse, Silly rabbit, Myasuda, Headbomb, West Brom 4ever, SalvNaut, Alu042, Leyo, Voorlandt, Geometry guy, Addbot, DOI bot, Luckas-bot, Fizyxnrd, Kilom691, Xqbot, Point-set topologist, Sławomir Biały, Citation bot 1, Rausch, Slawekb, Mgvongoeden, Helpful Pixie Bot and Anonymous: 19

- **Graviton** *Source:* https://en.wikipedia.org/wiki/Graviton?oldid=718050735 *Contributors:* CYD, Bryan Derksen, Timo Honkasalo, XJaM, Fubar Obfusco, Maury Markowitz, Kaczor~enwiki, Jketola, TakuyaMurata, Eric119, Looxix~enwiki, Glenn, Cyan, Wooster, Charles Matthews, Timwi, Wik, BenRG, Donarreiskoffer, Scott McNay, Stephan Schulz, Arkuat, Chris Roy, Merovingian, David19999, Giftlite, Xerxes314, Jason Quinn, Matt Crypto, CryptoDerk, Amaier, RetiredUser2, Icairns, Zfr, Lumidek, Ukexpat, Urvabara, Discospinster, Pjacobi, Vapour, Brian0918, El C, Joanjoc~enwiki, Dalf, Army1987, Mpvdm, La goutte de pluie, Physicistjedi, Daniel Arteaga~enwiki, Zenosparadox, Dethtron5000, Keenan Pepper, Viridian, SidP, Falcorian, Skeejay, Simetrical, Dr Archeville, Mpatel, Kyleca, Tmassey, Christopher Thomas, Tevatron~enwiki, Kbdank71, Drbogdan, Nightscream, Koavf, Mike Peel, Ems57fcva, FlaBot, RexNL, Chobot, DVdm, Roboto de Ajvol, Spacepotato, Anonymous editor, SnoopY~enwiki, Salsb, Bachrach44, Hyperbrand, NickBush24, Pnrj, RL0919, EEMIV, IslandGyrl, Bota47, C h fleming, Petri Krohn, Mario23, Alias Flood, Tim314, Teply, GrinBot~enwiki, SmackBot, Amcbride, Melchoir, Eskimbot, Gilliam, Skizzik, Timneu22, Complexica, Villarinho, Colonies Chris, Vladislav, Chlewbot, Xyzzyplugh, Jmnbatista, Fuhghettaboutit, Sadi Carnot, Yevgeny Kats, TenPoundHammer, Lambiam, Zaphraud, JorisvS, Mr Stephen, Ramuman, Quasar Jarosz, Lottamiata, Firewall62, Kurtan~enwiki, CmdrObot, BeenAroundAWhile, WeggeBot, Shultz IV, UncleBubba, Michael C Price, Anthmoo, Thijs!bot, Epbr123, Headbomb, KevinS06, Opelio, Spartaz, JAnDbot, Xoneca, SHCarter, BatteryIncluded, Pikazilla, Robin S, STBot, Kostisl, J.delanoy, Tarotcards, Coppertwig, Wesino, Sava ankit2006, Tygrrr, Idioma-bot, Sheliak, JoAnneThrax, TXiKiBoT, WilliamSommerwerck, Hqb, Anonymous Dissident, Antixt, SieBot, Flyer22 Reborn, Henry Delforn (old), ClueBot, Ergn, Niceguyedc, Darkicebot, DenverRedhead, Addbot, Eric Drexler, Uruk2008, DOI bot, BrianBop, PJonDevelopment, F Notebook, Legobot, Yobot, Picturesofnothing, Dov Henis, Alfredschrader, Eric-Wester, AnomieBOT, VanishedUser sdu9aya9fasdsopa, Jim1138, Materialscientist, Citation bot, Vayvor, Tomflaherty, Gap9551, ProtectionTaggingBot, Waleswatcher, FrescoBot, Juto20, LucienBOT, Paine Ellsworth, I dream of horses, Tom.Reding, RedBot, Omar.tigereyes, IVAN3MAN, Ashish.kotwal, Ale And Quail, Michael9422, D0wnfalle, Earthandmoon, Onel5969, EmausBot, Octaazacubane, 8digits, Slightsmile, K6ka, Thecheesykid, User10 5, Mattedia, Rcsprinter123, Orbjeeples, Puffin, Herk1955, ClueBot NG, Raidr, Masssly, Helpful Pixie Bot, Bibcode Bot, BG19bot, Shapoopy178, ServiceAT, PhnomPencil, Trevayne08, Brainssturm, Tjamcclain2, ChrisGualtieri, Ariscod, TheUyulala, LightandDark2000, Jessybun, Makecat-bot, Kryomaxim, JRYon, Andyhowlett, Mark viking, Yorsh07, CensoredScribe, Master Lenman, WPratiwi, Monkbot, Horseless Headman, Bryan Paul Senior, Dr.Begich, Nompynuthead, Eat me, I'm an azuki, Jacobflarsen, Gopal padma kumar, Roger Peartree, WilliamJennings1989, Brianjaythomas, DavidMiller2015 and Anonymous: 216

- **Superpartner** *Source:* https://en.wikipedia.org/wiki/Superpartner?oldid=699009083 *Contributors:* Roadrunner, SimonP, Phys, Donarreiskoffer, Giftlite, 4pq1injbok, Kocio, Alai, Duncan.france, Mpatel, Rjwilmsi, R.e.b., Drrngrvy, FlaBot, KFP, Conscious, SCZenz, SmackBot, Reedy, Dauto, Jgwacker, Thijs!bot, Headbomb, Maliz, Hans Dunkelberg, LovroZitnik, Agharo, Antixt, AlleborgoBot, Madacs, Bobathon71, Niceguyedc, Alexbot, SilvonenBot, SkyLined, Addbot, Barak Sh, Luckas-bot, ArthurBot, Xqbot, Erik9bot, Carlog3, Paine Ellsworth, Haeinous, Cracrunch, RedBot, EmausBot, Hydroxonium, Flloater, ClueBot NG, Bibcode Bot, Hrttu523, Rolf h nelson, Akro7 and Anonymous: 15

- **Gravitino** *Source:* https://en.wikipedia.org/wiki/Gravitino?oldid=685058282 *Contributors:* EddEdmondson, Schneelocke, Maximus Rex, BenRG, Donarreiskoffer, Icairns, Lumidek, Urvabara, Army1987, Jag123, Physicistjedi, Pauli133, Jef-Infojef, Kbdank71, Rjwilmsi, FlaBot, Ewlyahoocom, Conscious, Pigman, DavidConrad, Martinwilke1980, Vald, Bluebot, Vladislav, T-borg, Yevgeny Kats, Dan Gluck, Mssgill, DJBullfish, Thijs!bot, Headbomb, Spartaz, Shambolic Entity, Hair Commodore, Tarotcards, Telecomtom, Cuzkatzimhut, Antixt, AnonyScientist, SkyLined, Addbot, F Notebook, Luckas-bot, ArthurBot, Acky69, FrescoBot, Three887, StringTheory11, Ebrambot, Suslindisambiguator, Trevayne08 and Anonymous: 17

- **Cosmological constant** *Source:* https://en.wikipedia.org/wiki/Cosmological_constant?oldid=712605773 *Contributors:* AxelBoldt, Magnus Manske, Vicki Rosenzweig, Bryan Derksen, The Anome, Ed Poor, Enchanter, William Avery, Roadrunner, Schewek, Hephaestos, Boud, Bcrowell, Lquilter, TakuyaMurata, Minesweeper, Stevenj, Kimiko, Samw, Timwi, Reddi, Asar~enwiki, Dogface, Bevo, Anupamsr, Johnleemk, BenRG, Phil Boswell, Robbot, Goethean, Wereon, Giftlite, Bobblewik, Jonel, Rjpetti, Icairns, Rgrg, Burschik, JimJast, 4pq1injbok, Pjacobi, Vsmith, StephanKetz, Pavel Vozenilek, Dmr2, Bender235, RJHall, Pt, El C, Frankenschulz, RoyBoy, Rbj, I9Q79oL78KiL0QTFHgyc, Knucmo2, Jumbuck, Falcorian, Angr, OwenX, Linas, StradivariusTV, Kzollman, Mpatel, Joke137, Wisq, Christopher Thomas, Rnt20, Ashmoo, Coneslayer, Rjwilmsi, Coemgenus, Nightscream, RE, Itinerant1, Srleffler, Chobot, PointedEars, YurikBot, Hillman, RussBot, Ytrottier, SpuriousQ, Gaius Cornelius, Salsb, Sir48, Muu-karhu, DeadEyeArrow, Helge Rosé, Petri Krohn, KasugaHuang, SmackBot, Incnis Mrsi, WilyD,

Dan Gluck, Michael C Price, Alaibot, Opheicus, Headbomb, Appraiser, AlphaEta, Hans Dunkelberg, HowardFrampton, Red Act, Likebox, WikiLaurent, Sabri76, Ssaco, Truthnlove, Addbot, Tassedethe, OlEnglish, HerculeBot, Citation bot, Omnipaedista, Edited0001, Patchy1, Knowandgive, Citation bot 1, Tom.Reding, Foobarnix, ClueBot NG, Frietjes, Helpful Pixie Bot, Bibcode Bot, AHusain314, Rmlkcl, Polytope24 and Anonymous: 27

• **String theory** *Source:* https://en.wikipedia.org/wiki/String_theory?oldid=717496071 *Contributors:* AxelBoldt, Sodium, Mav, Bryan Derksen, Zundark, The Anome, Tarquin, Taw, Eean, Malcolm Farmer, Hephaestos, Olivier, Drseudo, Stevertigo, Spiff~enwiki, Edward, PhilipMW, Michael Hardy, Bewildebeast, Dante Alighieri, Gabbe, Graue, Tgeorgescu, Mcarling, CesarB, Looxix~enwiki, Ahoerstemeier, Theresa knott, Suisui, Angela, Den fjättrade ankan~enwiki, Jdforrester, Julesd, Salsa Shark, Schneelocke, Charles Matthews, Timwi, Bemoeial, Jitse Niesen, 4lex, Greenrd, ErikStewart, Furrykef, Saltine, Phys, Omegatron, Bevo, Topbanana, Trent, Nufy8, Robbot, Craig Stuntz, Fredrik, Chris 73, R3m0t, COGDEN, Mirv, Wjhonson, Sverdrup, Academic Challenger, DHN, Hadal, Khlo, ElBenevolente, HaeB, Xanzzibar, Tea2min, Giftlite, DocWatson42, Christopher Parham, Awolf002, Mporter, Amorim Parga, Mikez, Harp, Kim Bruning, Tom harrison, Ferkelparade, Leflyman, Fropuff, No Guru, Anville, Moyogo, Curps, Pashute, Nomad~enwiki, Mboverload, Solipsist, SWAdair, DemonThing, Wmahan, Btphelps, MSTCrow, Decoy, Chowbok, Gadfium, Steuard, Pgan002, Quadell, Carandol~enwiki, Antandrus, Beland, JoJan, Khaosworks, Tothebarricades.tk, Thincat, Tomruen, Shidobu, Icairns, Lumidek, NoPetrol, Avihu, Fanghong~enwiki, Trevor MacInnis, Lacrimosus, Zro, Mike Rosoft, D6, Urvabara, Felix Wan, Jkl, Discospinster, ElTyrant, Rich Farmbrough, Rhobite, Pjacobi, Alien life form, Vapour, Silence, Kzzl, LindsayH, Manil, Pavel Vozenilek, Paul August, Bender235, Kjoonlee, Mashford, Kelvinc, Perlman10s, Panu~enwiki, Brian0918, Dpotter, Livajo, El C, Laurascudder, Shanes, Zegoma beach, RoyBoy, Causa sui, Bobo192, Directorstratton, Janna Isabot, Smalljim, John Vandenberg, Flxmghvgvk, I9Q79oL78KiL0QTFHgyc, Physicistjedi, Bongoo, 4v4l0n42, Merope, Geschichte, Linuxlad, Phils, Merenta, Alansohn, Gary, JYolkowski, Enirac Sum, Ryanmcdaniel, Arthena, Borisblue, Rd232, Plumbago, Axl, R Calvete, Lightdarkness, Kocio, Bart133, Wtmitchell, Isaac, Tycho, Cal 1234, Fadereu, CloudNine, Sciurinæ, Computerjoe, Kusma, DV8 2XL, Pwqn, Gene Nygaard, Ringbang, Ceyockey, Falcorian, Bobrayner, Joriki, Mel Etitis, Linas, BillC, Jacobolus, HFarmer, Before My Ken, Netdragon, MONGO, GeorgeOrr, Mpatel, Bbatsell, GregorB, 阿部偉哉, Joke137, Christopher Thomas, Dysepsion, GSlicer, Jan.bannister, Graham87, Magister Mathematicae, Hillbrand, BD2412, Elvey, Galwhaa, Raymond Hill, JIP, RxS, Athelwulf, Edison, Sjakkalle, Rjwilmsi, Xgamer4, Jake Wartenberg, Arabani, MarSch, TheRingess, Jmcc150, Aero66, Crazynas, Juan Marquez, R.e.b., Bubba73, DoubleBlue, Zelos, AlisonW, Asafavi, Lionelbrits, Conorific, Zunz, Mathbot, Crazycomputers, RexNL, Gurch, Algri, TeaDrinker, Zifnabxar, XAXISx, Erik4, Phoenix2~enwiki, Antimatter15, Ggb667, Chobot, Visor, DVdm, Mhking, VolatileChemical, Bgwhite, Algebraist, Ben Tibbetts, YurikBot, Ugha, Wavelength, Borgx, NuclearFusion~enwiki, Angus Lepper, Hairy Dude, Jimp, Hillman, Cyferx, Wolfmankurd, Pip2andahalf, RussBot, Moronoman, Crazytales, Pippo2001, Bhny, Pigman, SpuriousQ, Branman515, Stephenb, Gaius Cornelius, Eleassar, Rsrikanth05, Bovineone, Cheesus, Shanel, NawlinWiki, Tong~enwiki, Mike18xx, SCZenz, Cleared as filed, Bdiah, Pym98, SColombo, Haemo, FF2010, Closedmouth, Reyk, Brina700, Chris Brennan, Vicarious, Brianlucas, Geoffrey.landis, Hitchhiker89, Spliffy, Pred, ArielGold, Roy Fultun, Ilmari Karonen, Katieh5584, Pentasyllabic, Lunch, DVD R W, WikiFew, That Guy, From That Show!, Street Scholar, AndrewWTaylor, QSquared, Sardanaphalus, Vanka5, MacsBug, Hvitlys, SmackBot, Kurochka, Zazaban, Tom Lougheed, Prodego, KnowledgeOfSelf, Hydrogen Iodide, Melchoir, Vald, Skrewtape, Atomota, Canthusus, GaeusOctavius, Cool3, Andyvn22, Gilliam, Skizzik, RobertM525, Dauto, Bluebot, SSJ 5, Keegan, Aidan Croft, Thumperward, Oli Filth, Silly rabbit, Timneu22, SchfiftyThree, Moshe Constantine Hassan Al-Silverburg, Complexica, Rediahs, RayAYang, Aero77, Adamstevenson, Ikiroid, Epastore, Baronnet, Ned Scott, Sbharris, Colonies Chris, Konstable, Sct72, Scwlong, Can't sleep, clown will eat me, Timothy Clemans, Onorem, Neilanderson, EvelinaB, TKD, KerathFreeman, Addshore, UU, The tooth, Pepsidrinka, Somebody2292, --=The Doctor=--, Fuhghettaboutit, Cybercobra, Irish Souffle, Nakon, Jdlambert, James McNally, MichaelBillington, Lostart, Insineratehymn, Drphilharmonic, SpiderJon, DMacks, Ihatetoregister, Where, Michael IFA, Yevgeny Kats, Vasiliy Faronov, Byelf2007, Angela26, Visium, Rory096, Zymurgy, Harryboyles, Mdl53711, T-dot, Titus III, Ergative rlt, MagnaMopus, UberCryxic, Vgy7ujm, Lazylaces, Linnell, Mgiganteus1, Nonsuch, IronGargoyle, Ckatz, DoItAgain, AstroGod, Kirbytime, Jimbo Mahoney, FredrickS, Invisifan, Ryulong, Ryanjunk, MathStuf, Mike Doughney, Norm mit, Hindol, Dan Gluck, Huntscorpio, Iridescent, K, Sunoco, You? Me? Us?, CzarB, Rabinzkaman, JoeBot, Lottamiata, Tony Fox, Vrkaul, Torrazzo, Gil Gamesh, Areldyb, Courcelles, Tawkerbot2, Gebrah, Shamvil, Fdssdf, DKqwerty, Lbr123, Harold f, Heqs, Devourer09, Duduong, Sarvagnya, Dewayne76, JForget, Cg-realms, InvisibleK, CRGreathouse, CmdrObot, Earthlyreason, Van helsing, Olaf Davis, CBM, Rawling, Jibal, Witten Is God, Nunquam Dormio, Giko, KnightLago, Thubsch, Leujohn, SlashDot, TheTito, Karenjc, Myasuda, Emarv, Cydebot, Gmusser, Gogo Dodo, Jkokavec, Kahananite, Quajafrie, Michael C Price, Doug Weller, DumbBOT, Narayanese, AlphaNumeric, SRoughsedge, Vanished User jdksfajlasd, Woland37, Zalgo, Daniel Olsen, UberScienceNerd, Bkazaz, DJBullfish, Thijs!bot, Epbr123, Rwmnau, Babemachine, Pimpin101, Mbell, O, Faigl.ladislav, Ucanlookitup, Andyjsmith, Headbomb, Tcturner2002, Marek69, Brahmajnani, Arthurcprado~enwiki, Y.t., D3gtrd, Babemonkey, Dark dude, Duncan McB, EdJohnston, MichaelMaggs, Ancientanubis, Natalie Erin, Hempfel, Jomoal99, Mmortal03, Mentifisto, Geekdom04, AntiVandalBot, Luna Santin, Seaphoto, Ed270791, Opelio, Doc Tropics, David136a, NithinBekal, Dotdotdotdash, Helicoptor, Poshzombie, MontanNito, Dylan Lake, Maximilian77, Shlomi Hillel, Db63376, SamIAmNot, Knotwork, Res2216firestar, Superior IQ Genius, MER-C, Andonic, Sitethief, 100110100, TallulahBelle, Nestamachine, Savant13, Daynightrader, Goldenglove, Charibdis, Acroterion, Ophion, Aigisthos, Editmyhandman, Aruben537, Magioladitis, WolfmanSF, Bongwarrior, VoABot II, Yandman, JamesBWatson, باسم, Qutt, Jespinos, Kevinmon, Aka042, Froid, DAGwyn, Catgut, Panser Born, Ensign beedrill, Perspectival, JJ Harrison, Dirac66, Justanother, Aziz1005, Cpl Syx, ChazBeckett, Teardrop onthefire, WLU, Stephen shenker, Robin S, SkepticVK, Joshua Davis, Mkroh, B9 hummingbird hovering, S3000, Hdt83, MartinBot, FlieGerFaUstMe262, Ytomem, Shimwell, Arjun01, KrishSundaresan, Anaxial, Jay Litman, Alexcalamaro, Andrej.westermann, Smokizzy, LedgendGamer, Cyrus Andiron, Peteryoung144, Tgeairn, Artaxiad, HEL, AlphaEta, J.delanoy, AstroHurricane001, Maurice Carbonaro, Yonidebot, Morris729, M C Y 1008, 69gangsta420, It Is Me Here, Shawn in Montreal, Janus Shadowsong, Bailo26, Fredsie, Madagaskar07, Duchesserin, AntiSpamBot, CHIAGEHYANG, Chiswick Chap, Watsup1313, Belovedfreak, HaloInverse, NewEnglandYankee, Scott1329m, Thesis4Eva, Policron, Jrcla2, KylieTastic, WJBscribe, Rnricklefs, Jamesofur, Eyelidlessness, Jonnyk aus, Kvdveer, JavierMC, Izno, Xiahou, CardinalDan, Sheliak, HamatoKameko, Malik Shabazz, Concertmusic, JohnBlackburne, JustinHagstrom, Fences and windows, Wooba doob, Philip Trueman, DoorsAjar, HowardFrampton, TXiKiBoT, Zidonuke, Red Act, Kriak, Calwiki, Technopat, Hqb, Andrius.v, Anonymous Dissident, Crohnie, AlysTarr, Qxz, Vanished user ikijeirw34iuaeolaseriffic, Impunv, Seraphim, Martin451, Don4of4, ABigGreenHippo, Huperphuff, LeaveSleaves, Kaenneth, StringyGuy, Maxim, Erth64net, Meters, Lamro, Rickstauduhar, Enviroboy, Turgan, Anna512, PhysPhD, Northfox, NPguy, Matthew Sanders, Luke Walkerson, Newbyguesses, MissMJ, SieBot, Escher26, J.A.Ireland, BA (IHPST), 4wajzkd02, Robdunst, Dreamafter, Pallab1234, Dbelange, MTHarden, Lemonflash, Kylemew, Yintan, GlassCobra, Discrete, Bentogoa, Likebox, Flyer22 Reborn, Exert, ProGeek314, Arbor to SJ, Babawhitemoose, Caidh, Dhatfield, Audree, Oxymoron83, Pretty Green, Weaselstomp, Manway, Alex.muller, Taco Manipulator, Tschach, Manheat84, Anchor Link Bot, Mikebernstein, ImperialismGo, Nergaal, Ionfield, Ayleuss, Sh4wz0r, Naturespace, ImageRemovalBot, Martarius, Phyte, ClueBot, The Thing That Should Not Be, String4d, Illusion96, Polyamorph, Mpd1989, Alexdeburca18, Wiggl3sLimited, Excirial, Kjramesh, Jusdafax, Resoru, WikiZorro, Eeekster, Verum~enwiki, Tamaratrouts, Gtstricky, Humanino, Brews ohare, NuclearWarfare, Cenarium, Arjayay, Razorflame, Scoobey, BOTarate, Sideswiper, Thingg, Capudo, BVBede, Versus22, Introductory adverb clause, Melon-Bot, SoxBot III, Egmontaz, Notpayingthepsychiatrist, DumZiBoT, BahTab, TimothyRias, Aj00200, Reaperfromhell, Dunkaroo207, XLinkBot, AlexGWU, Impshum, Saeed.Veradi, Little Mountain 5, Guy392, David424, Truthnlove, Qweeveen, Tayste, Addbot, Steven66s, Denali134, Elemented9, Varrey280303, Eric Drexler, Some jerk on the Internet, Fizzycyst, Uruk2008, DOI bot, Jojhutton, AngryBacon, Non-dropframe, Captain-tucker, Auspex1729, Kongr43gpen, Fgnievinski, Rhetoric Of A Sophist, Ronhjones, CanadianLinuxUser, Cst17, Download, Glane23, Bassbonerocks, Chzz, Favonian, Kronix35, LinkFA-Bot, Udugunit, Aktsu, Tassedethe, Numbo3-bot, Anpecota, Tide rolls, HerpesVirus, SDJ, OlEnglish, Scourge of God, Davidmedlar, Couldbenoway66, Yobot, Maxdamantus, Terrisknickers, Kartano, TaBOT-zerem, Julia W, Unique and proud of it, FireMouseHQ, Terrifictriffid, ArchonMagnus, CinchBug, Synchronism, AnomieBOT, Cleeseheb, 1exec1, Charlesvi, Bigdaddy4x4, Gitman4, Jim1138, IRP, Mintrick, Drweetmola, Ornamentalone, M00npirate, Gautam10, Csigabi, Poli-Psy, Materialscientist, 90 Auto, Citation bot, Teleprinter Sleuth, Vuerqex, Twri, Frankenpuppy, Fuzzy Bob Saget, DirlBot, Georgepowell2008, Heidisql, Cureden, Ekwos, Capricorn42, Gensanders, NFD9001, Anna Frodesiak, Tomwsulcer, A23649, Pra1998, Coretheapple, RadiX, Jagbag2, Vandalism destroyer, Ab1, Omnipaedista, Bandit5005, Shirik, RibotBOT, Waleswatcher, Saalstin, Amaury, Aaron35510, Caz34, Doulos Christos, Sewblon, Born Gay, Capricorn24, SchnitzelMannGreek, A. di M., SpacePyjamas, Kierkkadon, A.amitkumar, Dougofborg, StringLove, Nobelprizewinner, Astiburg, FrescoBot, Fortdj33, Paine Ellsworth, Goodbye Galaxy, HJ Mitchell, Steve Quinn, Vhann, Kwiki, Xhaoz, Citation bot 1, Batong, Gil987, Pinethicket, I dream of horses, Tallboyhoops1991, Three887, Steveo27five, RedBot, Sardinita, Serols, Vhsatheeshkumar, Swisstingle, DeletionUK, Reconsider the static, IVAN3MAN, Remingtonhill1, Orenburg1, Coltonhs, Angus Guilherme, Smamaret, Bethovenn, Dinamik-bot, Dc987, Oswaldo Zapata, Egemont, Syebo, Alaithiran, Reaper Eternal, Seahorseruler, Ybungalobill, Quaker phil, Specs112, Dr. Aakash Patel, Tbhotch, StormbringerUK, Minimac, Mathgenius3141592, Keegscee, Omgwaffels, Mick le pick, Solancel, Aznhero3793, Dwielark, Afteread, Enauspeaker, EmausBot, MaooaM, Immunize, Az29, Milkocookie, Faolin42, Fotoni, RA0808, RenamedUser01302013, 8digits, Yukiseaside, Slightsmile, Tommy2010, Winner 42, Wikipelli, Dcirovic, JonezyKiDx, Joe Gazz84, ZéroBot, Timeitsways, John Cline, Cogiati, Quaqa, Chrispaps2413, Nasulikid, Vollrath2323, Benjamin1414141414141414, Arbnos, Green Lane, A930913, Bamyers99, Azeraphale, H3llBot, Encyclopadia, Danga1988, Ollainen, PoisonGM, Wayne Slam, OnePt618, Knome335, L Kensington, Lulzprotuns, Kranix, Rpcappello, Maschen, Vastly~enwiki, Donner60, CatFiggy, CountMacula, Orange Suede Sofa, Etov, M1k3 101, Bill william compton, Wakabaloola, TERBAFAN, Nickslspride34, NeuralLotus, Isocliff, Brechbill123, Xanchester, ClueBot NG, Martti Muukkonen, KagakuKyouju, Jeff Song, This lousy T-shirt, Satellizer, Name Omitted, Marcdean123, Wiki incorp, Frietjes, O.Koslowski, Alexdamaino9, Dream of Nyx, Blackhall616, Widr, Sashhere, WikiPuppies, Stu181, T00g00d96, Pluma, Storm.sarup, Helpful Pixie Bot, Manzeet, Waffleboy36, HMSSolent, Mikeshelton1, Bibcode Bot, 2001:db8, Phillip.phillipson, Hoaxinator, Lowercase sigmabot, Thor cherubim, BG19bot, Mrshabam, Nishch, Flowerhat15, AvocatoBot, Housegeek224, MahRanch, Benzband, Altaïr, Benhenchdickthomas, Shreyakstring, Sweaty maori sphincter, DaFalk, Dsabo74, Ratanmaitra, MM4EVAH, Steven.w.kowalski, Minsbot, JGallardo2600, Dylanlatham, Myfriendganesha, OCCullens, Likeaboss189, Sean271293, LinusE8, BattyBot, Several Pending, Aldrich2122, CommanderMoka, Cyberbot II, The Illusive Man, ChrisGualtieri, KoalamaN2, Trevorkid45, Catsloveit07, Alex Modzz, Rustyjamsen, Goh ryangoh, Dexbot, Exolius, Hilander316, Alman1234321, SuperCalzer, LightandDark2000, MeekMelange, BQND, Cdarrai1, Kephir, TheMonkeyboy524, Michael Anon, TwoTwoHello, Mattfat8, Lugia2453, Anruy, Rachel weld, Jamesx12345, AHusain314, BossEditors, Hillbillyholiday, Joeinwiki, Mattninja, Theshadow444, Asaa82, Jakemarz197, Kzhang1025, Epicgenius, Spongbob456789, ⬚, TestMaster, Ianreisterariola, GrapperJ, Makeitnasty, Moemajdi, I am One of Many, NualaIvy, BAZINGASS, St3fanPC, Eyesnore, Isaac grozd, Jordanissexyaf1999, Baruch6525, Mosbruckercj, Ihatedirac2k13, Jonamithy121314, 123physicsquantum, Jt198, RaphaelQS, HeyJude70, AParker628, DimReg, A.k.blaze1, Joshuk, Zenibus, Nianoobasik, Ihelpapplen, Gamo To Apoel, SacredLabyrinth, Ginsuloft, Vampre1122, Dimension10, Howard Wolowitz, AddWittyNameHere, Polytope24, Elysion, Tutun12$, Longerboats7, SimonWombat8, Konveyor Belt, Vtank54, Micheal545, Hck24, Caliae19, Hexafish, Simpick, TheRealTheKoi, Bballbro62, Monkbot, ArmyPath, Gabero.88, TheQ Editor, Jtsmith098, Joshmiller1, Hanseer360, XXvPIEvXx, Dbennett 24, Ghikpenos, Nick65633, Saundra03, Thehippothatknows, Sewwgers, Teelaskeletor, Cirksena, Balockaye1234, PloppyDoo, Yesufu29, Lumpy2k14, Podayeruma, Abstract92, Sbenfiel, Monkman2k4, Swegwegdgfyetkfoffkkfkfkv, John95541234, Poopman224, ScrapIronIV, Tetra quark, GeneralizationsAreBad, Shivansh2014n, KasparBot, SHUCKYLUCKY, Fabiotheoto, FartGoblin, Joca potato, Joshcool246, Theoretical Physisist4444, JanetTom55, Reg7d88, CHANDLER MERRILL, Baking Soda, FklfjDKFd bfl, Rajputclann, Entranced98, Jahziahk, Mjhog, Strong81, WikiTikiDude007, ILoveShukli, Qwerty2345B, A1D1A2D2 and Anonymous: 1594

- **String duality** *Source:* https://en.wikipedia.org/wiki/String_duality?oldid=703639606 *Contributors:* Kingturtle, Fropuff, Mpatel, SmackBot, Dan Gluck, Thadius856, PhysPhD, Addbot, Bte99, Lightbot, Xqbot, Omnipaedista, Sa'y, CaroleHenson, ChrisGualtieri, Yikkayaya and Anonymous: 3

- **D-brane** *Source:* https://en.wikipedia.org/wiki/D-brane?oldid=696680667 *Contributors:* Zundark, TakuyaMurata, Karada, JWSchmidt, AugPi, Smack, Schneelocke, Gandalf61, Michael Snow, Fropuff, Anville, Just Another Dan, Phe, Lumidek, Rgrg, H0riz0n, El C, Constantine, I9Q79oL78KiL0QTFHgyc, FlaBot, Bhny, Nick, Zwobot, Sardanaphalus, KnightRider~enwiki, Teemu Ruskeepää, Colonies Chris, Scwlong, QFT, Eric Olson, Fuhghettaboutit, Vampus, JarahE, Twyder, Eewild, 345Kai, Cydebot, Headbomb, J. W. Love, Nick Number, Magioladitis, Jpod2, STBot, HEL, VolkovBot, TXiKiBoT, PhysPhD, Jonathanroxhead, Excirial, Alexbot, ResidueOfDesign, Addbot, LaaknorBot, Tassedethe, Lightbot, Luckas-bot, Yobot, Amirobot, Azcolvin429, Royote, Citation bot, Twri, ChristopherKingChemist, Omnipaedista, Galaktiker, Mentibot, Wakabaloola, Petrb, Frietjes, Luizpuodzius, OCCullens, Polytope24 and Anonymous: 30

- **Brane** *Source:* https://en.wikipedia.org/wiki/Brane?oldid=714337937 *Contributors:* Bth, Michael Hardy, DIG~enwiki, Samuelsen, JWSchmidt, Silverfish, Wetman, Bcorr, Blainster, Fropuff, Just Another Dan, D3, Lumidek, Yuriz, Rhobite, H0riz0n, Ben Standeven, RoyBoy, Mairi, Constantine, GatesPlusPlus, Kocio, Agquarx, Mpatel, Liface, BD2412, Quiddity, R.e.b., Mathbot, BradBeattie, Metropolitan90, YurikBot, Wavelength, NawlinWiki, Wknight94, Closedmouth, SmackBot, Kurochka, Jwestbrook, Autarch, Seanor32, Silly rabbit, Colonies Chris, Nsmith4658, Mesons, Monotonehell, TheVikingRaider, Yevgeny Kats, Spiritia, PaddyM, Czoller, Calmargulis, BeenAroundAWhile, Adailton, Julius M-D, J. W. Love, Julia Rossi, Chrisjj3, MER-C, Steveprutz, Just H, N.Nahber, Urco, Alexrussell101, Cyborg Ninja, Idioma-bot, VolkovBot, Anonymous Dissident, Michael Frind, Paucabot, Drschawrz, SieBot, Tresiden, OKBot, ClueBot, The Thing That Should Not Be, SilvonenBot, NonvocalScream, Addbot, Jujutsuka, Royote, LilHelpa, Patmethenyfan, Omnipaedista, Nagualdesign, Kgrad, Tbhotch, Tesseract2, EmausBot, MathMaven, ClueBot NG, Mikeflem, Gilderien, Baseball Watcher, Frietjes, DBigXray, Lowercase sigmabot, BG19bot, Solomon7968, OCCullens, BattyBot, Brirush, E8xE8, Dimension10, Polytope24, Ihsanturk, Eno Lirpa and Anonymous: 50

- **Perturbation theory** *Source:* https://en.wikipedia.org/wiki/Perturbation_theory?oldid=714444479 *Contributors:* CYD, FlorianMarquardt, Stevertigo, Michael Hardy, Kku, AugPi, Ideyal, Jitse Niesen, Phys, Robbot, Lowellian, Tea2min, Giftlite, BenFrantzDale, Neilc, Karol Langner,

The Land, Igorivanov~enwiki, MuDavid, Bender235, Cmdrjameson, Haham hanuka, Keenan Pepper, RJFJR, Count Iblis, Dirac1933, Mattbrundage, Djsasso, Oleg Alexandrov, Linas, Yansa, SeventyThree, Bubba73, Mathbot, ChrisChiasson, YurikBot, Piet Delport, Tong~enwiki, Joel7687, Dhollm, Tony1, DerHannes, Artemisfowl3rd, SmackBot, Mmernex, Tom Lougheed, Mcld, Chris the speller, Bduke, Complexica, Nbarth, Colonies Chris, MaxSem, Ohconfucius, Nishkid64, Harryboyles, Tomatoman, JorisvS, Frokor, Hiiiiiiiiiiiiiiiiiiii, Charles Baynham, Chetvorno, Khromegnome, CBM, Myasuda, Cydebot, Tawkerbot4, Roy W. Wright, Headbomb, Ben pcc, Engelbaet, David Eppstein, Alexei Kopylov, P.wormer, Cuzkatzimhut, Maghnus, Bphillab, Lechatjaune, EverGreg, Vsst, SieBot, JerroldPease-Atlanta, Nancy, Yhkhoo, ClueBot, Warbler271, PtolemyGalen, Mild Bill Hiccup, Zllvette, CohesionBot, Guiermo, Lacce, Crowsnest, DumZiBoT, Terry0051, Queenmomcat, Download, Yobot, TaBOT-zerem, AnomieBOT, Pownuk, Nfr-Maat, J04n, Resident Mario, Pradameinhoff, FrescoBot, Lotje, Mhilferink, Yger, Qweilun, Mattedia, Zfeinst, ClueBot NG, Wcherowi, LPOG1, Helpful Pixie Bot, Vlos2008, PhnomPencil, Anylai, Dexbot, Andyhowlett, Pde-calculus, Hctrmycss and Anonymous: 78

- **UV completion** *Source:* https://en.wikipedia.org/wiki/UV_completion?oldid=609369095 *Contributors:* The Anome, Phys, Hugo~enwiki, Lumidek, Tweet Tweet, Lockley, Conscious, Aaron Schulz, SmackBot, QFT, ShelfSkewed, Erik9bot, Trinitresque and Anonymous: 3

- **N=8 Supergravity** *Source:* https://en.wikipedia.org/wiki/N%3D8_Supergravity?oldid=703897466 *Contributors:* Michael Hardy, Axl, King of Hearts, Wavelength, Magioladitis, Katharineamy, Drschawrz, Yobot, AnomieBOT, Bj norge, CaroleHenson, Reddogsix, Oudheusa, Shedoblyde, Coal scuttle, Tyslaugengifbhef, JohnStarling and Anonymous: 6

- **Higher-dimensional supergravity** *Source:* https://en.wikipedia.org/wiki/Higher-dimensional_supergravity?oldid=711883658 *Contributors:* Michael Hardy, Rich Farmbrough, Zazaban, Colonies Chris, R'n'B, Student7, Telecomtom, Davehi1, Yobot, AnomieBOT, LilHelpa, Omnipaedista, Hep thinker, Paine Ellsworth, I dream of horses, Maschen, RockMagnetist, Snotbot, CaroleHenson, Miszatomic and Anonymous: 7

25.10.2   Images

- **File:080998_Universe_Content_240_after_Planck.jpg** *Source:* https://upload.wikimedia.org/wikipedia/commons/b/b6/080998_Universe_Content_240_after_Planck.jpg *License:* Public domain *Contributors:* http://map.gsfc.nasa.gov/media/080998/index.html updated data from http://www.nasa.gov/mission_pages/planck/news/planck20130321.html *Original artist:* NASA, Modified by User:▨▨

- **File:3-orbifold-voronoi-animated.gif** *Source:* https://upload.wikimedia.org/wikipedia/commons/c/c3/3-orbifold-voronoi-animated.gif *License:* CC BY-SA 4.0 *Contributors:* Own work *Original artist:* PureJadeKid

- **File:AdS3.svg** *Source:* https://upload.wikimedia.org/wikipedia/commons/4/47/AdS3.svg *License:* CC BY-SA 3.0 *Contributors:* This file was derived from: AdS3 (new).png
Original artist:

- derivative work: Alex Dunkel (Maky)

- **File:Albert_Einstein_portrait.jpg** *Source:* https://upload.wikimedia.org/wikipedia/en/f/f7/Albert_Einstein_portrait.jpg *License:* PD-US *Contributors:*
http://images.google.com/hosted/life/628e99cf2e26233d.html *Original artist:*
E. O. Hoppe. (1878-1972) Published on LIFE

- **File:Ambox_important.svg** *Source:* https://upload.wikimedia.org/wikipedia/commons/b/b4/Ambox_important.svg *License:* Public domain *Contributors:* Own work, based off of Image:Ambox scales.svg *Original artist:* Dsmurat (talk · contribs)

- **File:BBH_gravitational_lensing_of_gw150914.webm** *Source:* https://upload.wikimedia.org/wikipedia/commons/a/a4/BBH_gravitational_lensing_of_gw150914.webm *License:* CC BY-SA 4.0 *Contributors:* https://www.ligo.caltech.edu/video/ligo20160211v3 (video link); see also http://www.black-holes.org/gw150914 *Original artist:* Simulating eXtreme Spacetimes Lensing

- **File:Bundesarchiv_Bild183-R57262,_Werner_Heisenberg.jpg** *Source:* https://upload.wikimedia.org/wikipedia/commons/f/f8/Bundesarchiv_Bild183-R57262%2C_Werner_Heisenberg.jpg *License:* CC BY-SA 3.0 de *Contributors:* This image was provided to Wikimedia Commons by the German Federal Archive (Deutsches Bundesarchiv) as part of a cooperation project. The German Federal Archive guarantees an authentic representation only using the originals (negative and/or positive), resp. the digitalization of the originals as provided by the Digital Image Archive. *Original artist:* Unknown<a href='//www.wikidata.org/wiki/Q4233718' title='wikidata:Q4233718'><img alt='wikidata:Q4233718' src='https://upload.wikimedia.org/wikipedia/commons/thumb/f/ff/Wikidata-logo.svg/20px-Wikidata-logo.svg.png' width='20' height='11' srcset='https://upload.wikimedia.org/wikipedia/commons/thumb/f/ff/Wikidata-logo.svg/30px-Wikidata-logo.svg.png 1.5x, https://upload.wikimedia.org/wikipedia/commons/thumb/f/ff/Wikidata-logo.svg/40px-Wikidata-logo.svg.png 2x' data-file-width='1050' data-file-height='590' /></a>

- **File:Calabi-Yau-alternate.png** *Source:* https://upload.wikimedia.org/wikipedia/commons/5/55/Calabi-Yau-alternate.png *License:* CC BY-SA 2.5 *Contributors:* Transferred from en.wikipedia to Commons by Lunch. *Original artist:* The original uploader was Lunch at English Wikipedia

- **File:Calabi_yau.jpg** *Source:* https://upload.wikimedia.org/wikipedia/commons/f/f3/Calabi_yau.jpg *License:* Public domain *Contributors:* Mathematica output, created by author *Original artist:* Jbourjai

- **File:Clebsch_Cublic.png** *Source:* https://upload.wikimedia.org/wikipedia/commons/7/7c/Clebsch_Cublic.png *License:* CC BY-SA 3.0 *Contributors:* I created this on my own computer using the free software Surfer *Original artist:* Fly by Night

- **File:Commons-logo.svg** *Source:* https://upload.wikimedia.org/wikipedia/en/4/4a/Commons-logo.svg *License:* CC-BY-SA-3.0 *Contributors:* ? *Original artist:* ?

- **File:Compactification_example.svg** *Source:* https://upload.wikimedia.org/wikipedia/commons/f/f5/Compactification_example.svg *License:* CC BY-SA 4.0 *Contributors:* Brian Greene (2004). The Elegant Universe (DVD). Part II (String's the thing): WGBH Boston Video. Event occurs at 43:55. OCLC 54019786 *Original artist:* Alex Dunkel (Maky)

- **File:Crab_Nebula.jpg** *Source:* https://upload.wikimedia.org/wikipedia/commons/0/00/Crab_Nebula.jpg *License:* Public domain *Contributors:* HubbleSite: gallery, release. *Original artist:* NASA, ESA, J. Hester and A. Loll (Arizona State University)
- **File:Cyclic_group.svg** *Source:* https://upload.wikimedia.org/wikipedia/commons/5/5f/Cyclic_group.svg *License:* CC BY-SA 3.0 *Contributors:*
- Cyclic_group.png *Original artist:*
- derivative work: Pbroks13 (talk)
- **File:D3-brane_et_D2-brane.PNG** *Source:* https://upload.wikimedia.org/wikipedia/commons/8/88/D3-brane_et_D2-brane.PNG *License:* Public domain *Contributors:* Image:D-brane.PNG, oeuvre personnelle. *Original artist:* <a href='//commons.wikimedia.org/w/index.php?title=User:Rogilbert,,&,,action=edit,,&,,redlink=1' class='new' title='User:Rogilbert (page does not exist)'>Rogilbert</a>
- **File:Edit-clear.svg** *Source:* https://upload.wikimedia.org/wikipedia/en/f/f2/Edit-clear.svg *License:* Public domain *Contributors:* The *Tango! Desktop Project. Original artist:*

  The people from the Tango! project. And according to the meta-data in the file, specifically: "Andreas Nilsson, and Jakub Steiner (although minimally)."
- **File:Edward_Witten.jpg** *Source:* https://upload.wikimedia.org/wikipedia/commons/9/97/Edward_Witten.jpg *License:* Public domain *Contributors:* Own work *Original artist:* Ojan
- **File:Einstein_cross.jpg** *Source:* https://upload.wikimedia.org/wikipedia/commons/c/c8/Einstein_cross.jpg *License:* Public domain *Contributors:* http://hubblesite.org/newscenter/archive/releases/1990/20/image/a/ *Original artist:* NASA, ESA, and STScI
- **File:Elevator_gravity.svg** *Source:* https://upload.wikimedia.org/wikipedia/commons/1/11/Elevator_gravity.svg *License:* CC BY-SA 3.0 *Contributors:*
- Elevator_gravity2.png *Original artist:*
- derivative work: Pbroks13 (talk)
- **File:Em_dipoles.svg** *Source:* https://upload.wikimedia.org/wikipedia/commons/f/f0/Em_dipoles.svg *License:* CC0 *Contributors:* Own work *Original artist:* Maschen
- **File:Em_monopoles.svg** *Source:* https://upload.wikimedia.org/wikipedia/commons/2/2f/Em_monopoles.svg *License:* CC0 *Contributors:* Own work *Original artist:* Maschen
- **File:Ergosphere.svg** *Source:* https://upload.wikimedia.org/wikipedia/commons/0/0c/Ergosphere.svg *License:* CC-BY-SA-3.0 *Contributors:* own work based on the graphic uploaded by IMeowbot *Original artist:* MesserWoland
- **File:Fano_plane.svg** *Source:* https://upload.wikimedia.org/wikipedia/commons/a/af/Fano_plane.svg *License:* Public domain *Contributors:* created using vi *Original artist:* de:User:Gunther
- **File:Folder_Hexagonal_Icon.svg** *Source:* https://upload.wikimedia.org/wikipedia/en/4/48/Folder_Hexagonal_Icon.svg *License:* Cc-by-sa-3.0 *Contributors:* ? *Original artist:* ?
- **File:GabrieleVeneziano.jpg** *Source:* https://upload.wikimedia.org/wikipedia/commons/9/95/GabrieleVeneziano.jpg *License:* CC BY-SA 2.5 *Contributors:* Taken by Betsythedevine *Original artist:* The original uploader was Betsythedevine at English Wikipedia
- **File:Gravitational_red-shifting.png** *Source:* https://upload.wikimedia.org/wikipedia/commons/5/5c/Gravitational_red-shifting.png *License:* CC-BY-SA-3.0 *Contributors:* ? *Original artist:* ?
- **File:Gravwav.gif** *Source:* https://upload.wikimedia.org/wikipedia/commons/5/5c/Gravwav.gif *License:* CC-BY-SA-3.0 *Contributors:* self-made, using standard (TT-gauge) description of linearized sinusoidal gravitational wave *Original artist:* Mapos
- **File:Heawood_graph_bipartite.svg.png** *Source:* https://upload.wikimedia.org/wikipedia/commons/7/72/Heawood_graph_bipartite.svg.png *License:* CC-BY-SA-3.0 *Contributors:* Transferred from en.wikipedia to Commons. *Original artist:* The original uploader was Mathsci at English Wikipedia
- **File:Henri_Poincare.jpg** *Source:* https://upload.wikimedia.org/wikipedia/commons/2/28/Henri_Poincare.jpg *License:* Public domain *Contributors:* ? *Original artist:* ?
- **File:Hqmc-vector.svg** *Source:* https://upload.wikimedia.org/wikipedia/commons/6/68/Hqmc-vector.svg *License:* CC BY 3.0 *Contributors:* Own work *Original artist:* VermillionBird
- **File:Ilc_9yr_moll4096.png** *Source:* https://upload.wikimedia.org/wikipedia/commons/3/3c/Ilc_9yr_moll4096.png *License:* Public domain *Contributors:* http://map.gsfc.nasa.gov/media/121238/ilc_9yr_moll4096.png *Original artist:* NASA / WMAP Science Team
- **File:Joseph_Polchinski.jpg** *Source:* https://upload.wikimedia.org/wikipedia/commons/c/cf/Joseph_Polchinski.jpg *License:* Public domain *Contributors:* Transferred from en.wikipedia to Commons by Magnus Manske using CommonsHelper. *Original artist:* The original uploader was Lumidek at English Wikipedia
- **File:JuanMaldacena.jpg** *Source:* https://upload.wikimedia.org/wikipedia/commons/b/bc/JuanMaldacena.jpg *License:* CC BY 3.0 *Contributors:* Own work by the original uploader *Original artist:* Lumidek at English Wikipedia
- **File:KleinInvariantJ.jpg** *Source:* https://upload.wikimedia.org/wikipedia/commons/3/37/KleinInvariantJ.jpg *License:* Public domain *Contributors:* made with mathematica, own work *Original artist:* Jan Homann
- **File:LISA.jpg** *Source:* https://upload.wikimedia.org/wikipedia/commons/b/b5/LISA.jpg *License:* Public domain *Contributors:* ? *Original artist:* ?
- **File:Labeled_Triangle_Reflections.svg** *Source:* https://upload.wikimedia.org/wikipedia/commons/3/38/Labeled_Triangle_Reflections.svg *License:* Public domain *Contributors:* Own work *Original artist:* Jim.belk

- **File:Lensshoe_hubble.jpg** *Source:* https://upload.wikimedia.org/wikipedia/commons/a/a9/Lensshoe_hubble.jpg *License:* Public domain *Contributors:* http://apod.nasa.gov/apod/image/1112/lensshoe_hubble_3235.jpg *Original artist:* ESA/Hubble & NASA

- **File:LeonardSusskindStanford2009_cropped.jpg** *Source:* https://upload.wikimedia.org/wikipedia/commons/f/f8/ LeonardSusskindStanford2009_cropped.jpg *License:* CC BY-SA 3.0 *Contributors:* File:LeonardSusskindStanford2009.jpg *Original artist:* Jonathan Maltz

- **File:Light_cone.svg** *Source:* https://upload.wikimedia.org/wikipedia/commons/2/27/Light_cone.svg *License:* Public domain *Contributors:* Own work *Original artist:* Sakurambo

- **File:Light_deflection.png** *Source:* https://upload.wikimedia.org/wikipedia/commons/c/c2/Light_deflection.png *License:* CC BY-SA 3.0 *Contributors:* self-made, using numerical integration methods to solve the geodetic equation for light near a spherical massive object (Schwarzschild metric) *Original artist:* Markus Poessel (Mapos)

- **File:Limits_of_M-theory.svg** *Source:* https://upload.wikimedia.org/wikipedia/commons/b/b8/Limits_of_M-theory.svg *License:* CC BY-SA 3.0 *Contributors:*
Limits of M-theory.png

  Original artist:

- derivative work: Alex Dunkel (Maky)

- **File:MSSM_Flavor_Changing.svg** *Source:* https://upload.wikimedia.org/wikipedia/commons/c/cb/MSSM_Flavor_Changing.svg *License:* CC BY-SA 3.0 *Contributors:* Transferred from en.wikipedia to Commons. *Original artist:* JabberWok at English Wikipedia

- **File:Max_Born.jpg** *Source:* https://upload.wikimedia.org/wikipedia/commons/f/f7/Max_Born.jpg *License:* Public domain *Contributors:* ? *Original artist:* ?

- **File:Meissner_effect_p1390048.jpg** *Source:* https://upload.wikimedia.org/wikipedia/commons/5/55/Meissner_effect_p1390048.jpg *License:* CC-BY-SA-3.0 *Contributors:* self photo *Original artist:* Mai-Linh Doan

- **File:Model_space_development.svg** *Source:* https://upload.wikimedia.org/wikipedia/en/0/09/Model_space_development.svg *License:* PD *Contributors:* ? *Original artist:* ?

- **File:MontreGousset001.jpg** *Source:* https://upload.wikimedia.org/wikipedia/commons/4/45/MontreGousset001.jpg *License:* CC-BY-SA-3.0 *Contributors:* Self-published work by ZA *Original artist:* Isabelle Grosjean ZA

- **File:Newtonian_gravity_field_(physics).svg** *Source:* https://upload.wikimedia.org/wikipedia/commons/2/22/Newtonian_gravity_field_ %28physics%29.svg *License:* CC0 *Contributors:* Own work *Original artist:* Maschen

- **File:Nuvola_apps_edu_mathematics_blue-p.svg** *Source:* https://upload.wikimedia.org/wikipedia/commons/3/3e/Nuvola_apps_edu_ mathematics_blue-p.svg *License:* GPL *Contributors:* Derivative work from Image:Nuvola apps edu mathematics.png and Image:Nuvola apps edu mathematics-p.svg *Original artist:* David Vignoni (original icon); Flamurai (SVG convertion); bayo (color)

- **File:Nuvola_apps_kalzium.svg** *Source:* https://upload.wikimedia.org/wikipedia/commons/8/8b/Nuvola_apps_kalzium.svg *License:* LGPL *Contributors:* Own work *Original artist:* David Vignoni, SVG version by Bobarino

- **File:Open_and_closed_strings.svg** *Source:* https://upload.wikimedia.org/wikipedia/commons/5/56/Open_and_closed_strings.svg *License:* Public domain *Contributors:* Own work *Original artist:* Xoneca

- **File:Pascual_Jordan_1920s.jpg** *Source:* https://upload.wikimedia.org/wikipedia/commons/a/a6/Pascual_ Jordan_1920s.jpg *License:* Public domain *Contributors:* http://www.gettyimages.co.uk/detail/news-photo/ the-austrian-physicist-paul-ehrenfest-posing-in-front-of-a-news-photo/141551561 *Original artist:* Unknown (Mondadori Publishers)

- **File:Penrose.svg** *Source:* https://upload.wikimedia.org/wikipedia/commons/a/a8/Penrose.svg *License:* Public domain *Contributors:* Transferred from en.wikipedia to Commons by Andrei Stroe using CommonsHelper. *Original artist:* Cronholm144 at English Wikipedia

- **File:Portal-puzzle.svg** *Source:* https://upload.wikimedia.org/wikipedia/en/f/fd/Portal-puzzle.svg *License:* Public domain *Contributors:* ? *Original artist:* ?

- **File:Psr1913+16-weisberg_en.png** *Source:* https://upload.wikimedia.org/wikipedia/commons/7/79/Psr1913%2B16-weisberg_en.png *License:* Public domain *Contributors:* M. Haynes et Lorimer (2001) (redrawn by Dantor as Image:Psr1913+16-weisberg.png, English labels added by mapos) *Original artist:* ?

- **File:Qcd_fields_field_(physics).svg** *Source:* https://upload.wikimedia.org/wikipedia/commons/4/41/Qcd_fields_field_%28physics%29.svg *License:* CC0 *Contributors:* Own work *Original artist:* Maschen

- **File:Question_book-new.svg** *Source:* https://upload.wikimedia.org/wikipedia/en/9/99/Question_book-new.svg *License:* Cc-by-sa-3.0 *Contributors:*
Created from scratch in Adobe Illustrator. Based on Image:Question book.png created by User:Equazcion *Original artist:*
Tkgd2007

- **File:Question_dropshade.png** *Source:* https://upload.wikimedia.org/wikipedia/commons/d/dd/Question_dropshade.png *License:* Public domain *Contributors:* Image created by JRM *Original artist:* JRM

- **File:Relativistic_gravity_field_(physics).svg** *Source:* https://upload.wikimedia.org/wikipedia/commons/7/79/Relativistic_gravity_field_ %28physics%29.svg *License:* CC0 *Contributors:* Own work *Original artist:* Maschen

- **File:Relativistic_precession.svg** *Source:* https://upload.wikimedia.org/wikipedia/commons/2/28/Relativistic_precession.svg *License:* CC-BY-SA-3.0 *Contributors:* Own work, self-made using gnuplot with manual alterations *Original artist:* KSmrq

25.10.3 Content license

www.ingramcontent.com/pod-product-compliance
Lightning Source LLC
Chambersburg PA
CBHW080655190526
45169CB00006B/2124